018871
QH
31
.W54
E8
Evans
William Morton Wheeler,
biologist

William Morton Wheeler, Biologist

MARY ALICE EVANS and
HOWARD ENSIGN EVANS

William Morton Wheeler, Biologist

Harvard University Press
Cambridge, Massachusetts
1970

Distributed in Great Britain by Oxford University Press, London

Library of Congress Catalog Card Number 76–129117
SBN 674–95330–4

Printed in the United States of America

To our teachers and to our students

Contents

Illustrations

Preface

We cannot with any thoroughness acknowledge the many people who have helped us in the preparation of this book, but we have appreciated and enjoyed sharing their time and memories. For family recollections we are especially indebted to Miss Adaline E. Wheeler and Dr. Ralph E. Wheeler. We also thank them for giving us access to the family papers and allowing us to quote liberally from their father's diaries and correspondence.

In addition, we would like to thank the following institutions for permission to consult and use manuscripts and letters on deposit in their archives and libraries: Widener Library and the Museum of Comparative Zoology at Harvard University, Cambridge, Massachusetts; Nils Yngve Wessell Library at Tufts University, Medford, Massachusetts; The Academy of Natural Sciences of Philadelphia; The American Museum of Natural History in New York; University of Chicago Library; University of Texas Library in Austin; Flower Memorial Library in Watertown, New York; Rochester Public Library in Rochester, New York; Milwaukee Public Library and the Milwaukee Public Museum; Marine Biological Laboratory at Woods Hole, Massachusetts; Clark University in Worcester, Massachusetts; State Historical Society of Wisconsin in Madison.

As well as to the many librarians and research assistants who have helped us, we are obliged to H. B. Jefferson, former president of Clark University; Dr. James A. Oliver, director of the American Museum of Natural History; Dr. Chester Kielman of the University of Texas; Mrs. Franziska Dittlinger Liebscher of New Braunfels, Texas; Mrs. E. N. Sandstrom of the Rochester Historical Society and Mr. Blake McKelvey, city historian of Rochester, New York;

Dr. Kenneth MacArthur and Mrs. Irene Hanson of the Milwaukee Public Museum; Mrs. G. W. Hoffmeister, Jr. of Wellesley College, Wellesley, Massachusetts; Miss Julia G. Bacon of Vassar College, Poughkeepsie, New York; Professor Charles H. Lyttle, formerly with the Meadville Theological School, Chicago; Dr. Russell E. Miller of Tufts University; and Dr. Robert Rosenthal of the University of Chicago.

We are indebted to Dr. and Mrs. Henry Dietrich of Cornell University and to Professor Edward O. Wilson and Dr. Mary Jane West Eberhard of Harvard University for reading the manuscript critically. Professor John A. Weir of the University of Kansas also reviewed the chapters dealing with the Bussey Institution.

Our research in Milwaukee, Chicago, and Watertown, New York was made possible by a grant from the Penrose Fund of the American Philosophical Society. Restoration of a number of old photographs was supported by the W. M. Wheeler Fund of the Museum of Comparative Zoology, contributed by Dr. Caryl P. Haskins.

The Society of Sigma Xi has graciously given us permission to reprint the prologue, which first appeared in *The American Scientist*.

William Morton Wheeler, Biologist

Prologue / Ants, Elephants, and Men: Strands from the Fabric of Natural History

When P. T. Barnum's Museum reopened on Broadway in New York in October 1860, its proprietor promised "a mysterious novelty never before seen in that establishment": no less than a cherry-colored cat. When his audience was primed, he pulled from the bag an ordinary black cat, explaining that it was, of course, "the color of black cherries." The audience roared, the newspaper reporters scuttled off with their stories, and Barnum had scored another triumph.

It was little wonder that when, a year later, Barnum ("There's a sucker born every minute") announced his newest exhibit, two rarely seen white whales, there were those who were skeptical. But as Barnum had launched a major expedition to the mouth of the St. Lawrence River to capture the whales, and had paid $4,000 to have a tank twenty-four feet square built to hold them, he was determined that they should be certified as genuine white whales. No less a person than Professor Louis Agassiz of Harvard University was induced to examine them. He announced that they were, indeed, white whales. Unfortunately, the whales soon died, but even this was turned into a triumph for Barnum. Agassiz's endorsement was published far and wide and displayed for many years in the museum, and the tank was shortly thereafter used for "the first and only genuine hippopotamus that had ever been seen in America."[1]

Barnum was to call on Professor Agassiz for help once again. This time the charge was cruelty to animals, leveled by Henry Bergh, one-time diplomat to Russia, founder of the American S.P.C.A., and a colorful crusader immortalized by Longfellow as "the friend of every friendless beast." Bergh himself had turned to Agassiz when seeking support for his campaign against cruelty to turtles. Agassiz, a turtle-lover, gladly complied with a testament on the suffering of turtles. But when Bergh stormed Barnum's Museum, loudly protesting the feeding of live rabbits to boa constrictors, it was Barnum who appealed to Agassiz. Agassiz was a snake-lover, too, and he pointed out that snakes will accept only living food. "I do not think the most active member of the Society," he wrote, "would object to eating lobster salad because the lobster was boiled alive . . . or raw oysters because they must be swallowed alive." Barnum did oblige Bergh by feeding the snakes at night, thus sparing the public a scene the S.P.C.A. president regarded as "cruel and demoralizing."[2]

While Barnum's showmanship was sometimes suspect, perhaps with cause, that of Louis Agassiz was without taint. Trained under some of Europe's finest zoologists, he had come to Boston in 1846 as a visiting lecturer, and he had stayed to charm a nation and to found a museum of rather different character from Barnum's. Like Barnum, Agassiz was sometimes embroiled in controversy. His theories of glaciation were not accepted by all, though history has shown him more right than wrong. Much of his later life was devoted to efforts to refute the Darwinian theory of evolution, and here history has not treated him kindly. Agassiz often chose popular magazines, such as *The Atlantic Monthly,* for his essays on these and other subjects, for to him science was not something merely to be pursued in the dark corners of his museum. "The time has come," he said, "when scientific truth must cease to be the property of the few, when it must be woven into the common life of the world; for we have reached the point where the results of science touch the very problem of existence." That was in 1862!

Agassiz was an imposing figure on the lecture platform. In the words of one contemporary, "He was strikingly handsome, with a dome-like head under flowing black locks, large, dark, mobile eyes set in features strong and comely." Horace Greeley heralded his arrival in New York on a lecture tour by remarking in the *Tribune* that "never have our citizens enjoyed the opportunity of acquiring so large a measure of knowledge of the laws of nature . . . as these lectures will afford them." Both scientist and layman were held spellbound as Agassiz discoursed on subjects as diverse as embryology, glaciation, and fossil fishes.[3]

When Agassiz was scheduled to lecture in Rochester, New York, in January of 1854, a lad of twenty named Henry A. Ward hiked thirty miles from school through mud and snow to see and hear the famous naturalist. Ward, who had been interested in rocks and fossils for at least half of his life, was overjoyed the next morning when he was asked to show Agassiz the local geological points of interest. Agassiz was delighted with his day's excursion with Ward, and he promptly invited Ward to study with him at Harvard. During his apprenticeship with Agassiz, Ward first learned the value of museums in the advancement of scientific teaching and research. He went on to devote his life to providing specimens for museums, including Agassiz's.[4]

In the late nineteenth century, ordinary traveling was not easy, but Ward's desire to secure unusual specimens to supply the museums took him to many remote parts of the world. His reports home, when he found time to make them, must have made some of his friends wonder if his imagination was not rivaling Barnum's. One story told of a collecting trip in 1859, when Ward was barely twenty-five, on the Niger River in West Africa. He was taken ill with blackwater fever, then the scourge of the white man in the tropics, and the superstitious crew of the river boat put him ashore and left him to die. A native woman rescued him and nursed him back to health, with the clear intention of making him her husband. Somehow Ward managed to slip away and eventually to appear at his home in Rochester. Sometime later a large packing case of gifts was left by a passing ship on the bank by the hut of his dusky benefactress. Meanwhile, at a party in London, David Livingstone was reported to have ignored many important guests so that he might converse with the young American who knew Africa so well.

Agassiz himself traveled widely, and his travels were closely followed by the press—usually in prose, but sometimes in rhyme. Oliver Wendell Holmes celebrated Agassiz's departure for Brazil with a long ode that contained the lines:

> May he find, with his apostles,
> That the land is full of fossils,

That the waters swarm with fishes
Shaped according to his wishes,
That every pool is fertile
In fancy kinds of turtle.[5]

Agassiz's ability to raise money for his many scientific projects became legend even in his time. In early 1873, he conceived the idea of starting a summer school "somewhere on the coast of Massachusetts where teachers from our schools and colleges could make their vacations serviceable by the direct study of nature." John Anderson, a wealthy New Yorker, read the details of Agassiz's plan in the *Tribune,* and immediately presented him with the island of Penikese, in Buzzard's Bay, along with $50,000 for the founding of the school. As one contemporary put it, Agassiz had only to desire, and desire became reality.

Agassiz, himself, felt that he was primarily a teacher and that his greatest triumph was his summer school at Penikese. It was a pioneering effort both as a coeducational summer school and as a seaside laboratory. Although the physical plant was incomplete when the school opened in July, Agassiz declared that never had a laboratory been better equipped. The sea was abounding in life, the shore had in addition to its living inhabitants a great diversity of pebbles all with interesting stories to be revealed. "Study nature, not books," was Agassiz's motto; "I will only help you to discover what you want to know yourself." In retrospect, David Starr Jordan (later to become the first president of Stanford University) said: "The summer went on through a succession of joyous mornings, beautiful days, and calm nights, with the Master always present, always ready to help and encourage, the contagious enthusiasm which surrounded him like an atmosphere never lacking."[6] The poet Whittier was moved to describe that remarkable summer in his poem beginning:

On the isle of Penikese
Ringed about by sapphire seas,
Fanned by breezes salt and cool,
Stood the Master with his school.[7]

This was to be the last summer of Agassiz's life; when his group met again the next summer, but without their famed teacher, the magic was gone, and the Penikese school was never to be again. But many of the students at Penikese were to become leaders in biology who would bring fresh insight to the constant search for new truths. Perhaps James Russell Lowell sensed this, for he wrote after Agassiz's death: "He was a teacher; why be grieved for him whose living word still stimulates the air?"

One of the students at Penikese was Charles Otis Whitman, a high school teacher of quiet manners who had, already at thirty, a shock of snow-white hair. Whitman was a Bowdoin College graduate, but it was not until his two summers at Penikese that he took up biology in earnest. Like most biologists of his time, he went to Europe to learn the latest theories and techniques, and like most of his contemporaries, he spent some time at the new biological station at Naples, Italy. Built by a devoted German scientist on the picturesque shore of the Bay of Naples, this gleaming white building, dedicated to science, has caused some to remark that not all temples are built for the love of a beautiful woman.

In 1882 Whitman came back to the quiet, dusty alcoves of Mr. Agassiz's museum at Harvard. This Mr. Agassiz was, of course, Alexander, the son of Louis. For four years Whitman assisted Agassiz with his studies on fish, but Whitman was destined to be much more than an assistant. Then in his forties, he was ready for an opportunity to exercise his independence of spirit and to develop his own ideals. That opportunity came when he was asked to direct a new research laboratory in the cultural center of the midwest, Milwaukee, Wisconsin.

Milwaukee had one of the best public schools in the country, a public library, and a public natural history museum. The latter, of course, owed its existence to Henry A. Ward of Rochester, New York. At the Milwaukee Exposition of 1883, Professor Ward had exhibited a portion of his tremendous collection of stuffed animals in an effort to persuade the city fathers that a great and influential metropolis such as theirs should have a public museum. It could be started easily, he suggested, by purchasing his exhibit and combining it with the small but good collections then at the local German-English Academy. Ward was successful, not only in improving Milwaukee, but also in saving his business, which needed this sale to keep from going bankrupt. Someone else profited from this venture, too: a local boy named William Morton Wheeler, a student at the German-English school. Wheeler, classically trained but with an unusual enthusiasm for things biological, had haunted the school museum for years, and knew every specimen. When Ward arrived, Wheeler was asked to help arrange the exhibits, and before long Ward had invited him to work in his Establishment—for nine dollars a week minus six for room and board at Ward's own house.

Young Will Wheeler's eagerness to work among the wonderful beasts in Rochester gave way to amazement when he saw Ward's Natural Science Establishment. Sixteen wooden buildings of various sizes, each with a gilded totem pole at its peak to show the nature of its contents, opened into a common courtyard. As Wheeler entered the court, he passed beneath a Gothic arch formed by the jaws of a whale and was confronted by a conspicuous placard:

THIS IS NOT A MUSEUM
But a Working Establishment
Where All are Very Busy.

Slightly to the left was the main building, and on the doorstep was a stuffed gorilla with an open, leering mouth, clinging to a naked tree trunk. A less determined young man might have fled. But in Ward's study, a room lined with books and littered with maps, sketches, and specimens, Wheeler was warmly welcomed by his host.

Wheeler began his work identifying and listing birds and mammals, and later shells, with pleasure, but unfortunately Professor Ward, whom he so much admired, spent more time traveling than he did in Rochester. But there were a number of young men around, and in spite of the dictum outside they enjoyed many informal times together. Wheeler was especially attracted to a young taxidermist brought up on a farm not far from Rochester, Carl Akeley. The two spent many evenings together, often with Wheeler reading aloud and Akeley working on his animals.[8]

By the spring of 1885, both Wheeler and Akeley felt that there were better opportunities elsewhere. Wheeler wanted to go back to his cultural Milwaukee. He suggested that if Akeley would come, too, he would tutor him as preparation for entering a scientific school, and that besides, there might be a job for him in the new municipal museum. Wheeler did return to Milwaukee, where he soon became involved in teaching and research, and where he shortly came under the influence of C. O. Whitman at his laboratory. But an unexpected challenge kept Akeley at Rochester another year.

At about nine o'clock in the evening on September 15, 1885 at St. Thomas, Ontario, P. T. Barnum's seven-ton elephant, Jumbo, was being led along the railroad tracks to his car. Out of the night roared an unscheduled freight train. The locomotive struck Jumbo with terriffic force, and within moments he was dead, the train derailed and badly damaged. The news was cabled all over the world, for Jumbo had been the greatest attraction Barnum had ever offered to the public. He had been bought at the astronomical price of $10,000 ($30,000 with carrying charges included!) from the Royal Zoological Gardens in London. Jumbo was a sensation in America, and Barnum reputedly got back his money in ten days.

Not wanting to lose this asset, especially with the added publicity of the killing, Barnum decided to have Jumbo mounted. He wired Henry A. Ward, and Ward, Akeley, and a number of others rushed to the scene of the wreckage. With the aid of several local butchers, Jumbo's remains were removed as quickly as possible. It fell to Akeley mainly to oversee the feat of preserving Jumbo, so that the elephant could be put back on exhibition almost as though he could still breathe. This was done so well that Jumbo's massive skin still stands with all its lifelike wrinkles, and his beady eyes still seem to stare out at those who come to see him in the foyer of the biology building at Tufts University. Jumbo's skeleton was mounted, too, and now is at the American Museum of Natural History in New York. His heart, which weighed forty pounds, was advertised for sale by Ward for forty dollars. Who, if anyone, bought it no one seems to know.

When Jumbo was finished, Akeley followed Wheeler to Milwaukee, where he set up his taxidermy shop in Wheeler's barn. Shortly thereafter Wheeler was made director of the Milwaukee Public Museum, and Akeley joined the staff. There, under Wheeler's liberal and sympathetic leadership, Akeley created his first habitat group—a natural scene of muskrats in a pond (which is still on exhibit). Wheeler helped Akeley start this project one cold October day in 1889 when the two walked shivering into a nearby swamp to gather the muskrat house and skins and a few typical birds. To these, Akeley applied his new techniques of taxidermy and added the artificial reeds and water that he had devised, creating not only a remarkable illusion of muskrat activity, but a whole new concept in museum displays and public education.

Though Carl Akeley's skill was evident in Milwaukee, it was not until he began his studies of African animals that his great genius emerged. First at the Field Museum in Chicago, and then at the American Museum of Natural History in New York, Akeley proved that he was much more than a mere taxidermist. To these museums he brought his knowledge of the outdoors and his ability to sculpture, creating for even the least informed audience a feeling for Africa's unique wilderness. Akeley was especially fascinated by elephants, though once he was almost crushed by one, and his dramatic group, "The Fighting Bulls," in Chicago was dwarfed only by his plans for his elephant group in New York. He wanted it to be a part of a great African Hall, but unfortunately he did not live to complete his plans. He died on safari and was buried on the slopes of Mount Mikeno in the Congo—a spot he once called the most beautiful in the world. But his ideas were completed by workers whom he had trained, and the African exhibit today is dominated by the tremendous herd of elephants collected earlier by Akeley and his friend Theodore Roosevelt. Thus, the Akeley Memorial African Hall stands as a tribute to a man of humble origins who cherished a dream to preserve for all to see an image of our dwindling wildlife in its natural setting.

The Milwaukee Public Museum and the nearby Allis Lake Laboratory were to provide the initial opportunity not only to Akeley and to Wheeler, but to several others who went on to distinguished careers. Although short-lived (seven years), the Lake Laboratory was a most unusual institution.[9] It was founded by a businessman, Edward P. Allis, Jr. Pressured by his family to forsake his childhood interest in natural history, Allis was educated as an engineer so that he could enter the family business, the manufacture of farm machinery (later to become Allis-Chalmers). When he had proved himself as a businessman, he again took up the study of zoology in the few spare hours that he could get away from the factory. In a large billiard room on the third floor of his aunt's home near Lake Michigan, he set up a laboratory and then proceeded to seek a young and promising biologist as a director. He proposed to organize the laboratory like a business, wherein a group of specialists would work on a particular problem as a body and not as individuals. But C. O. Whitman, the man he selected as director, had other ideas. Whitman believed that each worker should have absolute freedom of thought and action and should pursue his search for truth wherever it should lead. Whitman's ideas not only prevailed, but he soon persuaded Allis, who had had no interest in publishing his research findings, to help organize and finance the *Journal of Morphology,* which was to be a model of excellence for all scientific periodicals.

Whitman was a soft-spoken man of simple tastes, throughout his life completely absorbed in his teaching and research. Nevertheless, beneath his quiet and kindly exterior was a brilliant and original mind and a fiercely independent spirit. In 1888 he was appointed the first director of the Marine Biological Laboratory at Woods Hole, Massachusetts, which was in many ways a lineal descendant of Penikese and of Allis Lake Laboratory. There for almost every summer for twenty years, Whitman directed the Laboratory according to his ideals of cooperation of individuals and institutions, but with independence in government so that investigators could follow their own research with complete freedom. The stature of the Laboratory today may be attributed largely to Whitman's genius.

The challenge of founding a biological station might have been all an ordinary man could handle, but Dr. Whitman accepted another equally exciting offer at about the same time. In 1889 he was asked to direct the department of zoology of the new Clark University in Worcester, Massachusetts, which was designed solely as a graduate and research institution. There was little formal instruction, and each student was expected to work independently on his own particular problem. Yet no one was permitted to become a specialist completely. Dr. Whitman encouraged his students, as had Professor Agassiz, not to allow anything their animals did to escape attention. Thus Whitman's students were reported to greet each other not with "What is your special field?" but rather "What is your beast?"[10]

One of Dr. Whitman's most enthusiastic students was William Morton Wheeler. Wheeler had come under Whitman's influence first at the Allis Lake Laboratory and subsequently had followed him to Clark, where he had completed his graduate studies in 1892. He then left Clark, along with Whitman and others. Once more Whitman had been asked to organize a biology department, this time at the new University of Chicago. There Whitman remained, except for summers at Woods Hole, becoming one of the country's foremost students of animal behavior. His classic monograph on the behavior of pigeons was published posthumously in 1919.

If Charles Otis Whitman was a credit to the ability of his teacher, Louis Agassiz, then William Morton Wheeler was no less so for Whitman. In 1899 Wheeler left Chicago, after several years of teaching there, to direct the department of zoology at the University of Texas at Austin. After the years of rich academic associations and atmosphere, Texas seemed virtually a frontier. There were few books and little laboratory equipment available in the University, and Wheeler was faced with the formidable task of not only teaching the beginning course in zoology, but also most of the advanced ones. Discouraged at the prospect of so much to do and so little with which to do it, Wheeler took his lunch one day and went out to eat it in a dry stream bed near the campus. Passing near him in steady procession were long lines of ants carrying leaves. He watched them with puzzled fascination for a while, and then said to himself: these are worth a lifetime of study! Wheeler went on to become a world authority on ants, the most abundant of all animals on earth, as Wheeler himself often pointed out.

For four years Wheeler directed the zoological studies at the University of Texas, even organizing a short-lived marine laboratory on the Gulf coast. Then he was challenged anew at the prospect of reorganizing the invertebrate animal collections at the American Museum of National History. After five productive years in New York, he was appointed to a professorship at Harvard, where he spent the remainder of his life.

Wheeler was a brilliant writer and lecturer, a man of encyclopedic learning. He could and did discourse on the most abstruse philosophical matters, yet some of his essays find their way regularly into the anthologies of popular natural history. In one of these papers he wrote, "Natural history constitutes the perennial rootstock or stolon of biological science . . . because it satisfies some of our most fundamental and vital interests in organisms as living individuals more or less like ourselves." In the same essay he staunchly supported the amateur in a characteristic Wheelerism: "The truth is that the amateur naturalist radiates interest and enthusiasm as easily and copiously as the professor radiates dry-rot." A few lines later he excluded "dear old, mellow, disinfected professors of the type of Louis Agassiz . . . [who could] enter at once into sympathetic rapport with the humblest amateur." Wheeler had made his point—one worth reiterating in these times, when the amateur is so frequently barred from the halls of science.[11]

Like Agassiz, Wheeler was many things to many people. To Alfred North Whitehead, the Harvard philosopher, he was the only man he had ever known "who would have been both worthy and able to sustain a conversation with Aristotle."[12] William Beebe characterized him as "an outstanding example of one who fulfilled every requirement of scientist, investigator, naturalist and literary master."[13] To one of Wheeler's neighbors at his summer home in Colebrook, Connecticut: "Oh yes, there's Professor Wheeler. He comes up here every summer and turns over all the stones in the pasture and then he goes back to Boston."[14] In his later years, Wheeler became a regular inhabitant of the Agassiz Museum, and his ant collection, comprising more than a million specimens from all parts of the world, is one of the most valued treasures of that institution.

Thus the wheel is come full circle—or rather several circles. And who can trace fully the many strands of thought which connected these men and their students? All were individualists, often of very different temperament; yet through them all runs a common strand, a fascination with animals as living beings. In these days of specialization and depersonalization, in these times of the two cultures and the scientific elite, we look back upon these expansive and independent spirits with incredulity, sometimes with amusement. Animals were to each of them the touchstone, and their lives were intertwined with those of ants and elephants and other living things. Today biology, like every other human activity, is caught up in a whirlwind. Personally, we would hope never quite to lose sight of men such as Agassiz discoursing on fossil fishes to an audience of spellbound laymen, Whitman surrounded by his pigeons, Wheeler watching a column of ants—yes, even Barnum perpetrating some new extravaganza and Ward offering Jumbo's heart for forty dollars. By the way, what *did* become of Jumbo's heart?

Men such as these form the warp and woof of science. Their personalities, their times, their ideas are fated to become increasingly lost to sight except for those bits and pieces we preserve and periodically re-examine. Louis Agassiz has had his biographers—and we hope he will have more—and so have P. T. Barnum and Henry Ward. Perhaps some day Carl Akeley and C. O. Whitman will receive the tributes they deserve. We have always felt an affinity for William Morton Wheeler, and a desire to see his place in science more fully chronicled. This book is an attempt to place on record a fuller account than has been available of the life and accomplishments of this unusual man. Some of the many persons who influenced Wheeler or were influenced by him also appear in these pages, and we have not hesitated to let such persons divert us from our main theme from time to time. Wheeler's scientific work spanned more than fifty years and reflected many of the trends and some of the turmoil in what was surely one of biology's most exciting half-centuries.

1 / Bright Beginnings

It was a typical fall day in Milwaukee, with a cold wind coming off the lake. The city, seemingly trying to hug the curve of Lake Michigan's western shore to find a little warmth that the summer might have left in the deep blue depths, was bustling far more than the chilling air required. But the gaiety and excitement of the booming metropolis did not enter into the quiet interior of the small museum. Nor was it evident in the two men working there. Hardly speaking, they unpacked with care the contents of the large boxes, removed the specimens and carefully dusted and brushed them as they prepared them for display.

Henry A. Ward worked with the skill and assurance of a man with thirty years' experience in such things; William Morton Wheeler, though still in school, contributed a youthful enthusiasm that was to mature into a contagious zest for living things. These two men were preparing an exhibit which they hoped would be of such interest to the public that it would demand a larger and better public natural history museum. The small municipal museum had existed, at least technically, since 1882, when the city had accepted the natural history collections of the German-English Academy. The museum at the Academy had become so crowded that it was almost inaccessible to interested students. Consequently, after being donated to the city, it had been moved to larger quarters at the Exposition Building not far away.

Milwaukee had long ago healed over its scars from the Civil War and was entering an era of great prosperity. New industries, railroads, commercial shipping on the Great Lakes—all were coming to Milwaukee. The business interests of the city had financed an exposition building to promote further commerce by exhibiting Milwaukee's industrial accomplishments. Not only was this goal achieved, but the Exposition itself was a financial success. To this scene, in 1883, came Henry A. Ward of Ward's Natural Science Establishment of Rochester, New York, a man whose reputation as a founder of museums was well established. This was reflected in the following paragraphs written by Carl Doerflinger, the first director of the Public Museum, and included in its first *Annual Report:*

A NOTABLE EVENT

A notable event that may have more or less bearing on the future increase, popularity and usefulness of this establishment, is the exhibition at the "Milwaukee Industrial Exposition" of Prof. Ward's great collection of natural objects and casts. It seems to be the chief center of attraction at the Exposition and will do much to create an interest in natural history among the people. Teachers and their pupils look upon it with especial favor, and it is at this moment contemplated by citizens having the advancement of the educational system of the city at heart, to raise a subscription for the purpose of acquiring the collection and donating the same to the Public Museum.

I earnestly hope that this particular scheme will be successful. The acquisition of the Ward museum would at once place the Public Museum near the level of the more important American institutions of the kind; it would offer the teacher and student superior opportunities, become more attractive for the people in general and be apt to draw unto itself new additions. If you can do anything to facilitate the achievement of the desired end it will be a lasting credit to your administration.

The material of the Ward cabinet was partly selected with reference to what specimens would be most desirable for a display at Milwaukee; among the specimens there are many that are typical for those divisions of the animal kingdom not represented or not well represented in our museum.

Thus, the two collections are to a certain degree complementary to each other, and if united would present good synoptical series for the study of various branches of the science.

Small wonder that young Will Wheeler had been eagerly anticipating the arrival of Professor Ward and his boxes of specimens. Wheeler, a student at the German-English Academy, had always been attracted to natural history and had often visited the Academy museum. It was natural, then, that his request to act as Ward's assistant be granted. So it was that young Wheeler and Professor Ward spent several evenings together unpacking and arranging specimens in an exhibit which in turn would play an important part in the life of many a future naturalist, including William Morton Wheeler.

Will Wheeler's future in 1884 was promising and exciting, as he himself wrote, but of his past, his childhood, he left little record. Nor do we know much of his family. His grandfather, Simon Wheeler, first appears in the official records in northern New York State in the early 1800s. Presumably he came from New England, his parents probably from eastern Massachusetts, to which many Wheelers had come as English immigrants to the colonies. The usual route of migration west was through New Hampshire or Vermont, and the histories of those states include a great deal of general information on this mass exodus west. The lure of new and untried places was probably paramount, but also important was a series of misfortunes beginning in 1808 with President Jefferson's embargo against trade with Montreal and continuing with cloudbursts and floods in the summer of 1811, the collapse of the Vermont State Bank during the following winter, the unpopular War of 1812, and finally the cold season of 1816, when there was frost every month of the growing season. And so, for whatever reason, Simon Wheeler came to the eastern shore of Lake Ontario in Jefferson County to farm several acres of land. We do not know if he brought his first wife with him, or if he married locally, but in the 1820 federal census we find him living in Ellisburgh (about twenty miles south of Watertown) with a wife, four sons, and one daughter. Five years later he owned eight acres in the nearby town of Lorraine, where he lived with his wife, Jemima, and four sons and two daughters. In 1835, Simon was still in Lorraine, owned thirty-five acres, had a wife named Lucy, and had two children living at home, one daughter under sixteen and one son not yet one. The older four boys and the oldest girl were evidently gone.

Julius Morton Wheeler (William Morton's father) was born in Watertown on November 16, 1817. His brother Thomas Wilson, who was born about 1812 in New York State (Watertown?), later married Emma Andrus, whose father came from Connecticut and was one of the first settlers in Rutland, New York. Two other brothers were reported to be named Simon and Asa, but they left no records. There were two sisters, Clarinda, whose birth date we do not know, and Chloe, who was born about 1830. Chloe married James Avery, who was a cattle drover, or a butcher, or perhaps both at various times; they had at least three children and presumably lived out their lives in or near Watertown. Clarinda married S. A. Kimball, had four children (one named Morton), and lived in Milwaukee. When The Milwaukee Hide and Leather Company, manufacturers of harness and specialists in collar leather, was officially incorporated in 1867, Mr. Kimball was one of the original four stock owners. In 1881, when the company employed forty men and was turning out about 15,000 hides per year, he was president.

Julius Morton married Sophie H. Ingles, and their first son, Herbert Julius, was born in Watertown on July 21, 1853. Their second son, Marcus, was born January 24, 1857 in Milwaukee. Why and when did Julius Morton migrate to Milwaukee? Did Thomas Wilson or any of the rest of the family accompany him? Presumably Simon was no longer alive (though we know little more about the year or place of his death than we do of his birth) and thus could not influence his sons. But here the records of deeds suggest some interesting possibilities. Julius and his brother Thomas Wilson, though claiming to be butchers, were speculating in land. Between 1844 and 1857 they bought and sold a great number of parcels of land in and around the village of Watertown, and in the 1850 census of Watertown, both are credited with considerable real estate: T. Wilson with land valued at $20,000, and Julius with $10,000. But in 1857 some jointly owned property was put up at auction and sold by proxy to pay the mortgage, and Julius was in Milwaukee. We do not know the whereabouts of Thomas Wilson in 1857, but interestingly enough, Loveland Paddock, a local banker, bought the property at auction, and later Thomas Wilson's daughter, Olive, married the banker's nephew, Edwin L. Paddock. The Watertown histories credit Thomas Wilson as builder of the American block, an important landmark in the center of the city (now replaced by the Woolworth building). Olive Paddock eventually willed her mansion and grounds to the Historical Society to use as a museum in which to preserve mementos of the city and surrounding county. As she had no children of her own, she left a considerable amount of cash to her companion, and to the children of her two brothers, Frank and Abraham, who had long ago left Watertown for Chicago. There was no indication that she even knew of her Milwaukee cousins.[1]

Julius Morton's presence in Milwaukee is first noted in 1857, when he is listed in the City Directory as a land speculator. Perhaps Julius did try buying land in or near Milwaukee, but more likely Milwaukee was already too large a city and land too costly for Julius, so that he went back to his old interest in animal hides. We do not know that he ever practiced as a butcher, but he did work as a tanner and at one time was a stock owner in The Milwaukee Hide and Leather Company, of which, as previously noted, his brother-in-law S. A. Kimball was president. Fire was always a problem in the era of frame buildings, and especially to businesses requiring the use of fire. Family tradition says that Julius owned a tannery on the east side of town, that it burned about 1883, and that Julius never recovered emotionally, if financially, from this loss.

A year or two after arriving in Milwaukee, Julius' wife, Sophie, died, and in 1864 he married Caroline G. Anderson. Carrie, as she was called, not only cared for her two stepsons, but had children of her own, of whom William Morton was the first (March 19, 1865) and the only son to reach maturity. There were three girls: Minnie, born in 1867, Clara, in 1872, and Olive, in 1876. Minnie seems to have been the closest to Will, both in years and understanding, at least in their youth. However, she married John Grenvis when she was barely 18, and her children and other family matters soon separated the two. Clara was sickly most of her life and the object of her mother's care. Olive, the baby, though at some distance from Will chronologically, was probably most like him in ability and independence. She mentioned in a letter to her brother in 1893 that she hoped to present the valedictory at her high school. And for six years prior to her marriage to J. Potter Whittren in late 1904, she worked at the Milwaukee Public Museum, officially as an attendant, but unofficially doing many skilled jobs, such as painting backgrounds in water color for the whaling exhibit "showing the harpooning, lancing and 'cutting in' of a whale," giving special attention to the care of the live specimens (plants as well as animals, and often considerable in number) used in the very popular exhibits on local flora and fauna, and curating the shell collections.[2]

So much for what is known of the Wheeler family. Will left no written impressions of his father, other than noting the time of his death; nor did he or any of his brothers or sisters put a marker on the graves of their parents in the Forest Home Cemetery (though cemetery records do note the burial plots). But, fortunately, the

Milwaukee Journal carried the following story on the lower half of its front page on January 30, 1884:

Death of an Old Settler
Julius M. Wheeler Joins the Silent Majority—
Sketch of His Life

Julius M. Wheeler, an old and well-known citizen, died at his residence, 590 Island Avenue, at an early hour this morning, aged 67 years. The deceased was a native of Watertown, N.Y., coming to Milwaukee thirty-one years ago. During his life his quiet, unobtrusive ways combined with rare sympathy and kindly feeling, made a large circle of friends, both in a social and business way, who will long remember and regret him. Among his former acquaintances this morning, there was an expression of sorrow at the demise. Mr. Wheeler was at one time interested with the Milwaukee leather company, but in later years had withdrawn from any connection. A widow and six children—three sons and three daughters are left. Two of the boys are located in business in Montana, while the third and youngest was about to leave the city for Rochester, N. Y., to connect himself with the Ward museum there. The cause of Mr. Wheeler's death is said to be internal injuries suffered from a fall on the ice at the corner of Grand Avenue and West Water street, three weeks ago. The funeral will be Sunday afternoon, it being delayed to that time to give the two sons in Montana time to get to Milwaukee.

As students of genealogy must know, colorful ancestors are not common, and when found should be treasured and appreciated. The Anderson family seems to have provided William Morton Wheeler with that enjoyment, for more information about his mother's family has survived in the family papers. One ancestor of Welsh descent, a Tomas Jones (whose sons were all over six feet tall) fought in the American Revolution—on the side of the English. Another, a Rebecca Bunny, though born near Bunker Hill on the outskirts of Boston, Massachusetts, was reputed not to be "a very refined woman," and was put in the work house from time to time, her husband taking pecuniary charge of her until she died—of cancer of the tongue. But Grandfather John Anderson must really have been the delight and exasperation of his family. Physically he suggested his Scandinavian-Scotch descent, for he was a commanding figure with blond hair, a dark brown beard, a rich and rosy complexion, and brilliant blue eyes. As a boy he had worked in his father's blacksmith shop and had had no time for a formal schooling.

However, he was ambitious and physically and mentally strong. While a young man he became a dissenting Methodist circuit preacher, and, after working all week in his father's shop, walked to neighboring towns (often to a distance of nineteen miles) to preach on Sundays. This dissension subsided, however, when through his own efforts and ability he became a successful engineer. In Rugby, England he lived with his wife and children as a prosperous owner of a foundry and a respected member of the community, including the Church of England. But tranquility was never long part of Grandfather Anderson's life. In his early forties, he developed a pulmonary disease, and his English physician, knowing nothing of the United States, suggested that he live in a drier climate, perhaps Wisconsin. So the family migrated in three groups to Racine, Wisconsin, where it was finally reunited in 1858. Here, in what was in fact a harsher climate, John Anderson's true nature reasserted itself. He had not only trained himself for a vocation, he had also sharpened his mind, and now, no longer lulled by complacency, he could use it. He had become a skillful mathematician and draftsman and applied these skills in making many inventions; he had become an excellent scholar and had published a number of articles, mostly of a professional nature; but he also liked poetry and music. He was bold, outspoken, and argumentative, and as an intense Abolitionist, he cast his first vote in this country for Abraham Lincoln. He also "relapsed" into the Methodist faith, joined a local church, and then withdrew because he was "disgusted with the hypocrisy of his neighbors." Finally, he joined the Congregational Church—perhaps to please his worried family.

The family of John Anderson was large. His wife, Mary Ann (granddaughter of Tomas Jones previously mentioned), had borne him ten children, and all had migrated to Wisconsin. The oldest, Caroline Georgianna, who later married Julius M. Wheeler, had been twenty-two when she came with the first group of her family to Wisconsin. William E., who was probably William Morton's favorite uncle, had been only thirteen, and had come with the second group. Until he was seventeen, William E. worked with his father and the rest of the family at what must have been extremely hard work: farming the rugged Wisconsin countryside. His experiences in the Civil War are not known, except that he was on General Sherman's famous march to the sea. Following the war, he first worked three years in a machine shop, then studied to become a teacher. After several years of teaching in Waukesha, Wisconsin, he became a principal there. From 1883 to 1892, he was superintendent of the schools in Milwaukee, an important position in a city that was proud of its public educational system.

Superintendent Anderson's administration in the schools was considered to be "a positive one," for during those nine years teachers' salaries were raised, several new school buildings went up, the teachers' association became more professional, and a number of interesting curriculum changes were made. Mr. Anderson was personally interested in natural science; he introduced experimental physics into the eighth grade, and wrote a textbook to help in promoting the course. He also reintroduced German, to make the schools once more bilingual, remarking that so long as there was no German spoken in the public schools, the German parents kept their children home and thus denied them a chance to learn English. In the 1943 annual report of the Superintendent of the Milwaukee Public Schools, entitled "Our Roots Grow Deep," the author said: "It is worth noting that Superintendent Anderson, who was not of German descent, made a defense of the inclusion of the teaching of the language on these realistic grounds and himself as an adult began his German studies with a private tutor."

Throughout his life, William E. Anderson never ceased his energetic interest in education and public service. To continue from the annual report: "After leaving the superintendent's office Mr. Anderson prepared and installed the Wisconsin Educational Exhibition at the Chicago World's Fair. In 1894 he was elected City Clerk of Milwaukee, and reelected in 1896. In 1898 he joined the staff of the State Superintendent of Schools and conducted teachers' institutes throughout the state. Later he organized a school supply business. He was the inventor of an improved blackboard paint, an adjustable desk, and a pendant school room globe. He served as President of the Wisconsin Educational Association and was a member of the Board of Regents of the Wisconsin Normal Schools."

Unlike his sister Carrie Wheeler, who lived to be seventy-seven, Will Anderson died in 1903, a comparatively young man of fifty-eight. But he had left a memorable imprint on the affairs of Milwaukee, as well as upon his namesake, Will Wheeler. In fact, young Will credited his uncle with first interesting him in nature lore when at the age of nine, he saw a spider ensnare a moth in its web.

The explanation of this incident by his nature-loving uncle made a deep and lasting impression on the boy.

Although the Andersons and the Wheelers were not of German origin, Milwaukee was primarily a transported German city, built mainly by the intellectual refugees from an unsuccessful rebellion against the German monarchies in 1848.[3] There had been a fairly cohesive German community in Milwaukee before the impact of the Forty-eighters, but the new arrivals produced an intellectual and cultural ferment, especially in the fields of music, education, and liberal social thought. The *Musikverein*, formed in 1850, encouraged music in all forms: instrumental and vocal, onstage or among local groups, and even in the local beer halls. Milwaukeeans loved music, operatic arias or drinking songs. Amateur theatricals, frequently followed by dancing, made the formation of an active professional theater inevitable. Public readings and speeches were well attended, and the leading names of the day graced Milwaukee's billboards as celebrities of every kind came to town: Ralph Waldo Emerson, James Russell Lowell, James Whitcomb Riley, and Ella Wheeler Wilcox, to name only a few. The six German-language newspapers kept readers informed not only of the lively events in town, but also in other states and abroad, especially Germany. The political ferment which was to bring the socialist Robert La Follette to power at the turn of the century was beginning to appear in the seventies and eighties. The people of this city of free-thinkers were searching for "the better life." In this atmosphere, William Morton Wheeler grew to manhood.

Because of an accident of birth in a city where education was important to its citizens, Will Wheeler was able to obtain one of the best formal educations available to any young person in the country, and one commensurate with his exceptional ability. Prior to the emigration of the Forty-eighters, the city schools were poor because Milwaukee had been composed primarily of young men without families and with small means. The Germans of 1848, however, had had good educations, and they wanted the same for their children. Thus, in 1851 the German-English Academy was founded to "grant to children, regardless of nationality and religious belief, an opportunity to acquire a thorough elementary and scientific education at a very small expense." Peter Engelmann was hired as principal. His qualifications were high, for he had been educated

The German-English Academy, Milwaukee, as it appeared when Wheeler attended. (Redrawn from a woodcut in *Engelmann's Schule: Seminar,* 1871)

in natural science at Heidelberg and at the University of Berlin. Because of his excellent academic record and in spite of his unorthodox religious views and liberal political ideas, he was given a teaching certificate. He taught in Germany for about six years, until his revolutionary activities finally made it necessary for him to flee for his life. He organized his Milwaukee school along the lines of the schools he had known in Germany, including a kindergarten in the elementary division, and scientific studies as well as classical subjects in high school.[4] Courses offered to the first high school class included higher equations, mechanics, astronomy, electricity, analytical chemistry, mineralogy, French (Moliere), Latin (Ovid and Orations of Cicero), Greek (Homer), history of the Middle Ages, history of German and English literature, translation into German and English of the best writers of both languages, drawings from nature, and German and English rhetoric. In addition, young ladies had fancy sewing, embroidery, and similar household arts. All good schools, especially if they are expanding, need more good teachers, and so a three-year teacher's seminary, as it was then called, was soon appended to the Academy.

As if the curriculum offerings were not enough, the "Engelmann Schule," thanks mainly to the inspiration of its principal, also acquired a natural history museum, and extensive collections were made so that object lessons in natural history and science could be given to the classes in the Academy. The collections, however, were not limited to the use of the students in the school. In 1871, in a report entitled, "The German and English Academy; its History and Present Condition," the following paragraph was included:

The Museum, in the new building of the German and English Academy, is without doubt one of the best in the Northwest. Owing to the untiring zeal and energy of its few founders, it has grown in a few years so rapidly, that its collection of specimens, in every department of Natural History now numbers 8,500 objects, all properly classified and well arranged. In the Zoological department are upward of 2,400 specimens; in the Mineralogical, about 2,600; the Botanical department numbers 1,500 objects; the Ethnological 1,600; and the Library contains 340 volumes. Particularly the departments of Ornithology and Mineralogy are well supplied with excellent specimens. There is also a fair collection of the North American quadrupeds, conspicuous among which is a large specimen of the Black Bear. The various other departments are also well filled with interesting and instructive objects. The Museum is open to the public twice a week, admission free, and affords a splendid opportunity to persons wishing to obtain instruction in Natural History, to do so without expense. The teachers of public and private schools throughout the city can take advantage of this arrangement, and by bringing their scholars to visit the Museum, awaken in their minds an interest in the study of Natural History.[5]

Also available, but not mentioned in the above paragraph, were the bimonthly meetings of The Natural History Society of Wisconsin, which had been founded with the Engelmann Museum. Lectures and discussions on a great variety of pertinent subjects were available to anyone interested in attending the meetings. The small group of enlightened men, mostly German, who belonged to the Society, then proceeded in 1882 to donate the collections and their accompanying library to the city.

Small wonder Henry A. Ward had great expectations for Milwaukee and its museum. He was not disappointed. When he returned to Rochester, not only was he better off financially (the citizens of Milwaukee had raised $12,000 to buy the Ward collections), but he also had the promise of the services of a very able worker, William Morton Wheeler. Ward had developed a genuine fondness for the boy as they had worked together in the Engelmann Museum, and consequently he had invited him to come and work for him in his Establishment. Will accepted, though not without regrets. How could he bear to leave the school where he had spent so much of his life? How could he part with his family and his friends, especially the Grobs, and particularly Elise? Sometimes Will must have wondered why he needed to wait, for he was not going to teach, so why should he finish the teacher training at the seminary? But, who could tell? The position with Ward might not work out, and then he would be glad he knew a little about teaching.

Late January was a sad time for the Wheeler family. Will's first diary (1884) began with this entry on Tuesday, January twenty-ninth: "Father departed this life, 11:30 at night." There is no other entry in January. And then five days of weeping relatives, sad farewells, and finally he escaped the bonds of childhood and home and became an employee of Ward's Natural Science Establishment.

Henry A. Ward (1834–1906), a dignified and commanding figure at fifty, had organized his Establishment in 1862, while he was teaching geology and related subjects at the University of Rochester. He had felt that American teachers were greatly handicapped without tangible representatives of extinct animals, and he went to Europe to make plaster casts of various important museum specimens there. Many of the original fossils were ones that he had previously collected and sold to museums and collectors all over Europe so that he might have enough money to finish his studies at the School of Mines in Paris. The demand for the plaster casts was enormous, and the illustrated, descriptive catalogues he made to advertise his casts were even used for textbooks by many teachers. Thus, Ward's little business on the side became his major concern, and in 1869 he gave up his professorship at the University. With Ward's full attention, the Establishment grew to meet the ever increasing demands. Some years later, one of the employees said in recollection of it: "In all this there is nothing that even suggests the curiosity shop or the dime museum. On double headed calves, monstrosities in general, and relics of all sorts, the law of the establishment has laid the grand taboo. There is enough to handle that is purely scientific and educational. The establishment consists of twelve separate and distinct scientific departments, housed in sixteen buildings, several of which are quite large. The working force usually consists of about twenty-five persons, the great majority of

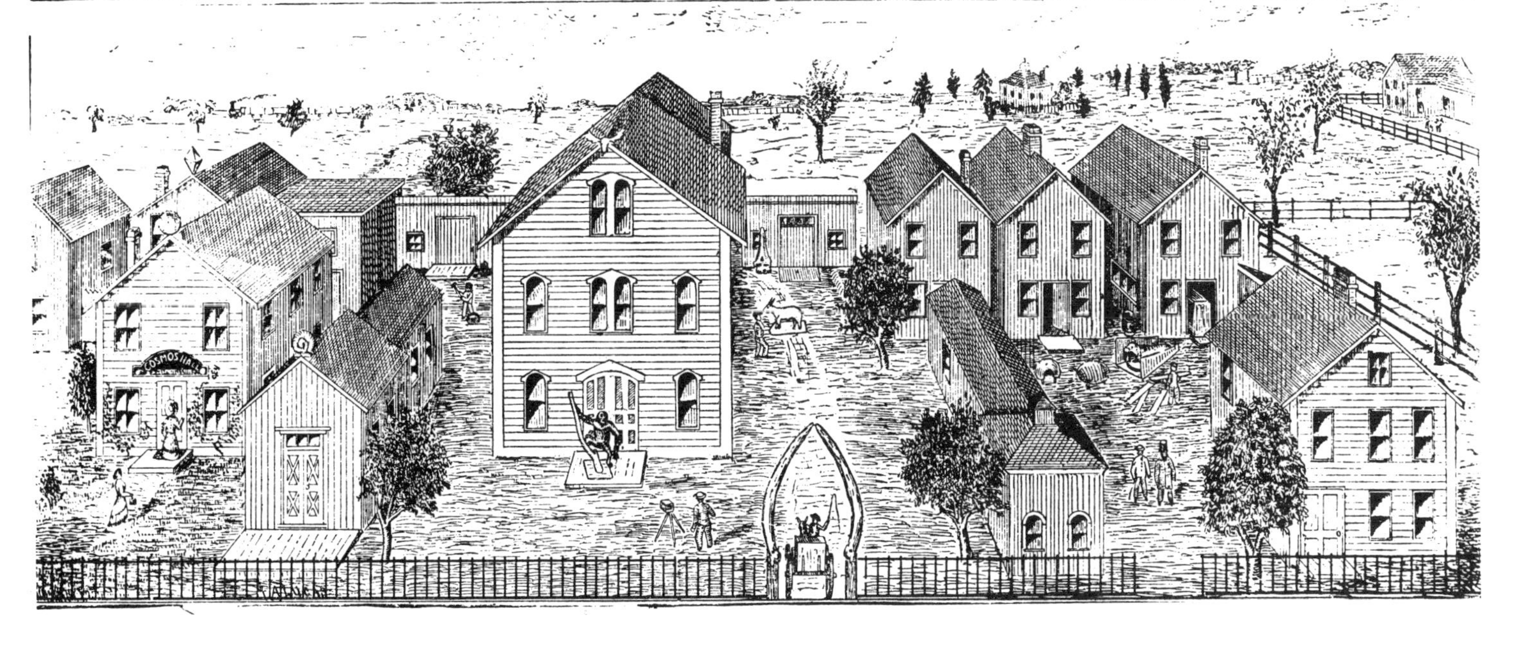

Ward's Natural Science Establishment, Rochester, in its early days. The gate is arched by the jawbones of a whale, and a stuffed gorilla stands on the threshold of the main building. This woodcut was used as the masthead of *Ward's Natural Science Bulletin* for many years. (Ward's Natural Science Establishment, Inc., Rochester, New York)

whom are trained experts. It requires nineteen printed catalogues . . . to adequately set forth to scientific students, educators and institutions the magnificent array of objects that are offered them for sale."[6] As previously mentioned, each building was devoted to one or more particular subjects; for example, the shell house, which contained shells, corals, starfishes, and such; the skeleton house; the building for skins; the geology building, and so forth.

Wheeler became a part of this unique business on February 7, 1884. Many years later (in 1927) he wrote of it: "My duties consisted in identifying, with the aid of a fair library, and listing birds and mammals. Later I was made a foreman and devoted most of my time to identifying and arranging the collections of shells, echinoderms, and sponges, and preparing catalogues and price lists of them for publication. Such is the present state of conchology that my shell catalogue is still used by collectors."[7]

Wheeler's arrival was noted in the *Bulletin* of the Establishment by the following comment: "Mr. William M. Wheeler, a young naturalist of merit from Milwaukee, has just joined our goodly ranks. We hereby intimate to him our intention to give him the editing of the next bulletin." This same number of the *Bulletin* also had an article on the new Milwaukee Public Museum.[8] Although the position of editor of the *Bulletin* apparently did not take up much of young Wheeler's time (its publication was a bit erratic in those days), he did have his first "scientific" paper published: "The Colugo and His Cousin." The date was July 1, 1884, and the paper described briefly some peculiar little East Indian mammals (sometimes also called flying lemurs, although they are not true lemurs). It did not represent any original research, and Wheeler never alluded to it in later life; nevertheless, it was the beginning of a long and outstanding career, which included a remarkable ability to gather pertinent facts and put them together into an informative and interesting article or book. He began in this way: "Glancing at the long series of living things, we frequently encounter species, which, besides having all the characteristics of their own order, possess other peculiarities observed only in widely different divisions of the animal kingdom. Thus the sea-horse (*Hippocampus*) combines some of the attributes of a monkey, a fish, a marsupial and a horse. It has the prehensile tail so common among the lesser Quadruma, the internal, and to a limited extent, the external structure of a fish. Its young are developed in a pouch as with the marsupials and its head has a decidedly equine appearance. Less peculiar, but no less interesting creatures, are the *Galeopitheci*."

Although Wheeler began his work at the Establishment with pleasure, he soon discovered, as young people often do in their first time away from home, that a new job is seldom ideal. Professor Ward, whom he so greatly admired, spent more time traveling in almost every corner of the world than he did at home in Rochester. "When other men are able to supply the specimens for one small department of a new scientific museum, his vast establishment can fill the entire museum, from the lowest depths of geology up to man himself, with every department reasonably complete." Such a promise required constant replenishments in the stock of the Establishment, and so Ward collected almost constantly.[9] On May 19, Wheeler noted in his diary that nineteen chests arrived from the Professor from Europe. Wheeler would have preferred Ward himself, of course, and always noted in his diary when the Professor arrived, adding, "He received me kindly."

Wheeler's other major problem was that he was homesick. He missed his mother and his sisters. He missed Lisa Grob and Mrs. Grob, and probably even Lisa's little sister, Amalie. Minnie, who was only two years younger, dutifully wrote to her brother, and Lisa and her mother were faithful correspondents. He answered all their letters immediately, and during the long evenings also wrote to many others, especially his old school friends. Lisa sent her picture, as well as a box of edible goodies; his family sent him some of his favorite books. Gradually, however, Wheeler became better acquainted with the other young men who worked around him. The first acquaintance was the Professor's younger son, Henry L. They had neighboring bedrooms and often chatted during the evenings. Later, Will tutored Henry in Latin. And at work, in spite of the dictum in the courtyard, the young men enjoyed many informal times together. They gossipped off and on during the day, and evenings and weekends went together to plays, musicals, dances, swimming, boat rides, and even various church services.

Rochester, in the mid-eighties, was enjoying a renaissance in its cultural and intellectual life. Previously conservative in its tastes for the arts and science, the city seemed to be rapidly awakening to the joys and pleasures which more advanced cities, such as Milwaukee, had known for many years. Various road companies, brought to Rochester almost under protest, met with surprising success (such as had Sarah Bernhardt in *Camille* in 1881, and the productions of Gilbert and Sullivan's musicals). The low prices of the summer theater brought out great crowds. Two theaters opened in the fall of 1884 offering for ten cents girls dressed in flesh-colored tights posing as nude paintings. These Dime Museums, as they were called, did not last long because they could not get permission to sell liquor, and they could not be said to have raised the city's cultural level (young Wheeler called them disgusting!), but they did contribute something. They brought down the prices at the other theaters, and that *was* appreciated by Wheeler and many others.[10]

Another aspect of Rochester's growth, which interested Wheeler and probably every other worker at the Establishment, including Professor Ward himself, was the rising debate between science and theology. This revolt from orthodoxy was led, not by any of the scientists at the University of Rochester or elsewhere, but by a local Unitarian minister, Dr. Newton M. Mann. He is thought to have been the first cleric in America to grasp the implications of Darwinism (in 1861). In essence, he said that all of life's functions were subject to evolution, and as man became increasingly conscious of his spiritual life, the evolution of his soul occurred. In 1885 Dr. Mann was still a dynamic speaker for his cause, and Wheeler wrote in his diary: "Took a walk with Carl [Akeley] and then went to church with him to hear Doctor Mann give a magnificent sermon." Mann not only pleased many laymen, he also convinced some of the other ministers in town—one of whom was eventually tried for heresy by his own church.[11]

There was much good fellowship at Ward's, and Wheeler enjoyed it just as did the other young men. But unlike most of the other young men, Will Wheeler did not devote himself solely to playing away his newly found freedom. He studied harder than he ever had in school, if that was possible. He studied algebra and geometry, and attended the meetings of the Geological Section of the Rochester Academy of Sciences (usually held at the University). In his 1884 diary, he listed seventy or so books which he had read that year. These included *The Iliad, Don Quixote,* various books by Herodotus and Rabelais, *Sartor Resartus,* Spenser's *Essay on the Nebular Hypothesis,* Ord's *Anatomy,* Huxley's *Anatomy of the Invertebrates,* and many others both scientific and classical. Furthermore, most were read in the language in which they were originally written. Much of Wheeler's reading was to increase his knowledge about

William Morton Wheeler at about the age of twenty. (Photograph taken in Milwaukee, about 1885)

the specimens with which he was working and to improve his handling of them. It was at Ward's that Wheeler seems to have first learned how to handle a microscope—through reading works on the microscope, and then by spending several evenings with the local society of microscopists. Incidentally, the Microscopical Section of the Academy of Science, led by Edward Bausch of the Bausch and Lomb Optical Company, was the largest and most active organization of its kind in the country.

Of the many new acquaintances made by Wheeler in Rochester, two men became especially close to him. One was Shelley G. Crump, who ran a prosperous grocery business in nearby Pittsford, but who was an amateur naturalist with a special interest in shells. The other was a young taxidermist, Carl Akeley, of whom we shall say more later. Crump and Wheeler met in the shell house, and they were soon spending much of their spare time together reading and studying shells and other natural history subjects. On Sunday, September 28, 1884, Wheeler wrote in his diary that he began the catalogue of shells at Mr. Crump's. Mr. Crump seems to have been a delightful person who not only found great personal satisfaction in his studies of the natural world, but who inspired and encouraged others to do so, too. In his diary (and 1927 paper), Wheeler described one of the weekends they spent together:

Sunday, May 23, 1885. From 10 to 12 worked with Professor Ward in the shell-house, labelling Echini—the last time I saw him [for many years]. In the afternoon Mr. Crump and his friend Dr. Dunning called on me. I walked with them to Brighton and thence took a train to Pittsford. We read together some recent papers on Pasteur by Tyndall and others and then walked along the Erie Canal bank where I collected two species of Valvata.

Monday, May 24, 1885. Rose late. Read some of Burrough's "Wake Robin" before breakfast. Then conversed with Dr. Dunning on Shakespeare's "Sonnets" [Dr. D. was blind and with the aid of his wife was preparing a volume on the sonnets]. At 9:20 took the train for Rochester, went to work in the shell-house, finishing the family Nassidae and part of the Volutidae.

The official list of publications of William Morton Wheeler begins in 1885 with the "Catalogue of Specimens of Mollusca and Brachiopoda for Sale at Ward's Natural Science Establishment." It includes "A List of Trees Found in the City of Milwaukee," although this was not actually published until 1886 in the *Proceedings* of the Wisconsin Pharmaceutical Association. Of how the latter paper came to be, we can find no explanation in Wheeler's writings. It is a continuation of an earlier list made by someone else, and one can probably safely assume that the floral study had been made while Wheeler was still a student at the German-English Academy or Seminary.

While Shelley G. Crump was an inspiration and guiding hand for young Wheeler, in Carl Akeley Will found a kindred spirit. "Of active, industrious young men," wrote Wheeler in later years, "there seem to be two types. One of them accepts a given environment and is not only satisfied with its routine and constantly recurring human contacts but prefers it to any change. These young men are apt to marry early and to become the conservative and contented *fond* of our society. Those of the other type, probably endowed with a more unstable if not more vivid imagination and with a peculiar defense reaction, or subconscious dread of being owned by people and things, soon exhaust the possibilities of their medium, like fungi that burn out their substratum, and become dissatisfied and restless till they can implant themselves in fresh conditions of growth. Akeley and I were of this latter type." Similar in temperament they may have been, but their backgrounds and general interests could hardly have been more diverse. Carl and his brother, Lewis, had grown up on a sixty-acre farm near Rochester. Carl was not fond of school, but he was fascinated with taxidermy. No one seems to know how such an interest had come about. He had studied and experimented on his own, and his one aim at nineteen was to be allowed to work for Professor Ward as a taxidermist. The privilege was granted at a weekly salary of $3.50, while his room and board cost $4.00 a week. It was hardly a paying proposition, except perhaps to understanding parents. From Wheeler's paper on Carl Akeley's early work we read: "He attached himself to William Critchley, a young and enthusiastic artisan, with the voice and physique of an Italian opera tenor, who had attained the highest proficiency in the taxidermic methods of the time, but did not seem to give promise of advancing the art. In the course of a year Akeley had more than mastered all that Critchley could teach him, and was longing for wider opportunities than could be offered by an establishment, which, after all, was neither an art school nor a scientific laboratory, but a business venture."

The two young friends, Carl and Will, soon outgrew Ward's Establishment, but not until after having many long, enjoyable evenings together. While Carl worked quietly, Will often read aloud, something he loved to do and had often done with Mrs. Grob and her daughters back in Milwaukee. Most of the reading was nonfiction and pretty heavy (apparently Will's choices), and Carl played the passive listener. Once, however, Will was interrupted in his reading of *Anna Karenina* by the need to make some drawings, and Carl took over the reading of this exciting novel. Many years later, after Carl's death, his brother, Lewis, wrote of their evenings together: "Wheeler was a big factor in Carl's early life. When people ask me how much education Carl had I tell them that he attended daily lectures by William Morton Wheeler for several years. I can remember the days when Wheeler and Carl worked in the same room . . . While their hands were busy their tongues were going. Wheeler's talk ranged over the whole field of science, and the whole range of human history. I never knew so versatile a man and I never heard or read a profounder thinker. And Carl had a mind of such quick and penetrating insight that he grasped everything. Any student of today who could have four years of such contact with Wheeler would consider that he had the most valuable kind of a post-graduate course."[12]

In the early spring of 1885, after a year at the Establishment, Will and Carl began discussing possible opportunities elsewhere. Will wanted to go back to Milwaukee, and he urged Carl to come, too, promising to tutor him as preparation for entering Sheffield Scientific School at Yale. On April 27, Wheeler received a letter from the Professor, who "acquiesced in my leaving the first of July." Carl expected to join Will in the fall, but an unexpected event kept him at Ward's for another year. That event was, of course, the death of Jumbo, P. T. Barnum's world-renowned elephant, in a tragic and bloody collision with a locomotive. Barnum desperately wanted to salvage Jumbo's hide and skeleton for exhibition, and what greater challenge could there possibly have been for a young taxidermist? But a year later, with Jumbo safely preserved for posterity, Akeley joined Wheeler in Milwaukee.

2 / The World of Science Opens

Milwaukee was still there, and still beautiful and promising to a young man who was hopeful of fulfilling his dreams and ambitions. Perhaps he, like so many others, felt a personal rapport with A. H. Bielfeld's poem, "Die Milwaukee Bai":

Erwacht, ihr Schläfer, aus dem Träumen!
Auf, schaut die Bai Milwaukee's an!
Die Vögelzwisschern auf den Bäumen—
Erwacht!—Die Sonne bricht sich Bahn.
Am Ufer deines Sees zu steh'n,
Und Dich nicht lieben, Deutsch Athen,
Wer das zu Stande bringen kann,
Ist ein beklagenswerther Mann.

Die Sonne sinkt.—Und traulich wieder
Beginnt der Mond die stille Bahn.
O, Phantasie, Du Gott der Lieder,
Verzeihe Du den süssen Wahn,
Wenn ich so denk' in meinem Sinn
Wohl über hundert Jahre hin,
Dann möcht ich wieder aufersteh'n,
Um die Milwaukee Bai zu seh'n.[1]

William Morton Wheeler was glad to be home. There was still the lure of the miles of shore and the deep blue waters of Lake Michigan. He liked to stand on the bluffs and watch the ships—

freighters, excursion steamers, little sailboats—come and go in the harbor. He often walked along the shore, or to the Menomonee Valley, or in Mann's woods near Bay View to collect beetles, which, as of old, he and Mr. Rauterberg would study at the museum. He had not seen Lisa since a brief trip home a year earlier, when he had bid her an affectionate farewell in the moonlight on the shore of the Lake. She had gone to teach in Texas, but was due home shortly. And then she was there for six long weeks, and life seemed full.

Now to do something about finding a position he liked. Prospects looked good. Uncle Will Anderson, then superintendent of schools, offered him a job in Sheboygan in July, but Will Wheeler was not about to leave Milwaukee so soon again. Dr. Heinrich Dorner, who had been his favorite science teacher at the academy and his mentor in the school museum, was interested in having his star pupil as one of his teachers. Thus in early September of 1885, Wheeler began teaching Natural Science and giving drawing lessons at the German-English Academy. Apparently he did not feel that his time was fully taken up, however, for three weeks later he also began to teach at the public high school. His subjects in the high school were biology (called physiology) and German.

Wheeler's tie with the high school was not through his Uncle Will, as one might suspect, but through a scientific friend who had long been an inspiration to the young man. This was George W. Peckham. Dr. Peckham (1845–1914) was one of those gifted men of the previous century who seemed to be able to do almost anything. He has been called "patriot, educator, scholar and scientist." He enlisted as a private during the Civil War and came out a lieutenant, having been commissioned and put in charge of a battery in an artillery regiment while he was still only nineteen. After the war he went back to school, receiving a law degree from the Albany Law School in Albany, New York, and then a medical degree from the University of Michigan. He was also an "amateur aeronaut," and made several balloon ascensions in and around Milwaukee during the 1860s and 1870s. Peckham never practiced law or medicine, as far as we know, but began his professional career in 1871 as a biology teacher in the public high school in Milwaukee. Later he was made principal, and in 1892 was elected superintendent of schools, succeeding William E. Anderson. Of the new superintendent, the *Milwaukee Journal* had the following to say:

Dr. George Peckham is the son of an early Milwaukee family and is kin to many of the old families of the city. Nearly everyone knows him personally, and those who do not, know him "by sight." In appearance, he is a character—of medium height with stooped shoulders he carries his head forward, wears gold bowed spectacles, and a hat several times too large for his head. His yellowish hair has been the least bit touched with gray. He is the ideal "old professor" so far as looks goes, though as a matter of fact, he is a young man in years and feeling. Sometimes he is induced to put on a dress-suit and then he is seen to be a handsome man. His manner is more than amiable, it is thoroughly friendly. He likes people, and people know it, and consequently he is a much liked man. That enables him to speak his mind frankly without giving offense. He thinks things, has pretty positive opinions on most subjects of human interest, expresses them freely, and effectively, and without incurring the resentment that almost anybody else would suffer. Nobody can spend ten minutes with him without saying that he is an unusual man, a good thinker, a close observer, and companionable in the highest degree. His pupils love him and perhaps the best work he has done in the schools has been to inspire, through his personal influence, an ambition for thoroughness. There are more young men and women in town who credit their success to his friendly interest and good advice than anybody can count in a day. He has grown up in this community and in the days when young men were doing things for the town he was "with them." He was active in the management of the Young Men's Association whose library was the substantial foundation of the present public library, and he married the most accomplished and charming librarian that ever was.

The article continued with a brief description of some of Dr. Peckham's scientific work on spiders, his support of Darwin's theories on evolution, and then a bit about his political views: "Of course Dr. Peckham thinks and says things about politics. He is a Democrat, a free-trader, a civil-service reformer and a Cleveland man. He circulated a petition, after Cleveland's inauguration, urging the retention of the Republican collector of internal revenue in this district, and was absolutely confident that Cleveland would not remove an efficient incumbent to make room for Mr. Wall. The Republican was removed, however, with promptness, but Dr. Peckham's confidence in the purity of Cleveland's 'intentions' never forsook him."[2]

Peckham was superintendent for a little more than six years, and during that time he not only worked on the curriculum, but he also, especially during the early years of his administration, devoted a great deal of time to improving the physical conditions that he considered essential for good work: ventilation, sanitation, and comfortable surroundings for the children. Many considered this a remarkable concern for a man mainly interested in theoretical matters. Dr. Peckham's love of books finally won out, however, and he resigned as superintendent to become director of the Public Library. But this was a position also concerned with public education, and he continued to work with the schools and the Public Museum until his retirement in 1910.

Dr. Peckham's scientific work was done during his leisure hours and in the company of his wife and three children. In 1880, he had married Miss Elizabeth M. Gifford, the previously mentioned librarian from the Young Men's Association who had graduated from Vassar College in 1876. Together they produced several very meticulous, yet very readable monographs, which are still studied and admired by scientists today.[3]

Peckham belonged to most of the scientific organizations of the day, and it must have been through the Wisconsin Natural History Society, which met at the German-English Academy, that William Morton became acquainted with him. Of him Wheeler later said:

Soon after my return to Milwaukee my old friend, Dr. George W. Peckham, who had long been making important contributions to arachnology and was beginning his well-known studies on the behavior of the solitary and social wasps, persuaded me to take a position as a teacher of German and physiology in the high school of which he was principal. Peckham was a very learned and charming man, deeply steeped in the evolutionary literature of the time and keenly alive to the possibilities of the new morphology that had been inaugurated by Huxley in England and a host of remarkable investigators in the laboratories of the German universities. Every year he most conscientiously read, as a devout priest might read his breviary, Darwin's *Origin* and *Animals and Plants under Domestication*. We became very intimate, and I find from my diaries that for some years I regularly spent my Sunday mornings in his house drawing the palpi and epigyna of spiders to illustrate the papers which he wrote in collaboration with his equally gifted and charming wife. I was privileged to collaborate with them in one paper (on the Lyssomanae) and to help them during the summers in their field work on the wasps at Pine Lake, Wisconsin. Under Peckham's management the biological work of the Milwaukee high school was carried far beyond that of any similar institution in the country. There were classes in embryology, with Foster as a text. We possessed a Jung microtome and the paraphernalia for staining sections and demonstrating the development of the chick, and, of course, the classes in physiology were required to master Huxley and Martin.[4]

We wish we knew more about this kind and scholarly man and his sympathetic wife, but at least we do know a little. About Dr. Heinrich Dorner we know almost nothing. He was a member of the botanical section of the Wisconsin Natural History Society, and seemingly a successful teacher. Apparently he wrote a textbook, or perhaps a workbook, on physiology, which was "embellished with anecdotes inculcating good hygienic habits." Wheeler, now an experienced man of twenty, must have felt, in comparing the two men, that teaching under Dr. Peckham was a greater challenge, for he resigned his position at the Academy at the end of November and gave his undivided attention to teaching in the public high school.

The Wisconsin Natural History Society met twice a month, and members gave lectures or listened to lectures by others. In the early days these were usually in German, as was each follow-up discussion. Generally the papers were later pubished in the *Proceedings* of the Society. Professor Peckham, as well as Dr. Dorner, participated in the Society's activities, and, of course, young Wheeler joined as soon as he was asked. He was formally elected to membership on November 14, 1885, and three weeks later he was elected corresponding secretary. With these honors, he immediately began to prepare his first paper for presentation to the Society. It was about beetles, a group he had been collecting and studying for many years under the able guidance of Herr Frederick Rauterberg (a mail carrier by vocation, a coleopterist by inclination), who had been an active member of the Deutscher Naturhistorischer Verein von Wisconsin before William Morton was born. When Wheeler had left home to go to Rochester, he had given his beetle collection of 2,600 specimens to the new Public Museum, and since that time he had contributed more beetles as well as his plant collections and other miscellaneous natural history items. As any scientist who works with biological specimens knows, collections are far more valuable when put in a reputable museum than if they are stored away to gather dust and decay in an unused garret.

Wheeler's paper on beetles was called "On the Distribution of Coleoptera Along the Lake Michigan Beach of Milwaukee County," and it was given (in English) on the evening of January 11, 1886, though not published until April 1887. From the title one might surmise that it was merely a list, but it was more than that, for Wheeler raised a number of questions regarding the distribution and abundance of these beetles. His paper begins:

Those who have frequently strolled along the Lake-beach north of our city, cannot fail to have noticed a long, almost unbroken, sinuous band of drift-wood and debris of every description extending along the sand. The more careful observer will also have seen lying here and there on the beach or clinging languidly to pieces of refuse a great number of insects of all orders. These too, he will conjecture, have been cast up by the water. Should the same observer have pushed aside a part of the drift-wood, he will have seen many active beetles and larvae run over damp sand which was previously protected from the sun's rays by the fragments of wood. Why are these insects under this isolated belt of chips? is a question, which naturally presents itself. These were phenomena which deeply interested me when, years ago, I began the study of insects. A few years of careful collecting among these piles of refuse revealed a great number of rare and beautiful Coleoptera. I then found at least 1,000 species of this one order, and my friend, Mr. Rauterberg, who had collected for many years previous, obtained several species which I have been unable to procure. Surely this is a remarkable littoral distribution. To what degree is this distribution accidental, i.e. to what extent influenced by the action of the lake water? What are the causes of this accidental distribution? To what degree is this distribution natural? These are the questions which I shall attempt to answer in the following paper.

Wheeler then proceeded to answer the questions he had raised, and especially perceptive was his paragraph, not far from the end of his paper, on some man-made problems affecting natural distribution:

Thus far we have considered only the littoral distribution of Coleoptera in space. It remains for us to consider this distribution in time. My attention was first directed to the distribution of Coleoptera in 1879. Mr. Rauterberg, whose accuracy as an observer and whose pains and diligence as a collector are well known to us all, states that previous to this date the number of Coleoptera found on the beach, even within the city limits, on that part of the shore since improved by the railroad, was truly wonderful. He had told me frequently, that he has filled bottle after bottle in a few minutes, that he could almost shovel the specimens up. From 1879 to 1883 we both observed the distribution and have noticed a steady decline in the number of specimens during these five years. Being absent from town during 1884, I was unable to continue my observations, but Mr. Rauterberg says that he could obtain but very few beetles. During the last summer, 1885, repeated visits to the beaches, which in former years had yielded such a rich harvest, were rewarded by nothing. How are we to explain this? We cannot consider this problem without regarding the distribution in time throughout the county. There seems to have been a slight decrease in the number of Coleoptera inland, but a decrease scarcely sufficient to explain the abrupt falling off in the littoral distribution. We are therefore obliged to ask the question: Can the building of embankments and piers of late years have anything to do with this problem? This question I am unable to answer. More lengthy and more thorough observations on the part of our county naturalists are required, both on this point and on many others touching this distribution. Perhaps an examination of the beach south of the city would aid in solving some of these problems.

In the spring of 1886, Wheeler continued much as he had the preceding fall. Daytimes he taught in the high school, evenings he often spent with Mrs. Grob and Amalie, either quietly reading at their home or sometimes at the Stadt Theater. Other evenings he spent playing cards with some of the other young lads, as he did to celebrate his twenty-first birthday. He took many walks—often botanizing in the daytime, but just exercising at night. His night walking, however, got him into trouble. On Tuesday evening, May 18, he was attacked by a gang of roughs, who blackened his eyes and gashed his cheeks. As a consequence, he was prevented from teaching for a week. Although he never indicated that he knew (or cared about) the cause of this beating, he apparently was a victim of labor riots, and there was no personal grudge involved.

Chicago and Milwaukee are only about ninety miles apart, and the history of their growth and development has been similar; although Chicago gradually won, there was some question in the early days as to which city would take the lead in economic growth. In the 1870s the working man, even when he could find a full time job, made barely enough money to feed and clothe his family. Hours were long, wages low, and working conditions usually miserable. A depression in 1879 and again in 1884 put even more men out of work. The leaders of industry continued to add to their wealth and to resist the attempts to form labor unions by using strike breakers and police to break up labor protests. Finally, on Tuesday evening, May 4, 1886, just a week before Wheeler was beaten, Chicago experienced the infamous Haymarket Riot, in which police and spectators alike were killed or hurt, either by the initial bomb thrown or by the ensuing bullets. The Riot was a disaster, and the activities of the so-called lawmakers which followed in the next few years were not much better, but out of these troubled times came the beginnings of reform for the working class.

Labor troubles seemed never to have touched William Morton very much except for that one unfortunate episode. In July his salary was raised to $800 per year. He fixed up a small laboratory at home to use during the summer. Part of August was spent with the Peckham family at their camp on Pine Lake. Wheeler was beginning to think more about his future, though, and made some tentative plans to go to Johns Hopkins University in 1887. But in November two events occurred which were to influence his future in ways which he did not suspect. Carl Akeley arrived from Rochester to take up his work and studies in Milwaukee, and Wheeler gave a talk to the Natural History Society. His talk, which he illustrated with drawings, included a résumé of the development of histology as a science and brief descriptions of the morphology of animal and vegetable cells. After his talk, as he noted in his diary, he "met Mr. Whitman, the embryologist, and a Mr. Patten." And at this point, for the second time, the magical wand of Louis Agassiz, the great naturalist and teacher, touched William Morton Wheeler. First it had been through Henry A. Ward, and now through Charles Otis Whitman.

Whitman, then a young high school teacher, had been one of the students at Penikese in 1873. After his death in 1910, it was said of him that "probably no teacher in zoology since Louis Agassiz has exerted so great an influence on young men." No other praise would have pleased him more. Whitman, like most of the young biologists in the 1800s, had gone to Europe to study, especially with Rudolf Leuckart, the famous zoologist at Leipzig, Germany. As Wheeler was later to do, Whitman also spent some time at the new biological station at Naples, Italy. In 1882 Whitman came back from Europe and the Orient (for two years he had taught zoology at the Imperial

University of Tokyo, Japan) to assist Alexander Agassiz for four years. Then he accepted a position as director of a new research laboratory in Milwaukee. This was the Allis Lake Laboratory.[5]

In 1876, when George W. Peckham was elected to membership in the Natural History Society of Wisconsin, another young man, Edward P. Allis, Jr., was also voted in. Edward P. Allis, Sr., was founder and leading proprietor of the Reliance Iron Works, one of Milwaukee's major industries, which later became part of Allis-Chalmers, noted mainly as manufacturers of farm machinery. Edward, Jr. (1851–1947) had become a success in his father's business, but a childhood interest in things biological never left him. He attended the meetings of the Natural History Society when he could; and then in 1883, twelve years after his graduation from M.I.T., when his business success provided him with a little more leisure time from the factory, he began to make plans to study biology seriously. First, he hired a young graduate student from Johns Hopkins University to come and tutor him in fundamentals. This young man was H. V. Wilson, who is now remembered for his classical studies on the regeneration of sponges (he devised experiments whereby he separated the cells of the little animals by pushing them through fine sieves, and then watched them regroup and reorganize themselves into whole animals again.)[6]

For a year Wilson and Allis worked together, whenever Allis could get away from the Company. After that time, Wilson went back to graduate school, convinced that he preferred pure science to medical studies. Allis then undertook to establish his own research laboratory. Although small and short-lived, the Allis Lake Laboratory was a pioneering institution of its kind. As originally conceived, we have Allis's own words:

> My plan was to establish a research laboratory on a business basis, that is, to engage at first one or two investigators to select some special or general problem; and then, with the necessary trained and specialized assistants, to attack the problem as a body and not as individuals. I felt certain that young men and women belonging to the mechanical and clerical classes could be found who would willingly undertake to do what I had come to consider, in my year's course of instruction under Mr. Wilson, as the purely mechanical part of the laboratory; and I felt equally certain that such trained assistants would do the work better, quicker and much more economically for the laboratory than professional men capable of research. My plan was, in fact, to apply to a

scientific laboratory what is now called in business matters "scientific management"; and as I did not expect for some time to be able myself to take any active part in the research work proper, my role was to be limited to that of business manager and general assistant. I expected to profit by association with the work going on.

But Allis's plans changed, because the man he was considering as director for his laboratory was Charles Otis Whitman, who had other ideas. In 1884 in Cambridge, Allis had an interview with Whitman, which Allis described as follows:

Whitman was immediately greatly interested in the project; but he thought that a laboratory devoted wholly to research, such as I had planned, could not succeed. Instruction of some sort, either by lectures or by tables offered to students, should, in his opinion, be combined with the other work of the laboratory. The business principles and methods proposed by me also did not appeal to him, excepting as they were limited to the maintenance of an ample supply of the sinews of war. Furthermore, he thought that an investigator should have absolute freedom of thought and action, even his general line of research not being in any way restricted or restrained. He also thought that no one but a competent investigator could properly prepare the material for investigation, who could properly do the somewhat mechanical work of dissecting, staining, sectioning, etc.; and that no one competent to undertake investigation could or would be limited to such work. In this I differed radically from Whitman but yielded to his opinion, insisting merely that someone would necessarily have to do most of such work for me, if I ever undertook research of any kind.

"If you ever undertake research of any kind," Whitman repeated, "why, you will begin such work at once."

Allis and Whitman continued their discussion as to possible organization of the laboratory, and Whitman said that although each worker would be free to investigate as he wished, when he published his findings he would give full credit to the Laboratory for the work done there. To continue in Allis's words:

I then asked how such work would be published and Whitman said it was usually sent to scientific periodicals or presented at the proceedings of scientific societies, but that sometimes it was published in a bulletin.

"In a bulletin!" I said. "Why, how can such work be published in a bulletin?," a bulletin being to me some sort of public and official notice of things interesting the general public.

Then Whitman explained that a bulletin would be the publication from time to time of the work done in the laboratory and asked me if I had considered such a publication.

I answered, "No," and added, "What would we do with a bulletin after it had been published?"

"Why, sell it, of course," he said, "unless you wished to distribute it free."

Here was something new and unexpected. Publish a bulletin and then offer it for sale! Not only let but even ask the general public to take cognizance of and interest in what was going on and being done in what I had considered up to that time as a wholly and strictly private and almost personal affair. A vague alarm began to take possession of me, and I felt myself embarking on a venture that seemed likely to lead me I knew not where.

"But," I said, "could such a bulletin have any value? I don't see what could be published in it that would induce any one to purchase it."

"There would not be much at first, of course," said Whitman. "But there would always be the work of those in the laboratory and then your own work."

"My work?" I said, still more alarmed. "But I shan't publish anything about my work."

"What will you do with it then?" said Whitman.

"Why, leave it, when I have got through with it, and take up something else."

It was now Whitman's turn to be surprised; and he said, "But if you are not going to publish your work, what are you doing it for?"

"For the love of it," I said. "I want something to do and some interest outside of business, and I have no wish to make it public in any way."

"But," he said, "if you do good work it will be a positive wrong not to give others the benefit of it."[7]

Allis went on to protest that he could not write up his research, even if he wanted to, because he had always had trouble, even in school, with composition. However, Whitman persuaded Allis to begin his research immediately, and Allis went on to publish his findings throughout his life (a total of eighty-one papers are listed in his bibliography). After hearing Whitman's explanation of the need for a good scientific journal in America, he helped organize and finance the *Journal of Morphology,* which is even today a leading journal in its field.

Unhappily, Allis, because of ill health, found it necessary to leave Milwaukee in 1889, and he took up residence in Mentone, France. During World War II his home and laboratory there were bombed and looted, and he had to flee to Nice. Nevertheless, despite privations, old age, and failing eyesight, he continued his scientific studies almost until his death in his ninety-fifth year.

In 1889, however, in the second volume of the *Journal of Morphology,* Allis published his first scientific paper, called "The Anatomy and Development of the Lateral Line System of *Amia calva.*" This paper, only five pages long, cannot begin to convey to the reader the many problems, often amusing as told later by Allis, which had to be overcome by Whitman and his pupil to complete the research. First, Allis had to learn what a bowfin or *Amia* was scientifically (although as a fisherman he was aware of its evil reputation); then he had to find out where and what was a lateral line system; and finally he had to procure his specimens to study. His studies were interrupted many times, particularly during the labor riots, when he could not leave his business for several days. But Allis persisted, and when he finally proceeded to the fine microscopical phase of his study, he was helped by a new laboratory recruit, William Patten.

Dr. Patten, who was later professor of zoology at Dartmouth College, is best remembered for his studies of ostracoderms (small, armored fossil fishes considered to be the earliest known vertebrates) and for his arachnid theory on the origin of vertebrates. It is interesting to note that although Patten still had his ostracoderm work in front of him when he came to Milwaukee in 1886, at the age of twenty-five, he had been studying insect development in Europe (his doctoral thesis under Leuckart at Leipzig dealt with "The Development of Phryganids, with a Preliminary Note on the Development of *Blatta germanica*"), and while at the Allis Lake Laboratory began to formulate his unusual theory on vertebrate origins. His first paper on the subject appeared in 1890.[8]

The meeting between William Morton Wheeler and "Mr. Whitman, the embryologist, and a Mr. Patten" proved to be a turning point in Wheeler's life. Early in 1887 Wheeler read Whitman's book *Methods of Research in Microscopical Anatomy and Embryology,* and within a month Dr. Whitman had made some tentative suggestions for a course of embryological studies that Wheeler might follow in his spare time from teaching. For the rest of the spring

Wheeler collected various developing eggs, made a few microscopic sections to study, and read voluminously, especially on insect and spider sense organs. And finally, on June 29, as Wheeler recorded in his diary, "Dr. Patten kindly helped me begin the embryology of *Blatta.*" Patten and Wheeler went their separate ways in 1890, but during the three years of their association, they seemed to have shared many profitable moments together, not only over the cockroach, but also at concerts and in stimulating evenings of discussion, with or without Whitman, the Peckhams, and other intellectual members of the community.

In 1887 Wheeler's world was changed in another way, too. Not only was he introduced to the world of scientific research, as practiced in the laboratory under the careful rules of the "new morphology," but he left the "narrow walls of a school room" to become director of a public museum. The museum was still housed in the Exposition Building, for there was no special building yet for the Public Museum. There he set to work immediately to improve the displays and to organize (with encouragement from the superintendent of schools, Uncle Will Anderson) lectures at the Museum

The Milwaukee Exposition Building, which housed the Public Museum during Wheeler's tenure as director. (Milwaukee Public Library)

for interested school classes. Thus, a happy balance of research, teaching, and museum work enabled Will Wheeler to make his last few years in his home town memorable ones.

The Museum ought not to appear [wrote Wheeler], as it doubtless does to many people, as a depository for cold and lifeless specimens intermingled with cards full of Latin and Greek names, nor as a place to which one can resort to spend an hour in idly looking at what is merely marvelous or beautiful, but the Museum should be a place where everyone can feel at home in learning something about the material objects of the wonderful planet on which we live. But the extracting of knowledge from objects requires skill in observing. So important is such skill that it is no exaggeration to say that the man who observes the things about him with the greatest accuracy will lead the most successful life, whether he be artisan or scientist. The ability of obtaining through the senses an accurate knowledge of the properties of material bodies is indispensable to every useful member of the community at the present day, and I feel confident that professors of paedagogy will agree with me that no better material can be found for training the powers of observation than natural history specimens. Such training must be one of the prime objects of our common schools.[9]

Though Peter Engelmann died while Wheeler was still a child, doubtless he would have been proud of this understanding young man who was a product of his school. Wheeler, at the end of his first year as custodian (that is, director) of the Milwaukee Public Museum, attempted to sum up the progress of his work and his hopes for the museum. He noted that the exhibits were crowded, and he tried to improve them by putting skins and skeletons of various animals together, by adding larger and simpler labels whenever feasible, and by overhauling the alcoholic and dried specimens. He also gave frequent talks to various classes from the public schools. Of course, he was also concerned with enlarging the collections, either by his own collecting (a month's visit with an aunt—his mother's sister, Julia—and uncle in Nebraska added 10,000 insects and about 300 plants), or by purchase or gifts to the museum. Lisa Grob sent a specimen from Texas, where she was teaching. Wheeler acknowledged this as follows:

Your letter came yesterday and the box containing the horrible reptile this morning. The express man came in carrying it as if it contained a caged bengal tiger and with a very peculiar, indescribable expression on his face. We placed the wild beast in a deep glass jar and have spent a portion of the morning making him angry. The specimen, which is a diamond rattlesnake (Crotalis diadematus) is not yet represented in the museum though we have the two other species, the prairie rattlesnake and the northern (Texas) rattlesnake. So you see that you have contributed a very valuable present to the museum. I thank you very much personally and will send you the thanks of our board in the near future. The specimen will be killed in the following manner: immersion in 35% alcohol for 5 hours, in 70% for one day, and will be kept in 95% after it has been nicely arranged on a white plaster slab in a fine large jar. On the label I will put "donated by Miss E. Grob."

He then added a few personal remarks:

Your mother and Amalie, whom I see every Sunday evening, tell me all the news about you, and I hear that you are very well and healthy . . . I have not wished to disturb your peace by writing. (This is a lie, the truth is I have been too lazy and neglectful, as of old, but I believe that when one is as bad as I am it is as impossible to try to improve one's character as to improve the shape of one's pug-nose.)

We are reading Sunday evenings . . . and miss you very much.

As museum director, Wheeler was especially interested in the insect collections, and in his report for 1889 he maintained that a museum should have both technical and popular collections. He then offered some interesting ideas for what a popular insect exhibit should contain:

Its object should be the imparting of knowledge to the general public . . . It is best to have a number of sub-collections, in the formation of which the larger, commoner, more beautiful and typical insects should play a prominent part. I suggest the following:

1. A collection to illustrate noxious insects. This collection should be arranged according to the plants infested and should give the specimens of the plants to show the manner in which they are destroyed by the insects.

2. A collection of beneficial insects, both of those which destroy noxious insects and do great service by playing the part of scavengers, as also of those which produce useful secretions. (Silkworm, cochineal, etc.) In this collection, too, the food of the insect should be present.

3. A small collection to illustrate the main facts in the external anatomy of insects. A large type of each order should be dissected out to show

the order and arrangement of the hard parts of the body and also to show the wonderful adherence to the main points of structure underlying infinite modification in detail.

4. A collection to illustrate insect development—both embryonic and postembryonic. The embryonic stages, being microscopic, must be represented by enlarged models and drawings, the postembryonic stages of a series of types from each order may be represented by the insects themselves.

5. A collection to illustrate insect architecture. The nests and other structures of insects of the different orders, notably of the Hymenoptera, should be exhibited together with their builders.

6. A collection to show sexual differences, in form, size and coloration.

7. A collection to show protective resemblance in insects—both the mimicry of other insects and the imitation of objects in the surroundings.

8. A collection to show dimorphism—both sexual and seasonal.

9. A collection to serve as an introduction to the classification of insects, including types of the orders and families, to show in a very general way the main results attained by systematic entomology.

10. A collection of insects noted for their great size, beauty, and for their figuring in history and literature. This collection would contain such forms as the gigantic beetles and butterflies, the sacred beetle of the Egyptians, the deathshead hawkmoth, the cicada of the Greeks, the noted phosphorescent species of the tropics; insects used as food by man; those infesting cattle, those forms which appear in immense swarms, like the locusts and ephemerids, etc.

The specimens in all the above suggested sub-collections should be accompanied by explanatory notes, and by all the sketches, photographs and diagrams necessary to the imparting of instruction. The arrangement should be simple and lucid, popular names should, as far as possible, replace technical expressions—in a word, the collection should impart valuable facts in an attractive guise and not be a monotonous arrangement of species hidden in a forest of labels covered with Greek and Latin.

Annual reports are generally dull, but the three that Wheeler wrote for the Milwaukee Public Museum are not. His biting sense of humor, which made him such a sought-after speaker and writer in later life, begins to show here, for example, in his report of 1890: "No great amount of time should be spent in a museum in arranging the legs and tarsi of insects and placing their antennae in uniform position. Such time had best be spent in studying the insects' external anatomy, their adolescent stages or looking up the literature. I would not be understood as wishing insects to be mounted carelessly, but Horace's *Est modus in rebus, sunt certi denique fines,* etc., holds good here as in so many other things. Nor am I finding fault with dillentanti. I have always had much tolerance and sympathy for these gentlemen for two reasons: first, because I know that scientific entomology owes them a great debt and, second, because I deem theirs both a more instructive and aesthetic amusement than the collecting of crockery, medals, etc."

Wheeler's appreciation of the importance of public education in natural history, formed in these years through the influence of Peckham, W. E. Anderson, and the German naturalists of the Milwaukee area, was to remain with him throughout his life. Although Whitman and Patten were to redirect Wheeler's efforts into morphology and embryology, the influence of these naturalists was profound. Many were amateurs and interested primarily in making lists of local species (Wheeler even helped Rauterberg write his long paper listing the local beetles). These men were also public-spirited citizens, and some of them were instrumental in raising interest and funds to start the public museum. One of the first curators they employed was a professional ornithologist, Thure Kumlien, who died in 1888, shortly after Wheeler became custodian. Wheeler wrote an obituary for the Museum on Mr. Kumlien, and it is interesting for what it tells us of both men—the old naturalist and the embryonic biologist:

Mr. Kumlien was essentially what we now call a naturalist of the old school. To him the accurate systematic comprehension of the species was the alpha and omega of biology. In my many talks with him, I well remember his looks of wonder whenever I touched on some of the embryological and morphological problems of the day. The advance of biology along a path so different from the one he had followed since he listened to the lectures in Upsala, never failed to astonish him. Though an admirer of the younger school he regretted, like Darwin, that its members were not in some respects more like the past generation of naturalists. I shall never forget his keen glance when he caught a younger naturalist showing in his knowledge of the classic tongues some flaw unpardonable in the days when the elder Fries wrote mellifluous Latin. He fondly loved the simple Linnaean names, which cling fast to the memory, and often wondered how some modern systematists could fall into such awkward and inapplicable nomenclature.

Too modest to think his own often very valuable observations sufficiently important to publish, he devoted many years of his life to helping other naturalists by sending them large and carefully prepared collections . . . He was very fond of the young and always ready to put at their disposal his long experience as a practical ornithologist and botanist. Such qualities are not to be underestimated in a naturalist for they are the means of charming the young and making good naturalists of youths who would be repelled by a cold exterior. Many of our rising botanists and zoologists owe much to Mr. Kumlien's warm and sympathetic enthusiasm, which was as contagious as hearty mirth.

The heritage of the past and the trends of the times were reshaping Wheeler's future. He did not hesitate to try new ideas, and Carl Akeley was a fortunate participant in Wheeler's life at this stage, as was Wheeler in Akeley's life. Carl Akeley had come to Milwaukee in November of 1886, and set up shop in the barn belonging to Will's mother at 139 Sherman Street ("C. E. Akeley's Studio of Scientific and Decorative Taxidermy. College and Museum Work a Specialty"). He was not hired full time at the Public Museum, however, until after Kumlien's death. Carl had been doing miscellaneous taxidermic jobs for the museum as well as for various private citizens, but on November 20, 1888, he was officially appointed full-time taxidermist by the museum's Board of Trustees.

This was to be a happy partnership, with Wheeler director and Akeley taxidermist, for Will was very sympathetic with Carl's visions of animal preparation. Taxidermy in 1888 was not an art in any sense, and Wheeler, almost forty years later, described it well:

The museum curators and their assistants throughout the greater part of the nineteenth century in France, Germany, England and the United States somehow managed to develop taxidermy to the stage in which it was vegetating when Akeley began his work. The duty of the poorly paid curator had always been to amass, hoard, name, describe, and label as many different defunct animals as possible, and the duty of his famulus the even more poorly paid taxidermist, was to impregnate them with lethal chemicals in sufficient quantity to discourage the museum pests and to try to give them a semblance of life. The result was pathetic when it was not ludicrous, because the taxidermist, as least in museums open to the public, was confronted with the stupendous problem of making dead hides thrilling to the common run of humanity, and the curator, if he was a scientist, necessarily pursued the method of all science, namely, that of abstraction, which has never been attractive to the great majority of our species. He was mainly interested in animals in isolation from their natural environment and behavior and reduced to so much fur, feathers, horns, hoofs, bones, etc., which he could measure and describe in an esoteric jargon intelligible only to other curators in other museums. Akeley, of course, hugely enjoyed the taxidermic exhibits of those days. I remember walking with him through a certain museum and coming upon a stuffed lynx. The creature had been upholstered to about four times its volume in life, its fur had long been a happy hunting ground for *Dermestes,* and one of its glass eyes had become dislocated, so that it was wall-eyed. Just then a sunbeam stole through the dusty pane of the case and fell on that unfortunate orb. The pathetic but fiery glance which it emitted and which seemed to concentrate within itself the whole tragedy of contemporary taxidermy, threw us both into convulsions of laughter.

From the beginning, Akeley clearly realized that any animal mounted for public exhibition can have neither educational nor aesthetic value merely as a stuffed hide, furnished with a pair of glass eyes, attached to a turned wooden pedicel, and provided with a label giving its Latin and vernacular names and the name of the locality in which it was slain. He was thoroughly convinced that an animal is meaningless, except to a hard-shelled zoologist, unless it is presented in such a manner as to convey something of its real character, or *ethos,* which is manifested by its specific motor behavior in a specific natural environment. The development of the taxidermic "group" follows naturally from such a conviction. At the present time, owing largely to Akeley's intensive study of mammalian habits and musculature and his achievements in animal sculpture and the construction of groups, no curator, in the United States at least, would dream of tolerating those indecent, not to say immoral, stuffed beasts which were lined up in the museums of the Victorian age. Furthermore, Akeley's conception was, in a sense, prophetic of a change which through the influence of the ethologists, behaviorists, physiologists and psychologists, has now pervaded the whole field of the biological sciences, so that we have come to see that an organism cannot be isolated, even conceptually, from the peculiar environment to which it has become adapted during aeons of geologic time, without a serious misunderstanding of its true nature.

Carl Akeley's first habitat group, a scene of muskrats in a pond, marked the beginning of a new era in museum displays and presaged Akeley's remarkable accomplishments later at the Field Museum in Chicago and the American Museum of Natural History in New York. This habitat group is still in excellent condition and has recently been moved to the second floor of the Milwaukee Public Museum's imposing new building.

Carl Akeley's first diorama. This historic display is still on exhibit, although it has been moved several times. (Milwaukee Public Museum)

Wheeler and Akeley were both to leave Milwaukee within a few years, and their paths crossed again infrequently. Akeley became fascinated with Africa and its remarkable animal life, and visited that continent many times. After Akeley died in Africa in 1926, Wheeler recalled their relationship in the pages of *Natural History,* including their final meeting: "The last time I saw him, before he left for Africa, never to return, he said, 'Will, I want you to go to Africa with me so that we may end our careers, as we began them, together.' This remark, I believe, was neither a premonition nor an utterance of what has been called the subconscious will to death, but the expression of a desire that we might journey together to some delightful spot in the land he so ardently loved and be reunited in our old age, as we had been united in youth, by our common interest in animal life."

Wheeler's main interest during his museum days was in entomology, but it was divided between taxonomy and embryology. The latter was encouraged, of course, by Whitman and Patten at the Lake Laboratory, but his feelings for taxonomy were the result, apparently, of a natural liking for the subject. Wheeler worked with Dr. and Mrs. Peckham on a monograph on the "Spiders of the Subfamily Lyssomanae," published in 1888. He also worked off and on over the next few years on the taxonomy of certain groups of flies, particularly the family Dolichopodidae (sometimes called the long-legged flies, although the legs are by no means as long as those of the crane flies). His 1890 paper on Dolichopodidae was purely descriptive and without plates or figures, but Wheeler seems to have known that an ability to illustrate his papers would be of great help. So, almost as soon as he started work in the museum he began to take water color lessons. He had had considerable training in school, had even taught drawing briefly at the Academy, and had for some years drawn figures for the Peckhams' papers on spiders. His art lessons continued sporadically for years—including some studies in landscapes, though these he apparently did not leave for posterity.

Although there may have been no connection between his water color studies and his interest in the phenomenon of mimicry, one thing is certain: both required understanding of color relationships. One of the first exhibits Wheeler prepared for the museum was a case to illustrate mimicry (finished August 17, 1888), and he later did another one. In his diary for February 17, 1890, he wrote: "Thought some about the evolution of metallic colors in insects." Also in 1890 in *Science,* Wheeler reviewed favorably a new book by Professor E. B. Poulton of Oxford, entitled *The Colours of Animals.* His only adverse criticism was a comment saying that he wished Professor Poulton had included more examples of mimicry. "Many startling cases of *Hymenoptera* mimicked by *Diptera* seem to have escaped the author's notice."

Will Wheeler had assumed his new duties at the museum on October 12, 1887, but already there had been some meetings in eastern Massachusetts which would eventually be instrumental in luring him away from Milwaukee. At a March meeting in the library of the Boston Society of Natural History, consideration had been made of the possibility of the formation of a permanant seaside laboratory (something like Agassiz's Penikese). A year later, the Marine Biological Laboratory, to be located at Woods Hole, Massachusetts, was formally incorporated, and in June of 1888, the corporation issued a statement of policies and named the director, Dr. Charles Otis Whitman of Milwaukee—a natural choice because of his association with the Allis Lake Laboratory. Dr. Whitman was not involved in the initial policy making of the Board of Trustees of the Marine Biological Laboratory, but as its director he developed those ideals far beyond the original hopes.[10]

Wheeler did not attend the first session at Woods Hole in 1888, but in May 1889, he was given a month's leave of absence from the museum, and he started east. He stopped first in Rochester to see the "boys" at the Establishment and to spend a couple of delightful days with Mr. Crump. He stopped in New York City to visit the American Museum of Natural History and to attend a meeting of the Brooklyn Entomological Society. He also visited Washington, D.C., where he met many of the entomologists whom he had known only by correspondence or by scientific publications. He also visited Baltimore and Johns Hopkins, Philadelphia with its Academy of Sciences, the University of Pennsylvania and the Zoological Gardens, and Boston and its Natural History Museum and Cambridge with its Museum of Comparative Zoology. Scattered along his route were old friends, too, from the German-English Academy and Seminary, and they often guided him around the local sights. The climax of his trip, however, was his stay at Woods Hole from June 15 to July 3. The formal summer season had not yet begun, but a few of the men were around. Wheeler spent long, pleasant hours collecting on the beach or working in the laboratory on various marine organisms. Evenings were quiet—perhaps a talk with Dr. Whitman, T. H. Morgan, or Howard Ayers (who would replace Whitman at the Allis Lake Laboratory). At other times Wheeler went with another young man, Hermon Carey Bumpus, for refreshments at the local ice cream parlor. The time was all too short.

Time was getting short for William Morton Wheeler in Milwaukee, too. The Museum work continued, but young Wheeler was getting restless. He continued his embryological studies and in the evenings went with Carl Akeley to see *Carmen, The Flying Dutchman, King Lear,* the humorists James Whitcomb Riley and Bill Nye, and others. Collecting was not neglected, even on his annual summer picnic with Mrs. Grob and Amalie. Changes were coming to the Lake Laboratory, however, for Whitman had resigned his directorship. As Will told Lisa Grob in a birthday message dated September 9, 1889: "I have just come from the depot where I said good bye to Dr. C. O. Whitman, now Prof. of Biology in the new Clark University, which is to throw in the shadow all the institutions of learning in the country. This morning I was surprised on entering my office to find a friend in my chair, Mr. Bumpus, late Prof. of Biology at Olivet, a very fine gentleman, whom I met while in the east. Till one o'clock I spent with him & Dr. W. Patten running about town. This afternoon I worked at the museum, so am quite fatigued."

Later on Wheeler added:

My moustache has made excellent progress . . . Mr. Akeley just returned from his home in Holly, N.Y., & has started tonight for northern Michigan . . . to collect for the museum. Your mother and Amalie send their love. Hoping that you will forgive this brief and stupid letter and wishing you every happiness and myself an answer, I remain,

Yours,
Willie

Dr. Patten left the Lake Laboratory soon after Whitman, and Wheeler missed the two. The new men who came, A. C. Eycleshymer and Howard Ayers, were, however, most congenial. Together with Dr. and Mrs. Peckham and Wheeler they formed a little "Biology Coterie," and spent winter evenings discussing current biological theories and research. But this was not enough. Wheeler wanted some formal scientific training at an outstanding university. The question was where: Johns Hopkins, Clark, Louvain or Munich? Should he or should he not give up a job which was not difficult and paid fairly well, too? Dr. Whitman wrote a long (for him) letter with a number of helpful suggestions, and an offer of a fellowship at Clark. H. C. Bumpus, who was a candidate for a Ph.D. at Clark, and whose company at Woods Hole Wheeler had so much enjoyed, also spoke well of the new institution. Then on March 23, 1890, Lisa Grob came home from Texas preparatory to getting married—to a young business man, H. Dittlinger, of New Braunfels, Texas. On March 25, Will Wheeler wrote to Dr. Whitman accepting the fellowship at Clark, and resigned his directorship to be effective at the end of the fiscal year (August 31).

3 / Two Universities Are Born—and a Career Is Launched

On October 2, 1890, William Morton Wheeler took up his studies, which he hoped would lead to the degree of doctor of philosophy, at Clark University in Worcester, Massachusetts. Wheeler had been fortunate to grow up in bustling Milwaukee, but this quiet town provided a more stimulating atmosphere than anything he had previously known. In his diary for 1890 he summarized his feelings simply by saying: "Much happier."

The atmosphere of Clark University was created by many people, not the least of whom was Jonas Clark, a local merchant who provided the money for its founding. But it was the first president, G. Stanley Hall, who had the insight and ability to attract the original outstanding faculty. In *Science,* June 14, 1889, a two-and a-half-page unsigned article, entitled "Clark University," began as follows:

> Clark University was founded by the munificence of a native of Worcester County, whose plans, conceived more than twenty years ago, have gradually grown with his fortune. His affairs have been so arranged as to allow long intervals for travel and study. During eight years thus spent, the leading foreign institutions of learning, old and new, were visited, and their records gathered and read. These studies centered about the means by which the highest culture of one generation is best transmitted to the ablest youths of the next, and especially about the external conditions most favorable for increasing the sum of human knowledge. To the

improvement of these means and the enlargement of these conditions, the new university will be devoted.

It is the strong and express desire of the founder that the highest possible academic standards be here forever maintained; that special opportunities and inducements be offered to research; that to this end instructors be not overburdened with teaching or examinations; that all available experience, both of older countries and our own, be freely utilized; and that new measures, and even innovations, if really helpful to the highest needs of modern science and culture, be no less freely adopted; in fine, that the great opportunities of a new foundation in this land and age be diligently explored and improved.

Several paragraphs later, the article continued:

In the spring of 1888, G. Stanley Hall, then a professor at Johns Hopkins University, was invited to the presidency. The official letter conveying this invitation contained the following well-considered and significant expression of the spirit animating the trustees: "They desire to impose on you no trammels; they have no friends for whom they wish to provide at the expense of the interests of the institution, no pet theories to press upon you in derogation of your judgment, no sectarian tests to apply, no guarantees to require save such as are implied by your acceptance of this trust. Their single desire is to fit men for the highest duties of life, and to that end, that this institution, in whatever branches of sound learning it may find itself engaged, may be a leader and a light."

The president was at once granted one year's leave of absence, with full salary, to visit universities in Europe . . . The plans of the university have now so far progressed that work will begin in October next, in mathematics, physics, chemistry, biology, and psychology.[1]

The rest of the paper in *Science* went on to describe the general organization of the university curriculum, the types of students that would be admitted, available and hoped-for scholarships, and the names and brief histories of the faculty hired so far. Of the six so named, four, including Dr. Hall, were coming from Johns Hopkins. Professor C. O. Whitman was not included, as he had not yet accepted a position at Clark.

When Clark University opened its doors on October 2, 1889, the hopes of everyone connected with the institution were high. That Clark University never developed over the years into the research institution and graduate school envisioned by Dr. Hall and the men first associated with him was indeed unfortunate and undoubtedly was a great loss to science and to higher education in general.[2] But more of this later. Dr. Hall was without doubt an excellent choice for the presidency. He had been born in Massachusetts in 1846 and had done his undergraduate work at Williams College (where in 1867 he had been chosen Class Poet and elected to Phi Beta Kappa) and at the Union Theological Seminary. He studied in Germany for several years, especially in the new science of experimental psychology, and also at Harvard University, where he received the degree of doctor of philosophy in 1878. In 1882 he was asked to organize a laboratory and teach psychology at Johns Hopkins University. One of his students was Edmund C. Sanford, who helped Dr. Hall found and edit *The American Journal of Psychology*. Sanford moved with him to Clark and later also took over some of the administrative work. Dr. Hall devoted his life to furthering his ideals, both in psychology and in educational reform. He read omnivorously, enjoyed lecturing, and wrote almost continuously, as his lengthy bibliography shows.[3] William James, the Harvard psychologist and a personal friend, said of him: "I never hear Hall speak in a small group or before a public audience but I marvel at his wonderful facility in extracting interesting facts from all sorts of out of the way places. He digs data from reports and blue books that simply astonish one. I wonder how he ever finds time to read so much as he does—but that is Hall."[4]

When William Morton Wheeler arrived at Clark in 1890, the faculty had been complete for some time, and activities were in full swing. Wheeler, along with the other students, attended lectures and seminars with zest. Wheeler's diary of those days indicates that in typical student fashion, professors were rated for their lecturing ability. President Hall's series on contemporary educational systems was good. Franklin P. Mall, Professor of Anatomy, was a nice man, but his lectures were dull. Others were so-so, except for Prof. Whitman and Dr. George Baur.

Whitman began his lectures with a survey of the history of research in animal morphology, especially embryology, because he felt that "history illuminates a field, gives breadth and depth to concepts, enriches and enlarges general views, makes us more generous, considerate and respectful towards predecessors and contemporaries." Whitman did not enjoy giving lectures, and later in life avoided them whenever he could, but when he gave a lecture it was always a masterpiece in exposition. Unfortunately, he did not always have time to finish delivering them, and apparently at his next lecture he usually had forgotten he had not previously finished and began on a new subject. Nor did he, during a semester, survey the field as completely as his students might have wished. Nevertheless, no one missed one of his lectures from choice.

Dr. Baur seems to have been the most stimulating, if not controversial, scientist at Clark. Baur's untimely death in 1898, when he was but thirty-nine, was a great loss. In the obituary written by Wheeler for *The American Naturalist* (1899), we learn much of what he meant to Wheeler and his associates, both at Clark and later at the University of Chicago. George Baur was born and educated in Germany, and came to the United States in 1884, only to become embroiled in the infamous Cope-Marsh feud. He had come to assist Professor O. C. Marsh in paleontology at Yale and left, as did most of Marsh's assistants, greatly disillusioned with the Professor. He came to Clark in 1890 as docent, or lecturer, in comparative osteology and paleontology. Wheeler said of him:

> The calm atmosphere of the investigation pervading that institution allayed the excitement into which he had worked himself on leaving Yale, and having entered on a position where free and independent investigation and publication were not merely tolerated but required, he began to plan several extensive works. One of these was an elaborate monograph of the North American tortoises, to be published by the National Museum as a companion volume to Cope's *Batrachia*. Another was the investigation of the faunas and floras of oceanic islands. During the two years that Dr. Baur held his position at Clark he made great progress in both of these undertakings. In 1891 . . . he was enabled to fit out an expedition to the Galapagos Islands. Accompanied by Mr. C. F. Adams, he left in May and returned in October, after visiting nearly all of the islands of the archipelago. The study of his extensive collections of the plants and animals of these islands has since occupied Dr. Baur and several zoologists and botanists both in this country and in Europe. The various reports, embodying descriptions of many new species, had been nearly all published, and just before his last illness Dr. Baur was planning a general work on the Galapagos Islands to include all the results of the expedition, together with an elaborate introductory chapter by himself. The valuable collections were recently purchased by the Tring Museum, which is undertaking a further study of the Galapagos fauna.

Although Baur did not speak English fluently, his lectures and seminars provoked many hours of discussion and debate among his students and fellow faculty members. His studies centered on reptiles, both living and fossil, and he used these facts as the basis for his theories on the origin of variations. Interestingly (in the light of some of Wheeler's own statements in later life), Baur was a Neo-Lamarckian. Again, to quote Wheeler:

> Dr. Baur's interest in the more general problems of morphology, such as the origin of variations, was first keenly aroused at Clark University at a time when Weismann's essays were the subjects of much general discussion. He always remained a steadfast Neo-Lamarckian, never forsaking the position he had taken in his inaugural dissertation. A perusal of Wagner's and Eimer's works convinced him that isolation and environment are potent factors in producing variation. A previous study of the giant land tortoises of the Galapagos, in connection with his turtle monograph, and further studies on a genus of lizards (Tropidurus) which has produced different so-called species on the various islands of the archipelago, led him to conclusions which his subsequent visit to the islands did not modify. It was perhaps fitting that one born in the year of publication of the *Origin of Species* should gain inspiration from the spot where the idea of the "Origin" was conceived.

Wheeler did not neglect his own research, nor his writing, though the latter was restricted to a few small papers, mainly book reviews. He prepared and delivered a lecture to the Biology Club on "Insect Metamorphosis," but apparently this was not overly successful, as he noted in his diary that it was too technical, and he and his "hearers" were disappointed. Another disappointment was the unexpected absence of Hermon Carey Bumpus from the campus. In the fall of 1890, Bumpus had been asked by the new president of Brown University at Providence, Rhode Island (where Bumpus had been an undergraduate) to reorganize the biology department and to develop a graduate program, perhaps not unlike the one materializing at Clark. He could not refuse the offer from Brown, but this prospect did not keep Bumpus from finishing his thesis on lobster development—an original and thorough piece of work for its day. Whenever Bumpus came to Clark, the day was devoted to long discussions on lobsters, as Wheeler noted in his diary, and then on May 1, 1891, Bumpus successfully took his final examination for the doctorate. Dr. Bumpus had the distinction of receiving the first

Ph.D. from Clark University. But the association between Wheeler and Bumpus was just beginning, and it was lifelong. They had first met at Woods Hole in 1889, then in Milwaukee, when Bumpus had stopped briefly at the Allis Lake Laboratory on his way to teach at Olivet College in Michigan, and now at Clark. During the previous Christmas holiday (1890), together they had attended the meetings of the Morphological Society and the American Society of Naturalists in Boston, toured the sights of Boston and Cambridge, and, along with E. O. Jordan, another student from Clark, called on that delightful old gentleman Oliver Wendell Holmes. Bumpus was even briefly at Woods Hole in the early summer, though Wheeler, and most of the rest of the biology faculty and students from Clark, spent the entire summer season at the Marine Biological Laboratory.

Wheeler was to spend the next two summers at Woods Hole. In 1891, he assisted Whitman with the laboratory sections of the elementary course, and in 1892 he did the same, except that Bumpus

The staff of the Marine Biological Laboratory in the summer of 1892. Front row, left to right: W. M. Wheeler, W. A. Setchell, C. O. Whitman, H. C. Bumpus, S. Watase; back row, left to right: P. A. Fish, J. Loeb, E. O. Jordan, C. L. Bristol, and E. G. Conklin. (Library, Marine Biological Laboratory, Woods Hole)

supervised the course instead of Whitman. In addition to the teaching, Wheeler continued his research and his studies. But far and above the routine was the sheer joy of a summer in this lovely place situated between Buzzards Bay and the open sea, and the simple, unsophisticated life which all the workers led when away from the restrictive social environments of the large cities. Except for the few hours in the morning that Wheeler usually had to devote to helping the beginning students (mostly public school teachers) in the laboratory, he was free to roam the beaches collecting, or to go in a boat on a field trip with others—either along the coast, to some of the islands, or just deep sea fishing. The five o'clock swim every afternoon was a particular favorite of "the boys," and the varied evening lectures were stimulating. These included J. P. McMurrich on the blastopore, Sho Watase on karyokinesis, J. S. Kingsley on the Dakota Badlands, H. F. Osborne on the evolution of mammals, H. V. Wilson on sponges, S. H. Scudder on butterfly wings, and E. G. Conklin on gastropods.

Sundays were of unusual interest to the workers at the Biological Laboratory, for classes were suspended, and most of the people there took part in the excursions in the Laboratory launch. On August 2, 1891, an especially important event occurred, historically speaking, for the excursion was to Penikese Island, where Wheeler and many others visited the old Agassiz Laboratory and took the mottos off the walls. In 1874, during the second and last session of the Penikese School, the students and faculty, as a tribute to their sorely missed leader, had written out, on pieces of linen, various bits of Louis Agassiz's philosophy, and hung these around on the walls of the laboratory. There they stayed for seventeen years. Today, in the entrance of the main building at the Marine Biological Laboratory at Woods Hole hangs the best known of these: "Study Nature Not Books." No one at Woods Hole today seems to be able to find the other mottos brought back that day, or even knows how many there were or what precisely they said. One can still hope, though, that they will be rediscovered. Meanwhile, as Wheeler noted in his diary, when a return visit was made to Penikese on August 10, 1892, the old laboratory was gone—destroyed by fire during the previous summer.

Wheeler's research in the summer of 1892 began to depart from insects, for he became interested in marine flat worms, some of which were commensals on the snail *Busycon* and others on the horseshoe crab, *Limulus.* But prior to that time, he had made some remarkable investigations into the embryology of insects—climaxing a five-year period with his doctoral thesis at Clark in the spring of 1892 (published in 1893 in Whitman's *Journal of Morphology*).

Twenty to thirty years earlier, embryological and other cytological studies were made possible in a much more detailed way by the invention of a tissue-slicing machine, the microtome. The development of the microscope some two hundred years earlier had initiated such studies, but until thin sections of tissues were possible, many details of specific structures were not clearly visible and thus difficult to understand. Following the perfection of the microtome and the development of skill in mounting sections on glass slides came the development of organic dyeing, particularly in Germany. Blue nuclei against pink cytoplasm not only added a touch of color to otherwise dingy laboratories, they also enlightened the investigators as to cellular structures and possible functionings (to be explored later as various techniques were improved).

Modern descriptive embryology is considered to have begun early in the nineteenth century with the work of two non-Germans in Würzburg, Germany: a Russian, Heinrich C. Pander, and an Estonian, Karl Ernst von Baer. Pander first recognized (1817) the germ layers in a chick embryo after twenty-four hours of incubation. Von Baer, inspired by his friend Pander, studied the development of many animals and formulated (1828) a law for the development of all animals. He said that during development, the egg or ovum divides into layers of tissue, and that each layer gives rise to the same organs throughout various groups of animals. Though von Baer's "germ-layer" theory offered a great deal of evidence to support the evolutionary doctrine as presented by Darwin some thirty years later, von Baer himself did not accept the doctrine of organic evolution, although he lived until 1876.

Today embryological research is seldom solely descriptive, probably because investigators find experimental embryology much more exciting, especially when chemistry has proved to be such a useful tool in analyzing cellular physiology. That insects are used in these biochemical, cellular studies is often only incidental, much as the little fruit fly *Drosophila* just happened to be *the* experimental animal in the pioneering research in genetics. Entomologists specializing in embryology are scarce, and the most comprehensive as well as standard textbook on insect embryology today is *Embryology of Insects and Myriapods,* by O. A. Johannsen and F. H. Butt, which was published in 1941. One wonders if Dr. Wheeler, when he abandoned embryological research and writing after a decade of work, felt that descriptive embryology, even in the 1890's, would soon be passé as a research field. Be that as it may, insect embryology was very active at the beginning of the 1800's, and Johannsen and Butt's summary of the field is interesting:

Among the earlier works dealing with the embryology of insects may be mentioned that of Herold (1815) on the Lepidoptera, of Hummel (1835) on the roach, and of Kölliker (1843) on a comparative study of the development of insects and vertebrates. These researches were followed by Weismann's (1863) work on the Diptera which represents for its time an outstanding contribution . . . In the year 1871 a paper on the embryology of worms and arthropods by Kowalewsky appeared, a work of great merit, in which the method of cutting sections of tissues, previously fixed and embedded in paraffin, was used. During the thirty years that followed, numerous and important contributions to the subject of arthropod embryology were made by both European and American investigators. Among prolific European writers of this period may be mentioned Carriere, Cholodkowsky, Graber, Heider, Heymons, Korschelt, and Nusbaum. For the same period the works of the American zoologists Ayers, Claypole, Knower, Packard, Patten, Ryder, Wheeler and Woodworth are noteworthy. Wheeler's papers on the development of the Orthoptera are especially outstanding.[5]

Wheeler's first embryological paper, published in 1889, was entitled "The Embryology of *Blatta germanica* and *Doryphora decemlineata*" and was the result of his association with William Patten at the Allis Lake Laboratory, as previously described. In the historical discussion of his first paper, Wheeler explained his choice of experimental animals:

The cockroaches have long been favorite objects of morphological study. Easily obtained at all seasons of the year, of convenient size for dissection, and being but slightly modified descendants of the oldest insects of geological time, they combine qualifications which make them especially interesting and valuable to the morphologist.

Doryphora decemlineata [the Colorado potato beetle] has not been investigated heretofore from an ontogenetic standpoint. It is surprising that so common an insect, and one whose eggs present such advantages for embryological study, should have been overlooked. The favored

Coleopteran of embryologists has always been *Hydrophilus,* and it is certain that the water-beetles (*Hydrophilidae* and *Dytiscidae*) are much less modified forms than the leaf-beetles (*Chrysomelidae*), to which *Doryphora* belongs.

Thus Wheeler chose two common insects, one belonging to a more primitive order, Orthoptera, and one to a more advanced group, the Coleoptera or beetles, so that hopefully he might be able to observe phylogenetic differences as well as ontogenetic development. Wheeler subsequently worked on insects representing several orders, but the greater part of his embryological studies remained with the Orthoptera—first the German cockroach, now called *Blattella germanica,* and in his thesis *Xiphidium ensiferum* Scudder, a meadow grasshopper now called *Conocephalus brevipennis.*

Wheeler began his work with *Conocephalus* in the outdoors by noting when, where, and how the females laid their eggs. Then he collected the eggs at various stages of development and transferred them to the laboratory. There he killed, sectioned and stained them, and mounted them on microscope slides. In the early days of his work in Milwaukee, he was often discouraged with his poor histological techniques, but he soon learned to make excellent preparations, as is evidenced by the clarity and beauty of his drawings made from those slides and included with his papers. So excellent are his slides that some are still being used three-quarters of a century later in the Biology of Insects course at Harvard. In the laboratory Wheeler studied the early development of the fertilized egg, noting the type of cleavage, the formation of the embryonic membranes, and the differentiation of the germ layers: the entoderm or inner layer, and the mesoderm or middle layer, from the original outer layer of cells (blastoderm) now called ectoderm. From these layers are derived the specific tissues and organs. Tracing the development of insect tissues from specific germ layers is no easy task; even today there is much argument as to the exact dermal origins of many of the tissues. Insect eggs are small and contain a large amount of yolk, and these factors tend to obscure cleavage and the formation of the blastoderm and ensuing layers. In 1941 Johannsen and Butt said that "the difficulties in interpreting the homologies of the germ layers, especially in insects, are so great that writers from time to time have questioned the validity of the germ-layer theory." Wheeler did not doubt its validity, however, and modern experimental embryology seems to be proving that he was correct. According to C. H. Waddington, writing in 1956, though developing insect eggs may seem to show variations and modifications from the eggs of other animals, studies are proving that the germ-layer theory is valid for all animals, including insects.[6]

Wheeler's "A Contribution to Insect Embryology" helped to clarify many disputed points and to standardize the terminology of the field. He named the indusium, a peculiar thickening in the blastoderm, as well as various stages in blastokinesis, the rotation of the embryo. He believed that rotation might serve to produce circulation and aeration in the yolk and to "bring fresh pabulum in contact with the assimilating cells of the embryo." He did not hesitate to criticize the views of other workers—including those of William Patten as well as his own as expressed in earlier papers. Now and then a little of the vitriol for which he later became well known shows through, as when he attacked Cholodkowsky's ideas on the origin of blood cells from the yolk as being the result of overtreatment of his preparations with a potent fixative, Perenyi's fluid. "The result of this heroic method," Wheeler remarked, "is apparent enough in the distortion of the tissues, but its effect on the yolk is quite remarkable."

Following his discussion of the embryology of the meadow grasshopper, Wheeler discussed in detail the embryonic envelopes, the yolk, and blastokinesis, incorporating facts from many groups of insects. Finally, he treated the development of the nervous system and the reproductive organs of insects in general terms. The result was the most comprehensive treatment of insect embryology available in English at that time. The publication of this monumental treatise in 1893 placed Wheeler in the forefront of his field. A few more years would see the rediscovery of Gregor Mendel's laws and swift, new advances in genetics and in cellular and developmental biology. That Wheeler would soon abandon this area and achieve fame in wholly different aspects of biology could not have been suspected by anyone who observed the young Ph.D. who had effectively mastered so difficult and disputatious a subject as insect embryology.

Of course, Wheeler was hardly in a position to plan further research until he had located a position. President Hall had offered him an opportunity to stay at Clark, and E. B. Wilson had approached him regarding a position at Columbia. But as a result of undercurrents at Clark which suddenly came to the surface just as

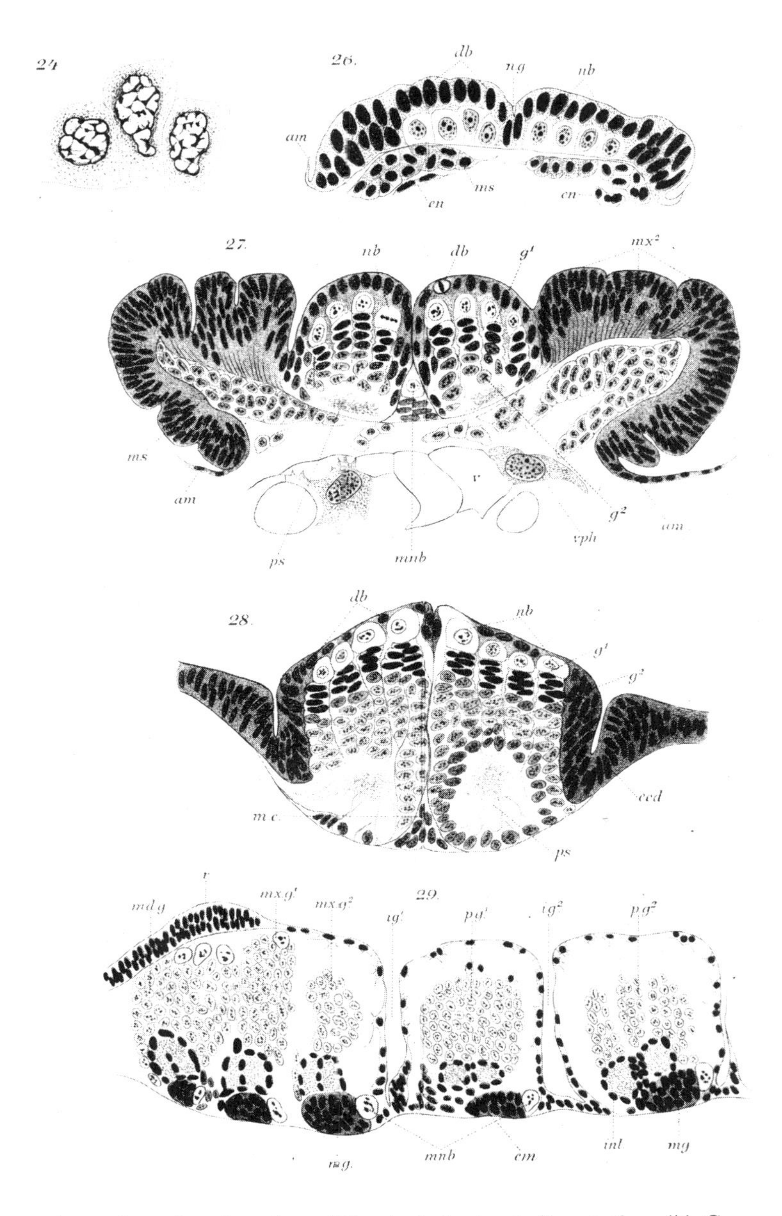

A portion of a plate from Wheeler's doctoral dissertation, "A Contribution to Insect Embryology," as published in *The Journal of Morphology* in 1893.

Wheeler was completing his studies, he was to be swept—along with much of the Clark faculty—to a still different institution: another university born of the munificence of wealthy businessmen and led by an idealistic and forceful president.

Over the past few years, G. Stanley Hall's dream of a utopian society of scholars and investigators seemed to be shattering for reasons over which he had very little control. Posterity has found him blameless, except, perhaps, for the fact that he did not take his faculty into his confidence. He did not explain to them his problems in dealing with Jonas Clark, the founder. Mr. Clark, a businessman, thought he could run a university like a business, wherein he as owner need not explain to employees (including the faculty) how he expected to use his money. When he received requests for specific equipment, he did not fill them as asked but instead sent his own ideas of the most useful sinks, shelves, etc. He carefully supervised all building and other activities as much as he could, and he saw no need to submit a proposed budget ahead of time, so that the president could have some ideas as to the amount of money he would have for salaries, books and equipment. In 1891 Clark's health broke, and gradually his mind faded so that he became more and more suspicious—thinking that people, and especially Dr. Hall, were trying to get his money from him. Eventually President Hall could not get any more money from Clark, but he kept his trouble to himself, and took the blame for the college deficits. A personal tragedy in 1890, the death of his first wife and eight-year-old daughter by gas suffocation in their home, had perhaps inclined Dr. Hall to avoid burdening others with his own troubles. The only other person who seems to have had any idea of the difficulties with Mr. Clark was the University clerk (or treasurer) L. H. Wilson, and in writing of it later (1914), he summarized the dilemma as follows: "But with a founder who could not understand these ideals and who gave no intimation of his real wealth; with a faculty of very earnest and very ambitious scientists, with an income that did not cover the salary list, serious difficulties and misunderstandings were inevitable."[7]

It is difficult to say which of the many outstanding members of the faculty at Clark were most troubled by the restraints they felt at the University that had once seemed so promising. Many meetings were held about the lack of understanding between the administration and the faculty, and finally, in late 1891, it became apparent to at least most of the biological and physical scientists that conditions would not improve. Consequently, on January 21, 1892, nine of the men handed in their resignations. Few records remain of these troubled times, but the two leaders of this "revolt" seem to have been C. O. Whitman, who had never hesitated to stand up for his ideals, and A. A. Michelson, head of the department of physics, and a similarly strong personality. Dr. Michelson already had made some of his outstanding experiments with light, namely, the discovery of a simple method for measuring the speed of light with great precision. His later experiments would inspire Albert Einstein in the promulgation of his theory of relativity, as well as bring to Michelson in 1907 the Nobel prize for physics—the first ever given to an American.

If the agitation and disillusionment of the faculty in the fall and winter of 1897 were known by the students, Wheeler, at least, made no mention of it in his diary. He talked only about some of the lectures he attended, how his laboratory studies were progressing, the activities of the Biology Club, the variety shows in the local theater, and even how he nearly fainted during an autopsy (the first he had ever seen) in the local hospital. Finally, however, on February 16, 1892, Wheeler wrote the following: "had a long talk with Dr. Whitman on the prospects in the new Chicago University." And in the next two months and after, the development of the new University of Chicago would come to be more and more important to the faculty and students of Clark.

In the latter part of the nineteenth century, the city of Chicago was showing a phenomenal growth (the population, estimated to be about 500,000 in 1880, had more than tripled by 1900). It had outstripped Milwaukee in size, and consequently its problems, though similar, were larger. Milwaukee had had some bad fires, but Chicago's great fire of 1871 (which made Mrs. O'Leary's cow famous) was more devastating. From it, however, emerged a more attractive and more permanent city of brick and stone to replace the ugly wooden city of earlier times. The difficulties of labor were not solved, though there was more hope among the immigrant laborers in 1892 when John Peter Altgeld, whose parents had emigrated from Germany when he was three months old, was elected governor of Illinois. After a long, careful study of the records of the trial of the Haymarket prisoners, Governor Altgeld pardoned them (something no former governor would even consider). Altgeld was a hope, but he was struggling against great odds, and poverty continued to be appalling, city politics a disgrace, and money and big business interests still in power. The many financiers, perhaps to ease their consciences, were spending some of their millions trying to bring "culture" to Chicago. In this era were founded the Chicago Art Institute (filled with paintings by Rembrandt, Frans Hals, Holbein, El Greco, and others), the Chicago Auditorium, the Chicago Symphony Orchestra (though its first conductor seems to have had difficulties persuading his audiences that the program should include a symphony), the Chicago Public Library, a natural history museum, and finally, a university.

The University of Chicago, though it had existed before the Great Fire, had not managed to get into the mainstream of the city's growth, and it had been allowed to languish. John D. Rockefeller made the first move to revive the University with a gift of $600,000. He soon added a great deal more, as did other Chicago tycoons, including Marshall Field, who gave a valuable 10-acre plot for the campus. But Rockefeller's greatest contribution was in helping to persuade William Rainey Harper to accept the first presidency. Dr. Harper was a professor of Semitic Languages at Yale University (where he had received his Ph.D. degree in 1875 when he was 18) and had established himself as an excellent teacher, a brilliant scholar, and an exceptional organizer. On July 1, 1891, he took up his official duties to create a "Great University" which would in ten years rival Yale and Harvard. After five years of the Harper administration, during which more than $9,000,000 had been given the University, some educational circles felt that the University of Chicago had even eclipsed the two or more centuries of growth made at Yale and Harvard.

William Rainey Harper was, in some ways, a man much like G. Stanley Hall, at least in his ideas of how to build a great University—with great men. He was lucky, too, for he was able to capitalize on much of the basic work done by Dr. Hall. President Harper's brother, Robert, had been a close friend of Franklin P. Mall (then professor of anatomy at Clark) since their student days together in Leipzig, and through this contact, President Harper heard about the trouble at Clark. Immediately, he pressed his advantages —one being attractive salaries, such as the unheard of sum of $7000 for a full or head professor. That money was not everything, however, is shown in a remarkable letter written by Professor Mall (in his not-too-easy-to-read longhand) to President Harper on January 27, 1892, from Worcester:

Dear President Harper—I have also constantly had my fears that the biological scheme might not develop. The amount he [C. O. Whitman] suggests is not great if all the departments are included; the physiological department alone at Columbia has a salary list of 15,000 and at Berlin over 50,000. Yet with these things clearly in view, I have constantly urged Prof. Whitman, and his enthusiasm has most of the time been the highest. When you wrote to him last he felt a little downhearted but the idea of making a biological department with various branches (but not full departments) represented seemed first to me and then to him a way out of the difficulty. Now I feel most hopeful and he tells me that he has written a hopeful letter to you.

You will pardon me if I tell you frankly what I think the reason is why men of Whitman's stamp are so attached to this place. They make of Science a religion and the highest duty of a scientist is to add to knowledge. This place is founded on such an idea and if I mistake not so is the Ogden School. In America, and in almost all countries, the biological sciences are taught from books and we all know that the . . . object is . . . "to study life." We perform our duty in the winter and in the summer when most people take vacations we do most of our conscientious work and attempt to reach our ideal. You see the difficulties. On account of much freedom here, in spite of our troubles (confidential), we cling to our ideals. You know of Whitman's organizing ability. I may add that his students idolize him.—I yet believe that if the ideals which biologists prize so much are again plainly laid before him that he will consider the place most favorably.

Assuring you that I have in the past and shall in the future do all in my power, I am, Most sincerely, F. Mall.[8]

Professor Mall's contribution to the University of Chicago was not inconsiderable, especially when he helped to bring Professor Whitman there and to help organize the department of biology. But Mall stayed only one year, for Johns Hopkins University asked him to come back and to organize its medical school. The decision to leave was not easy, for he had great hopes for the University of Chicago, but its plans for a medical school were still nebulous. So he went back to Baltimore, and the medical school there now bears witness to the idealism and ability of Professor Mall. He was much interested in research, and he felt that every doctor should have a strong background and experience in pure science before attempting any diagnosis or suggesting any treatment for disease. He believed medicine should be a science, not just a practical art.[9]

And so President Harper, with considerable help from Mall, persuaded Dr. Whitman to leave Clark for Chicago, and subsequently, mainly through Whitman, most of the rest of the science faculty. He even offered Dr. Hall the position as head of the psychology department, but Hall, a man with his own strong convictions, could not desert the institution for which he had worked so hard, especially when it needed strong leadership to see it through the crisis of 1892. On April 19 Wheeler, after an hour's interview with Harper in Whitman's house, accepted a position as an instructor in morphology for $1500. He also noted that "most of the men at Clark University have been engaged to go to Chicago." And, he might have added, so did most of the doctoral candidates—so much so, that the University of Chicago, at least in the departments of zoology, physics and chemistry, was almost a "pre-fab," complete with faculty and graduate students, thanks to Clark University.

While Clark still had its president in 1892, the same could not be said of a number of other institutions. Included on the Chicago faculty of one hundred and twenty (nine of whom were women) were eight former college or university presidents. For example, Albion W. Small had resigned as president of Colby College to found a sociology department at Chicago; T. C. Chamberlin left the presidency of the University of Wisconsin to organize the geology department; and in 1896 John Merle Coulter, formerly president of Indiana State University, came to head the botany department. All of these men left lasting impressions on their fields, not the least of which was the founding of important journals in their respective subjects. The dean of women was Alice Freeman Palmer, former president of Wellesley College. John Dewey, though not a former president of anything, came to Chicago in 1894 as chairman of the department of philosophy, psychology, and pedagogy. There, in his University Elementary School, he worked out his revolutionary ideas on education, saying, in essence, that a school should make a child's education relate directly to his everyday life. Nor was the University's athletic program given anything less than the best. President Harper engaged Amos Alonzo Stagg as director of physical culture and athletics, and Stagg, the best known college athlete of that day, was idolized by students, alumni, and most football fans in and out of Chicago. Ray Ginger, in describing the unbelievable speed with which Chicago became a cultural and intellectual center, wrote of Harper as follows: "William Rainey Harper, chief promoter and first president, had determined that the graduate schools should not be neglected in favor of the college, and he recruited for the faculty several of the greatest scholars in the world. Harper's round body, round face, and small genial eyes made him look like a friendly pig. But where talent was concerned, he was a greedy hog."

Ginger paid further tribute to Harper:

> For any organization the first years are crucial. If key positions are filled with incompetents, they will choose as subordinates other incompetents, who rise through the ranks to the key positions. Mediocrity becomes the characteristic of the organization, and it breeds itself. The level of quality is especially hard to raise in a university, where entrenched deadheads are protected in their jobs, not only by their power within the organization, but also by the tenure regulations peculiar to academic life. At Chicago, the start was right. The top men at the beginning were good, and they knew how to identify and attract their equals or superiors. (When Robert A. Millikan finished his postdoctoral work at Berlin and Gottingen in 1896, Michelson brought him to Chicago as an assistant in physics. He stayed for twenty-five years. Two years after he left, he won the Nobel Prize.) The University of Chicago started life as a great university: both storehouse and workshop. Harper's vision was sound, and he made it a reality.[10]

"If the first faculty of the University of Chicago had met in a tent," commented Robert M. Hutchins in his presidential inaugural address at the University in 1929, "this would still have been a great university." Such was the atmosphere into which William Morton Wheeler came in the fall of 1892. That the physical plant had not caught up with the intellectual acquisitions was obvious, too. Most of the campus was still in its natural state. The west and southeast sides were covered with young oaks, but the center, being low, was a morass in the spring and had standing water most of the year. There were three buildings which, although not finished, could be used when the University opened officially on October first. One was the general recitation, Cobb Lecture Hall (entered via a precarious board in lieu of steps), the other two dormitories for men. "Science Hall" was an apartment building which had been taken over for a year to house the sciences, and it was into this building that tons of equipment and books were moved, temporarily, so that lectures and laboratories could begin on that momentous Saturday morning of October. As Wheeler said when he first viewed the campus: "Everything is in great confusion!" Confusion, perhaps, but not in the minds of those who had come to teach and to learn, for instruction began on time and continued, in spite of sawing and hammering and dust and noise.[11]

Wheeler began his teaching immediately with a course in the embryology of vertebrates for four students: three young ladies and a gentleman. He wrote and delivered lectures mainly on the chick embryo and its development, but he also discussed current European research, particularly that of Dr. Theodor Boveri, who was making some important and exciting cytological discoveries. Though Dr. Boveri was working on invertebrate eggs, mainly those of worms, his conclusions seem to have wide applications throughout the animal kingdom. Wheeler's time was by no means taken up with his teaching, and he even noted in his diary that he really did not have too much to do, at least in the classroom. This, though Dr.

William Morton Wheeler as a young professor, probably at the University of Chicago, about 1898.

Wheeler may not have been aware of it, was one of President Harper's policies for his faculty: to give them a minimum number of teaching hours so that they would have plenty of time for research. Wheeler's research did not seem to take up much of his time, either, though he did continue at intervals to work on a problem he had started during the previous summer at Woods Hole concerning the description and naming of some little marine triclad flatworms or planarians. Wheeler attended some of the concerts (Paderewski moved him very much, he said) as well as lectures on campus (he especially mentioned enjoying several given by President Harper on "Job"). He took lessons in water color and gave a private entomology lesson each Thursday evening. But much of Wheeler's time during the fall quarter—another of Harper's innovations: a four-quarter system throughout the year—was taken up with the preparation of a paper on protective resemblance and mimicry for the Biology Club. The Biology Club was a favorite idea of Professor Whitman's, and he organized it at Chicago immediately upon arrival, just as he had done at Clark. Whitman allowed his graduate students to work pretty much independently, and he gave them advice only when they asked for it, except for the direction that no student should allow anything his animal did to escape attention. But the Biology Club served to bring these independent spirits, faculty and students, together once or twice a month, both to keep them informed of work going on within their group as well as elsewhere and to stimulate further research. Each speaker prepared and read a paper, and this was followed by considerable discussion, usually critical, sometimes vehement, but seldom dull. Wheeler did little original research in the area of protective resemblance, but he had long been interested in mimicry as a general biological problem. In 1892 and early 1893 he collected information and illustrations for his talk. In his short trips home to Milwaukee, he managed to discuss mimicry with the Peckhams and Mr. Rauterberg; he read Plateau's paper on protective resemblance and reread Bates' classic paper on mimicry. He borrowed specimens both in Milwaukee and Chicago, and on January 25, he gave his talk to the Biology Club. As to the exact contents of this lecture or of the discusssion following it, there is no record, except as Wheeler noted in his diary: had a discussion with Baur and Loeb. In light of what we know of both men, it would be interesting to speculate on what must have been a lively afternoon. George Baur was a Neo-

Lamarckian, interested in the causes of variation and well able to defend his convictions! Jacques Loeb was a no less vivacious and thought-provoking personality. Either you liked him or you did not, and obviously Wheeler did in those brief years they were together in Chicago. Of this relationship we shall speak further on a later page.

The list of personalities associated with the department of zoology during Wheeler's time (1892–1899) was impressive, especially when the summer associations at Woods Hole are included. Many of these men were of about the same age—late twenties and early thirties. Some of these were Henry H. Donaldson, F. P. Mall, Jacques Loeb, Sho Watase, Frank R. Lillie, A. S. Mead, E. O. Jordan, and, of course, Whitman and Baur. Add to these the early workers at Woods Hole: E. G. Conklin, E. B. Wilson, H. C. Bumpus, William Patten, T. H. Morgan, H. S. Jennings, Cornelia Clapp of Mt. Holyoke and Marcella O'Grady (later Mrs. Theodor Boveri) of Vassar, and so on. Of the major professors at Chicago, outside of Dr. Whitman, Professor Donaldson seemed to be the best liked. He was a gentleman of distinguished appearance, with a magnificent head, including, at least in later life, a beautiful crown of silver hair and a mustache and goatee. He was kind, understanding, and patient with students, and neat and orderly in his research and personal habits. Wheeler, and no doubt other young instructors and new students on campus, stayed in the Donaldson home until more permanent quarters were found. And on January 21, 1893, Wheeler and others (not specifically recorded) "after supper went over to Dr. Donaldson's where we started a club for the purpose of doing some study in the history of biology. Enjoyed ourselves very much." Such was the Donaldson hospitality!

Dr. Donaldson had been a student of G. Stanley Hall at Johns Hopkins, and had come with him to Clark. At Clark, Dr. Hall persuaded Donaldson to study the brain of Laura Bridgeman, a blind deaf mute who had learned to speak and had demonstrated considerable mental ability. This study led to an interest in the development of the brain and of the central nervous system in general, and this became Dr. Donaldson's life work. At Chicago, he was professor of neurology and dean of the Ogden Graduate School of Science and always active in his research and publication in spite of a tubercular infection of the hip joint which left him permanently lame. In 1906, he left Chicago to become professor of neurology and director of research at the newly established Wister Institute of Anatomy and Biology in Philadelphia.[12]

Of the younger men, Wheeler's closest friends at Chicago were E. O. Jordan, Shosoburo Watase, A. D. Mead, and A. C. Eycleshymer. All had come from Clark, and were all "Whitman" men. Jordan, along with Bumpus and Wheeler, had been one of the three who received Ph.D. degrees under Whitman at Clark. Eycleshymer had been at the Lake Laboratory in Milwaukee and followed Whitman to Clark and then to Chicago, where he received his doctorate in 1894. At Eycleshymer's death in 1925 at the age of fifty-eight, he was Dean of the College of Medicine at the University of Illinois. Watase was a cellular biologist with a doctorate from Johns Hopkins who had come under Whitman's spell at Clark, and had moved with him to Chicago. He was a capable research worker but not a good lecturer, mainly because he had trouble pronouncing English consonants, as, for example, making "copepods" sound like "coffee-pots" to his students. Watase left the University of Chicago in 1899, the same year as did Wheeler, and went to be professor of cellular biology in Tokyo. When Mead received his doctorate he went back to Brown University to teach comparative anatomy and to be with his old friend Bumpus.

Edwin Oakes Jordan and William Morton Wheeler must have been men of similar temperament, both serious and quiet, with brilliant minds and devoted to their respective fields of research, yet not without their lighter sides. The following quotation from a biographical memoir of Dr. Jordan could equally well have been written about Wheeler:

He went to Clark in 1890 to finish his formal training with Whitman and received his Ph.D. in 1892. The new environment was highly congenial in many ways. Clark University had been founded by G. Stanley Hall for the purpose of establishing a strictly graduate and research institution. Whitman's own mind had much in common with this ideal. Speaking of the graduate student, he had said, "He is recognized, not as an irresponsible school-boy, to be marked for absences, ranked for recitations, and rewarded, after a prescribed number of years of study and decent behavior, with a 'graduating degree'; but as a man who knows, or ought to know, his purpose, and who, if he ever expects to attain the distinction of a degree, must demonstrate his eligibility thereto by making some worthy contribution to the advancement of knowledge in his chosen field." This ideal, together with Whitman's strong conviction that a young man had best be given a problem and left largely to his own devices in working it out, were in complete accord with Jordan's own attitude. Despite the discrepancy in their ages—Jordan was twenty-four and Whitman fifty-eight—Jordan developed a great admiration for the older man and was undoubtedly greatly influenced by him. Both were interested in fundamental rather than superficial biological phenomena and shared the conviction that devotion to research was a prime means and chief end of higher education.

Jordan, a New Englander by birth, had been an undergraduate at M.I.T., where he had become interested in the new and developing science of bacteriology through a young professor there, W. T. Sedgwick. But, after a couple of summers at Woods Hole, Jordan's interests in biology enlarged, and for his doctoral thesis under Whitman he worked on the development of the newt. However, his old interests eventually won out, and the greater part of his life was devoted to microbiology, particularly as it related to the problems of public health. Jordan remained at Chicago throughout his life, and when he retired in 1933, he was Andrew McLeish Distinguished Service Professor of Bacteriology. His son bears the distinguished name of Henry Donaldson Jordan.[13]

One of the distractions from academic life for Wheeler and Jordan and others was the neighboring Columbian Exposition. In 1890, Congress had decided that there should be a mammoth celebration on the 400th anniversary of Columbus' arrival in the New World. Chicago, perhaps trying to prove to Boston and New York that it was a serious rival culturally as well as economically, bid for and received the honor to be the site for the great fair. In two and a half years, an incredible "White City" was built in Jackson Park on the shores of Lake Michigan—to the north and east of the University (the "Gray City"). Mud flats were converted to lawns, wooded islands, and lagoons, and dozens of buildings, small and large, and of every conceivable type of architecture, soon appeared over the twenty acres of previous swamp and wasteland. These buildings were finished in stucco (especially impractical in Chicago's climate) and painted white with gold trim. Over, around, and through these myriads of gleaming structures were hung hundreds of arc lamps, so that in the sunlight the White City blinded its visitors, at least physically if not mentally, and at night the lagoon could reflect this exotic scene to the crowds that were to come.

For come they did, the rich and important in carriages and the poor by foot. "Sell the cook stove if necessary and come. You *must* see the fair." So Hamlin Garland, then a young writer on his "way up," was said to have advised his parents in Dakota. Many people were reported to have used their burial money for this last fling. And unemployment, poverty, epidemics, and general misery were temporarily forgotten in an ephemeral haven of illusory beauty. Even the sophisticated Bostonian man of letters, William Dean Howells, was carried away when he said that the White City "gives men, as nowhere else on earth, a fore-taste of heaven."[14]

Although the Columbian Exposition did not open officially until May 1, 1893, the buildings were dedicated on October 21, 1892, and among the attendants was Dr. Wheeler. Nor did he miss the official dedication, when hordes of people and great excitement ran rampant over the area, and he even took Carl Akeley in early June to "see the sights." And let it be said that as far as is known Dr. Wheeler did not ever again let himself be hustled and bustled around in such a manner for such nonacademic reasons.

If Wheeler did not seem to be as dedicated to his academic pursuits as he had been, or would be, there may have been another reason, aside from being a young man with a new doctorate in a new position in a new university near an exciting world's fair. On September 18, a few days before he arrived in Chicago officially, he had had supper with Mrs. Grob and Amalie in Milwaukee, and then while boating up the Milwaukee River with Amalie, he had proposed marriage and she had accepted. Will Wheeler and Mrs. Grob and her daughters had been friends for many years, beginning when Will and Lisa were classmates at the German-English Academy and Seminary. Mrs. Grob seemed to give Will the intellectual stimulus and understanding he did not get at home, and her letters to him in German script were concluded with "Your motherly friend." When Will had lived in Milwaukee, he had spent many evenings at the Grobs', even after Lisa had gone, reading and talking. It is difficult to know exactly what the relationships were between the Wheelers and the Grobs (obviously they were all fond of each other), because Will's diaries are notably brief and lacking in emotional outbursts. He did not mention the proposal again in his diary until November 27, when he said that he had written to his mother "telling her of my desire to marry Amalie." And in December, to summarize his year's achievements, he wrote: "Another eventful year ended. Took my

Ph.D. and have a position in the Chicago University. Am engaged to be married. All important events!"

Although Will did not marry "Malie," as he called her, his year abroad studying, in 1893–94, was made a great deal happier by her "letters from home." She talked about Carl, about Will's mother and sisters, about Lisa's visits with her children (who asked after "Onkel Willie"); and her obvious interest in his work (though she probably did not understand it), his friends, and his sightseeing trips did much to ease his state of mind. As Malie was considerably younger than Will, she was perhaps his last tie to the world of his youth, a world which was rapidly fading for him, and that even she could not save.

President Harper gave Wheeler a year's leave of absence to spend in Europe, and on June 17, 1893, he boarded a steamer in New York City bound for England and the continent and a whole new world of study and travel.

4 / European Interlude

In late 1889 or early 1890, not long after he had arrived at Clark University, Dr. C. O. Whitman had written to his young friend Will Wheeler, who was still in Milwaukee: "Your decision to spend three years in Europe interests me much. I have been hoping that you might find your way here in some way next year." Whitman went on to describe what his hope for work at Clark was, but then he concluded his undated letter (Whitman never seemed to date his letters):

> Now as to Europe. I should rather study with Boveri and Boehm in Munich on karyokinesis than either Flemming or Carnoz. You would get very little attention from Strasburger. Van Beneden is probably the most brilliant and best trained man in Europe on this subject. Carnoz is nowhere in comparison.
>
> My second choice would be Weismann and Gruber in Freiburg.
>
> Have you seen Boveri's latest discovery? showing that the *male* nucleus of one sea-urchin may fertilize the enucleate fragment of the egg of another species of sea-urchin, and cause it to develop into the species from which the nucleus came. That seems to settle the question as to where the formative power resides, and to show a good reason for the existence of sexual reproduction.[1]

Thus was the seed planted, but not until January 7, 1893, did Wheeler write to Dr. Boveri asking to be allowed to work with him, and in a letter from Munich, dated January 25, Boveri replied: "Ihr Name und ihre schönen Arbeiten sind mir wohl bekannt." He then mentioned his projected move to Würzburg, and the work that he was beginning on the lamprey eel, *Petromyzon*. He concluded with "Ich bitte Sie, Professor Whitman und Professor Baur bestens von mir zu grüssen."[2] Wheeler probably knew that Whitman had tried to entice Boveri to Clark while the latter was doing post-doctoral work at the University of Munich.

Theodor Boveri (1862–1915) was born in Bamberg, in northern Bavaria, where his father was a surgeon. After graduating from high school in nearby Nürnberg, he went to the University of Munich to study history. However, he soon tired of this and took up natural history and medicine, and in 1885 published his first paper, "Beitrage zur Kenntnis der Nervenfasern." Shortly after this, the previously empty chair of Zoology and Comparative Anatomy was filled by a young man, Richard Hertwig, who had been a student of Ernst Haeckel in Jena. Years later, Boveri recounted his meeting with Professor Hertwig. Early in May 1885, a few days after Hertwig's arrival, he went into Hertwig's laboratory and announced that he was his assistant. Boveri then told Hertwig that there was no working space for an assistant, but that he would like to begin working as soon as possible. Hertwig got up from his worktable, told Boveri to grab the other end of the table, and they carried it into the next room. Boveri could begin immediately. Such was Hertwig's kindness and consideration for his students. To Wheeler, as to many others, Boveri proved an equally inspiring and sympathetic teacher and friend.

Boveri spent two years as an assistant for Professor Hertwig, and then left to become Professor of Zoology and Comparative Anatomy at Würzburg, where he remained until his death in 1915. Wheeler arrived in Würzburg not long after Boveri, on July 8, 1893, after a quick trip through Paris (where he visited the Louvre), Cologne, Bonn, and Heidelberg. At Heidelberg he found time to visit Baron von Osten Sacken and see his insect collection (kept in trunks), and to chat briefly with Professor Otto Bütschli, who was studying the chemical and physical properties of protoplasm. On the day after his arrival in Würzburg, he was assigned a place in Boveri's laboratory, in the evening had dinner with Boveri, and on July 10 began his work on the excretory system of *Petromyzon planeri*.

Although Charles Otis Whitman and Theodor Boveri may not have seemed to have had much in common, either in years or in background, Wheeler must have compared them often in his own mind and noted many similarities. Both were quiet, thoughtful, and completely dedicated to their work, and both were leaders in the scientific advance of their times. Neither restricted himself to studies of one aspect of biology or to one organism, though perhaps Whitman's observations were involved more with trying to understand the origin of species, and Boveri's with the origin of individuals. One of Boveri's students, Fritz Baltzer, wrote of his teacher:

> Almost the whole of Boveri's scientific work, as he himself once said, was devoted to the *investigation of those processes* "by means of which from the parental reproductive materials, a new individual with definite qualities arises." The basic problem of his work was thus, in the broadest sense, that of heredity and development. This includes three research fields: cytology, since the egg and the sperm are cells; embryology, since the definite qualities become visible only as the egg develops into an embryo; and finally, genetics, since the genes of the germ cells, in collaboration with the cytoplasm, determine the expression of these definite qualities. It was the triumph of the period to which Boveri belonged that the findings of these three fields came to be united into a conceptual whole; and precisely here, Boveri's theoretical views, for transcending his own data, were of pioneering significance.[3]

Before Wheeler had come to study with him, Boveri had demonstrated the importance of the chromosomes in fertilization—equal numbers of equal importance from the egg and sperm—thus paving the way for studies eventually proving that the chromosomes carried the determiners of hereditary traits. But in 1893 Boveri was interested in the development of the urogenital systems in various primitive chordates, including the lamprey eel, as he mentioned in his letter, and Amphioxus, as his publications indicate. These studies would certainly be classified today as "evolutionary biology," and Baltzer mentions a lecture entitled "Organisms as Historical Beings" as being of great interest because it showed Boveri's insights into nature. Wheeler remarked in his diary that he attended two lectures by Boveri in mid-November, the first on mimicry and protective coloration, the second on Darwinism and Lamarckism.

Not only was Dr. Boveri an outstanding scientist and teacher, but he also had considerable artistic talent, and this, along with some other remarks on Boveri as a person, is described by another of Boveri's students, Leopold von Ubisch:

> Theodor Boveri had a fertile imagination and a lively temperament, which was however held in check by strong self-control, so that he seemed rather reserved. He had much humor and could be quite sarcastic, and

when he was, he sometimes could be very sharp. He was very highly endowed artistically and would have liked to become a painter, which—fortunately, we may say—circumstances did not permit. But he painted and drew a great deal. He was very musical also, and played the piano well. His esthetic talent may be recognized in many of his scientific illustrations, which are not only clear but also beautiful. The drawings that he made on the blackboard with colored chalk during his lectures were masterly in their coloring and form, and he could devote himself freely to drawing them since he spoke without notes. The language and style of Boveri's works are also classical; he himself demanded that scientific works should be works of art.[4]

Wheeler's paper on his *Petromyzon* studies in Würzburg appeared in 1899 (the reason for the delay is unknown). It is a work of art, as his papers for Whitman had been. Boveri must have enjoyed his apt pupil.

William Morton Wheeler's six months in Germany were an education for him in many ways. Language was, of course, no barrier. Nor did he restrict himself to studying, but took advantage of many opportunities for sightseeing. Würzburg was an old and lovely city, known throughout the world for its baroque architecture and its "Stein" wines (the latter a favorite with the poet Goethe) as well as for its music and art. Wheeler often walked around the city, sometimes alone but usually with some of the other young men at the laboratory, enjoying the view, and examining the flowers and insects along the way. Some of the flowers he picked and pressed and sent back with his letters to Amalie Grob in Milwaukee. "Your dear letters are always such a treat to me," Amalie wrote to Will in an unusual letter dated August 23—unusual in that she wrote in English instead of the German script which was easier for her. "You always see something new again and tell me about it in such a nice way; I am thankful to you for so doing and the dear little flowers you sent along . . . I can well imagine that you feel lonesome and homesick at times. How gladly would I accompany you to some little village or other nice place in the neighborhood, or to the Steinberg—on one side of which you say so many different kinds of flowers grow, there we could pick a bunch of flowers and then arrange them into a pretty bouquet."[5]

Late in the summer Dr. Boveri left for Norway, and Wheeler was especially lonesome, as most of the students were away, too. So, after a week or two of drawing *Petromyzon* sections, Wheeler decided to take time off from his work and do a little touring. He went first to Nürnberg, where he visited the Burg and saw the royal apartments, torture chambers, and so forth. Then he went on to Munich, where he toured the Royal Palace and the art museum, visited the famous equestrian statue of Maximilian, and had a wonderful time at the opera *Siegfried*. He saw the falls on the Rhine at Neuhausen and marveled at the magnificent views of the Alps as he traveled to Zürich and Lucerne in Switzerland. Most of his trip was by train, but he took a steamer from Lucerne to Alpnach, where he again took a train (whose run was "the steepest in the world") to the summit of Mt. Pilatus (6998 feet high). The trip up took one and a quarter hours and was climaxed with dinner in a restaurant on top, and "a view which, though a little cloudy, was nice." On his way back to Würzburg, Wheeler stopped briefly at the University of Zürich to call on Arnold Lang, who was professor of zoology and comparative anatomy. In 1893 Lang was in the process of writing the *Text-Book of Comparative Anatomy* (the first of the two volumes was already out) covering invertebrates. Ernest Haeckel, in the preface to the first volume, said that Professor Lang, more than any other writer, had successfully utilized comparative embryology to explain comparative anatomy and furthermore had pointed out the phylogenetic significance of ontogenetic facts.

Wheeler arrived back at Würzburg early in September and took up his histological work again. In a letter dated September 7 and sent to his friend and co-worker E. O. Jordan, back in Chicago, Wheeler wrote:[6]

Boveri has been very kind to me and I have been pegging away most of the summer in the lab. trying to make out the development of the pronephros of *Petromyzon planeri*. Unfortunately I have not much material and each series of sections has to be made with extreme care. It would amuse you to see Boveri cut sections—the excessive care with which he draws the knife over the paraffin would certainly be ridiculous in a person of less celebrated technique. He is a thoroughly good fellow —though he knows nothing but morphology. He returns from Norway in about a week. Hope he will bring back *Myxine*. The assistant in the lab, Dr. Kathariner, is a *most consummate ass*. The docent, Dr. Schuberg, now thank goodness, gone to Karlsruhe as assistant in a technological institute, is a still more detestable specimen of humanity (by the way he was one of Ayer's teachers at Freiburg—which explains many things.)

There are three Germans in the lab. The toughest looking crowd you ever saw—"ganz verbiert." The youngest of the three, Herr Escherich, is a man of promise. He is working on the sexual organs of beetles. The others are hopeless cases. A Mr. Hill, a pupil of Ray Lankester's—a cousin of Harmer's of Polyzoan fame—a grand nephew of the Sir Somebody Hill who introduced the parcel post into England and who lies in Westminster—worked in the lab. most of the summer on the centrosome of Nephelis eggs. We became great friends and I have felt very lonesome since he returned to Oxford. We used to take long walks in the vicinity of Würzburg. (The valley of the Main is beautiful beyond description with the quiet little villages, and the long sloping vineyards!) The past two weeks I have been spending in Switzerland. I visited Nürnberg, Munich, Zürich, Luzern and a great number of smaller towns. Nürnberg is a gem, but Munich is divine. (Pardon the abortive effusion!) We could live in Munich all our days and never grow tired. I visited about 10 art galleries and the Hofbrau-haus—beer and painting, painting and beer everywhere.

Now I am back in Würzburg trying to get to work again, but my nervous system is still so excited with the things I have seen, that I find it difficult to concentrate my mind. When the winter semester opens I shall take Botany with Herr Geheimrat Prof. Dr. v. Sachs, who has kindly given me permission to attend his lectures—a great favor for you must know that he has grown very cranky. (It is rumored that he is addicted to the use of cocaine). He can't bear to see new faces—mine evidently seemed like an old one to him—for he was very pleasant during the visit which Boveri and I paid him! Boveri is almost his only friend here in Würzburg. I have seen Koelliker and Heidenhain several times but am not yet acquainted with them. Old Leydig lives here in Würzburg during the winter. I hope to have a chance to see the old boy, who has now grown somewhat childish, so I am told. Of course, you heard of Semper's death. He appears to have been insane and to have been troubled with some urinary disease. So much work has been turned out in this lab on the urogenital system that Semper could not very well die of anything else. The students here are curious animals. I cannot say that I admire them. Their social ideas would horrify an American. They spend most of the time drinking beer, duelling and running after girls. Thoroughly selfish and pig-headed (Mead would also add pot-bellied). They respect nothing which the ordinary American respects. And to see these animals enjoying the most superb health—rosy and hearty in every limb, while we poor Americans go about like living skeletons—because our ideals are so high!—Surely the lives of these students form a strange contrast to the struggling suffering existences of the poor people in this enlightened country. Still begging is forbidden by the law and beggars are subject to arrest if reported. People are so poor here! When I go to dinner I pass through the Stelzen Gasse by the Julius-Spital, a hospital founded by the prince-bishop Julius Echter von Mespelbrunn (in 15 hundred and something)—and hundreds of the most miserable mortals crowd around the door to receive their plate of soup from the good people in the hospital.

Apparently Dr. Wheeler had never been exposed to excessive poverty before, for he remarked on it many times, both in Germany and later in Italy, during this year of study and travel in Europe. It is difficult to surmise how much Wheeler was affected by this misery, or how aware he was of national or international economical problems. He mentioned in his diary that on September 21 he had received a letter from his mother telling him that he had lost $500 in a Milwaukee bank; no mention was made—perhaps he did not know—that his bank was only one of 600 banks and other financial institutions in the United States that had gone bankrupt in 1893. Only Chicago seemed immune to the pending financial crisis. "Tell it not in Goth," wrote E. O. Jordan to his friend in a letter dated October 13, "but it is a fact that the University has been very hard up this summer. Things are looking a bit brighter now owing to the great success of the Fair. The attendance last Monday—'Chicago Day'—was stupendous, 716,000 paid admissions!"

As indicated by his letters and his diary, Wheeler made a point of meeting as many of the leading biologists in Europe as he could. This was, of course, very exciting as well as potentially useful to him in his research and future publication. Some of the names he mentioned have escaped into oblivion, but the majority remain prominent in the literature and history of biology. Rudolph Albert von Kölliker (1817–1905) was a master teacher (at Würzburg for many years) and an indefatigable researcher in many aspects of zoology, though is probably remembered best for his investigations of spermatozoa (proving they were not parasites but a true sexual product) and his many fine, general histological studies. Franz von Leydig (1821–1908), noted for his work in comparative histology, had been an assistant of Kölliker and a long-time professional and personal friend, though most of his teaching and research had been at the University of Bonn. Professor Carl Semper (1832–1893), whom Wheeler did not have the good fortune to meet, had been a student of Kölliker, Leydig, and other morphologists at Würzburg, and during his life contributed much to embryology and the other morphological sciences, but today he is claimed by the ecologists. As a graduate student he had done some comparative studies, and these had left him with the feeling that he needed to know a lot more about many other forms, and in particular about the tropical fauna. So he spent most of his inheritance on a trip to the Philippines for several years of exploring, collecting, and study. He became a versatile and brilliant investigator, and encouraged his students to travel and to learn everything they could about their animals, both in the laboratory and in the field.

Julius Sachs (1832–1897), though old and difficult when Wheeler met him, had done much to modernize botany with his studies on photosynthesis, and his classic, *The History of Botany,* is still used as a source of much valuable information on the subject. Karl Escherich (1871–1951) went on with his entomological studies, publishing on various groups of insects, and especially on those which were of economic importance in forestry. Wheeler kept in touch with Escherich throughout his life, and in his correspondence files for 1934 is a handsome photograph of Dr. Escherich in his academic robes taken when he was made rector of the University of Munich. He had been director of the Institute for Applied Entomology at Munich for a number of years, coming there from the school of forestry at Tharandt.[7]

Wheeler finally settled down on his research in September and continued making sections of lamprey embryos. Once he was satisfied with these, he sectioned various salamander embryos, and even some ammocoetes (larval lampreys), tracing kidney development. After Dr. Boveri had returned from Norway, he looked over Wheeler's drawings and made various suggestions, as he did from time to time with all of his students. And when he became disgusted with his work, said Wheeler, Boveri cheered him up.

In mid-October Dr. Wheeler received from Washington the official notice of his appointment to the Smithsonian table at Naples (he had already been unofficially notified by the director of the zoological station, Dr. Anton Dohrn). So, in odd moments, Will studied Italian, too. Not long before the official letter, Wheeler's routine had been interrupted by the arrival in Würzburg of his old friend H. C. Bumpus, on leave from Brown University. Bumpus had come with the intention of studying at Würzburg, too, but for reasons not completely understood, went on to Munich after only a few days. In November Boveri and Sachs began a series of lectures which Wheeler attended and seemed to enjoy very much. November 22 was another notable date, for Wheeler saw his first duels—not one or two, but six, all fought in one evening. Evenings were usually more quietly spent however, studying or letter writing in his room, going now and then with the boys to the local variety show, or having a chat on zoology with Geheimrat Leydig, who "received me kindly."

Escherich and Wheeler continued their pleasant association, with Will helping Karl arrange the beetles in the collection of the Zoological Institute. Will accepted an invitation of Karl's to accompany him home for a week to his villa in Regensburg, with side-trips for sight-seeing and visiting along the way. Wheeler enjoyed the hospitality of Karl and his wife as well as the beauty of the Danube valley.

Karl Escherich finished his work with Dr. Boveri in mid-December, gave his "Doctor-Schmauss" dinner (excellent food and much drinking), and finally left for Regensburg on December 20. Two days later, Boveri and Wheeler had a quiet supper together. Boveri looked over the *Petromyzon* drawings, and they "talked on various topics until 12." The next day Boveri saw Wheeler off on the afternoon train to Munich—on the first leg of his trip to Italy. But Wheeler tarried in Munich to spend the Christmas holidays with Bumpus and his wife and boy. A "Prosit Neujahr" card he wrote to Jordan carried the following note: "Merry Xmas! Bumpus and I wish you were here in the Löwenbräukeller to drink beer with us and hear the fine music." Another card written a few days later, and mailed from Naples, said: "Have at last reached the 'Mecca' and hope to go to work tomorrow, when my place in the lab. will be ready for me. Naples is even more beautiful than I had imagined it to be. There is plenty of foliage on the trees and the weather is heavenly compared with what you are probably having in Chicago. Will write you a letter in a few days . . . Regards to the boys and to Prof. Whitman."

Like most north Europeans and Americans, Wheeler was both attracted to and repulsed by the city of Naples and its environs. The scenery was colorful, as was its palimpsest history. The luxurious villas of the wealthy stood side by side with the slums, with no apparent communication. The streets were filled with beggars and dirty children. Block after block of people lived in such poverty and misery that it was difficult for anyone but a Neapolitan to comprehend. Filth and amorality seemed everywhere, yet these people appeared content with their life. In a letter dated January 20, 1894, Wheeler wrote to Jordan:

I have now been in Naples just three weeks. The city is superb, to say the least, the people in the last stages of psychical, physical and moral degeneration. I never could learn to like [the Neapolitans] if I remained here the rest of my days. And worst of all, the charming weather with its endless sunshine and deep blue sky is telling on me. I feel like a lazzaroni and would rather be on a rug and eat macaroni and smoke cigarettes than work in the laboratory. The laboratory contains everything necessary for work, of course, and the library is superb. All the help about the establishment is well trained. Every morning I have two large jars of skimming to look over. They contain hosts of Appendiculariae, Medusae and Copepods, but very few Sagittae—the beasts I want to work on. I do not know whether I told you that I am going in for chromatin reductions, excited thereto by Boveri's last (not yet published) paper on *Ascaria*. To fill out the time till the Sagittae begin to lay (presumably sometime in Feb.) I am working on *Myzostoma*. We found some fine specimens the other day when we were out dredging. The species is *M. pulvinas* and occurs in the alimentary canal of the rather rare Crinoid, *Antedon phalangium*. The big females have little complemental males clinging to their backs—for what purpose I do not know, as I imagine that these so called females may be really hermaphrodites, as they are in some other Myzostomida.*

Bumpus will be down here in a few days, and then Eisig will give us a room together. For the present I am working in one of the curtained-off compartments of the general laboratory with a lot of pig-headed German privat-dozents, who make me very weary . . . There are two remarkably fine fellows here, Dr. Moore, one of Howes' students, working on spermatogenesis, young and iconoclastic, and a Mr. Fairchild, a botanist from the Smithsonian. We are together a great deal. Mayer and Eisig are both very kind and give us lots of assistance.

But Wheeler had more to say about the city and its people:

It is all that a young man's life is worth to go up the Via Roma in the evening. Females old and young, all exceedingly ugly, come up and take hold of one's arm, etc., etc. I was accosted at least 20 times in going the

* Myzostomida are marine annelid worms with flat, oval bodies having no external segmentation. They are usually placed with the free-living polychaetes, although they are parasitic on certain echinoderms.

distance of two blocks in the Via Roma the other night. When I walk through the villa with a cigar, a lot of little urchins follow me and beg for my cigar stump, and when I throw it away, they make for it like so many football players for the ball. I have seen men selling cigar and cigarette stumps in the street! You have heard the song Santa Lucia and have probably had pleasant dreams when you heard it, of sunny Italy and its charming (??) women, etc., etc., but you should sée the real harbor or street of Santa Lucia at Naples! Of all the filthy places! It is here where they sell the oysters with their gills full of cholera bacilli—and sea urchins—think of eating a sea urchin! and nasty little squids that have been out of water so long that they have deliquesced. It is interesting to study the faces of these people in Santa Lucia or, in fact, anywhere in Naples. Not a line that tells of any better feeling or a trace of intellectuality. The men look as if they were capable of stabbing anybody for a few liras. The women are no women at all. And then watch them when a few priests carry the consecrated host through the street! What a kneeling and bowing and taking off of hats!

But you will think I am very blue, and dissatisfied with Italy. I assure you that I am delighted with the country, i.e. with the scenery. To see the changing colors on the land and sea for a single evening more than pays me for coming here. I wish I could give you some idea of the wonderful tones that shift and melt on the sand and water and dull green trees, on the white houses and rocks where the sun sets off near Ischia in a glory of orange and vermilion. Last Sunday from the heights where the ruins of the Acropolis of Annae stand, I watched the sunset. I have never seen anything so beautiful. The sand was a delicate pink and the Mediterranean where it forms the lovely Bay of Circe was so blue, and the trees of the Royal Preserve were flushed with tones of purple and brown. This alone was worth a journey to Naples.

The zoological station of Naples is a remarkable institution, and Anton Dohrn, its founder and first director, was an amazing man. Dr. Dohrn (1840–1909) was born in Stettin in northern Germany, the son of Carl Dohrn, a wealthy sugar-refiner whose hobby was collecting beetles. The elder Dohrn was well-informed in his avocational interests, and encouraged his son in his pursuit of scientific studies. Anton Dohrn studied at Königsberg, and then Bonn, had a brief career in the army, but in 1866 he began his scientific studies in earnest when he went to Jena as docent in zoology to meet and work with Gegenbaur, Haeckel, Kleinenberg, Lankester, and many of the other leading European biologists of the day. In an obituary for Anton Dohrn, Professor E. Ray Lankester of Oxford wrote:

When I knew him at Jena, Anton had already made up his mind to do something really large and important for the progress of zoological science. Like others who had visited the Mediterranean in order to study its rich marine life, he had felt the difficulty of carrying on such work in lodgings, without apparatus, without library, and at the mercy of the fishermen whom it was necessary to employ and to conciliate . . . The plan took shape in Anton Dohrn's mind of establishing a larger and more completely equipped laboratory than these on the Mediterranean coast, and, but for the war between France and Germany, he would probably have carried out his first intention and placed his laboratory on the coast near Marseilles. When I knew him he had already thought out the scheme which he realised, and had determined to try to secure a site at Naples in the Villa Reale, which stretches along the shore. He had succeeded with no little difficulty in securing a certain sum of money from his father—his heritage, in fact—and he intended deliberately to risk this in his enterprise. His plan was to secure the cooperation of all European universities in building and maintaining the Naples laboratory, or "station," as he proposed to call it . . . But in order to obtain this support and cooperation he realised that it was necessary, at whatever effort and risk, to make a plunge—to start the "stazione," to erect a fine and imposing building, to demonstrate the convenience and excellence of its organisation, and thus to secure approval and unhesitating financial assistance. His plan was to sink his own fortune in that first step, and he did so. He obtained help from friends both at home and in this country as the building grew, and by tactful appeal and untiring effort—involving years of work given up to persuading statesmen, politicians, associations, professors, millionaires, and emperors of the value and importance of the great Naples "Stazione Zoologica"—he achieved for it a splendid and permanent position.

Professor Lankester then went on to describe some of the problems that faced Anton Dohrn in Naples: "An example of the innumerable difficulties which Dohrn had to surmount is the challenge to a duel brought to him by the representative of the Neapolitan architect whom he had agreed (in order to conciliate the Neapolitans) to employ for the design of the elevation. This gentleman considered himself insulted because Dohrn refused to promise him a ten per cent commission instead of the five per cent which is usual in northern Europe. I had to act as Dohrn's second, and conferred with the Neapolitan architect's friend. On my insisting that Dohrn was a soldier of the German Emperor, and a very deadly man with the sabre—and determined not to yield to any nonsense—

the challenge was withdrawn, and the insulted architect completed his task very satisfactorily."[8]

The Naples station was officially opened in 1874, thanks to a determined and dedicated man. But Anton Dohrn's talents were not limited to this one project. He was a great lover of classical music (Beethoven, Mendelssohn, and Brahms) and enjoyed Cicero, Horace, Shakespeare, and Goethe of classical literature. In biology, he was a strong pro-Darwinian, and encouraged the workers at his station to understand comparative physiology and morphology as it related to evolution. His research was mostly concerned with the probable ancestry of the vertebrates. He believed that Amphioxus and the Ascidians were not in the direct ancestral line, but degenerate forms of some earlier prevertebrate ancestor. He tried to connect the vertebrate stock with the chaetopod worms, and in his attempts to prove this hypothesis he made many interesting discoveries.

But Dohrn's greatest contribution to science was his zoological station. The building itself is picturesque with its white stone walls and red loggias gleaming like a true temple in the brilliant Naples sun. It contains a public aquarium on the first floor and research laboratories on the second floor, and of course has had to be enlarged to accommodate the ever increased demands for its facilities. The aquarium is a great tourist attraction, and contributes to the support of the station. Various universities and governments around the world also support the station by maintaining tables for investigators, such as the Smithsonian table Wheeler occupied.

The greatest attraction to biologists is, of course, the facility for research. Living material has always been supplied fresh every morning by local fishermen and can be maintained indefinitely in well-operated aquaria, and the library is excellent. During Wheeler's stay, Dohrn was assisted by Dr. Hugo Eisig, and the zoological department was presided over by Dr. Paul Mayer, who was always willing to help by teaching almost any technique, no matter how complicated, which might be needed for the then modern morphological investigations. Dr. Mayer also edited the publications of the station. But perhaps the most unusual staff member was Salvatore Lo Bianco, who as a local youngster of thirteen had joined the station as a laboratory boy, then went on to become, under Dr. Dohrn's tutelage and encouragement, "without the aid of schools or universities, a skilled linguist and biologist and an investigator of wide fame. His knowledge of the local and seasonal distribution of the fauna and flora of the Gulf of Naples has been one of the indispensable elements in the daily life and work of the station. His cheery greeting as he made the daily round of the laboratories to learn the needs of the three score and ten investigators, and his versatile talents, whether seen administering discipline—in Neapolitan—to some unruly member of his motley crew, or as the genial host at the trattoria on the Vico Pasquale, or singing 'John Brown' to the strumming of his guitar on an 'Ausflug' of the 'Johannes Muller,' endeared him to the biologists of many nations."[9]

Dr. Wheeler obviously enjoyed the station and his work and associates there. In early March he again wrote to Dr. Jordan:

> Your letter reached me today. Many thanks for enclosing the mail. Bumpus has been here more than a month now and we are working together in the same room at the Stazione . . . After much searching in the Auftrieb, which is brought me every morning, I fail to find any sexually mature *Sagittae,* so that I am unable to work on the subject. Some weeks ago I looked about for a new form, and finally settled on *Myzostoma.* This curious beast has not been studied embryologically since Beard's superficial attempt in '84. Lo Bianco supplies me with all the Crinoids *(Comatula)* I need and I have only to pick off the Myzostomes and fertilize their eggs artificially to get any of the early stages I want. I have made a curious discovery; *Myzostoma cirriferum, M. glabrum and M. pulvinar,* and probably all the other species of the group are males while very young, hermaphrodites when somewhat older, and females in their old age. Beard's interpretation of the sexual relations in these animals is quite erroneous.
>
> Last week Prof. Dohrn took us out in the steamer launch to Baja. A fine lunch was served on board and we had a most delightful time. After landing at Baja we walked over to Fusaro, catching lizards by the way. (You should have seen Carey Bumpus go for the poor little things!) Little urchins came out with orange branches which they had stolen from the neighboring orchards and sold them to us. From Baja we returned along Posillipo to the Station. The almond trees along the shore were pink with blossoms.

The scenery and the flora and fauna of the Bay of Naples were cause for many excursions from the laboratory. Naples is on the north side of the Bay, which faces southwest, and behind the city rise many old volcanic hills, capped by smoking Vesuvius to the east. In the Bay are a number of islands, including Ischia, which is nearest Naples and Capri, probably the best known. "Yesterday," wrote Wheeler to Jordan, "Bumpus and I went over to Capri in the steamer launch of the lab. Prof. Dohrn's son conducted us over. We visited the Blue Grotto and sailed around the island. We dredged near the Faraglioni—bringing up lots of sea-urchins (*Cidaris*), Holothurians, Chitons, etc. We landed at the Grande Marina and bought some coral of the charming native women. It was great fun to see the sweet little Capri boys dive to the bottom of the harbor to get the 5 centesimi pieces we threw in. Nothing can be more charming than Capri. The day was perfect and the sea magnificently blue."

Sometimes alone, but often with Bumpus and sometimes Mrs. Bumpus and their son, Carey, Wheeler walked around the city of Naples looking at the many chapels and churches, the palaces and the slums. With Dr. Moore, and later with Bumpus, Wheeler visited Pompeii, and on another occasion, Wheeler, Moore, and David Fairchild climbed Vesuvius, admired the view, and then "dined together at the 'Sirena' and discussed—women." The longest trip was a five-day steamer and train excursion Wheeler and Moore took to Taormina in Sicily. On a card dated March 27, 1894, Wheeler wrote to Jordan: "Am spending a few days here in Taormina with Mr. Moore (spermatogenesis man). Aetna is covered with snow but the oranges are in blossom under the hotel window. We kill time crawling about among the huge cacti and over the magnificent ruins of the old Greek theatre, where, by the way, one of Aeschylus' tragedies was performed. The foot-lights are badly damaged and the scenery of the theatre is mostly gone—but otherwise it is fairly well preserved. Hope to return to Naples in a few days."

In his book *The World Was My Garden,* published in 1938, David Fairchild also commented on those idyllic days at Naples in the spring of 1894:

My friendship with William Morton Wheeler, the other American at the Biological Station, continued until his death and grew with the years. But in those early days of our acquaintance, he suspected me, as a product of the Western plains, of knowing little about marine biology. One day in Naples he put me in my place quite brutally when I mistook some tiny rotifers, which had gorged themselves on chlorophyll granules, for swarm spores of my pet Valonia.

"Oh, rats!" he said. And I fled in disgrace.

But Wheeler's teaching instinct was too great to allow him to remain scornful. The next day he called me in to show me a sea urchin's egg lying in a watch crystal under his microscope, and explained the phenomenon of fertilization in the life of the sea urchin. It was an enormous egg surrounded by the flagellated swimming sperms, a sight which every school child should see, the beginning of an individual existence, a marvel which never fails to thrill me to this day.

Wheeler knew Dante almost by heart and also introduced me to the melancholy poems of Count Giacomo Leopardi, noted for their perfection of style. We climbed Vesuvius together, and frequented the curious cafes which used to line the Via Roma in those days.

A true spirit of research and peace presided over the Stazione Zoologica. We had a complete library, efficient servants, and a devoted staff. Indeed, it was a scientific monastery, the only one in which it has been my good fortune to occupy a cell. No click of typewriters or noisy conversation disturbed the quiet of the laboratories.[10]

The Neopolitan interlude came to an end for Wheeler in mid-April, when he started north toward Liège, Belgium, and the last two months of his European year of study. Enroute, he toured Rome (not missing St. Peter's or the Vatican art) and Florence (especially the art galleries), climbed the leaning tower of Pisa, and visited Genoa, Milan, again Lucerne and Heidelberg—finally reaching Liège on April 30, where he was to work at the Zoological Institute under Professor Edouard van Beneden. In a letter writter in Liège on May 2, Wheeler described some of his latest activities.

I arrived here Monday last after a most delightful trip down the Rhine. Stopped in Freiburg, but did not see Weismann as he had gone to England to deliver the Crooman lecture. Val. Haecker conducted me over the Zool. Inst. which is rather small and hardly as elegant as Boveri's at Würzburg. Stopped a few days in Heidelberg, charming as usual —but quite overrun and being rapidly spoiled by the English. (The English finally succeed in spoiling all the charming spots in Europe. In 10 years Heidelberg will be nearly as bad as Naples.) After taking a last look at the Cologne cathedral I started on to Liège.

Liège seems to be a hybrid between Worcester, Mass., and Paris. It reminds me very much of our American cities. The people "hustle" and the air contains a great deal of unconsumed carbon. The streets are not as clean as those of the north German cities. On the other hand, the boulevards with their horse-chestnut trees and the endless cafés full of blithe young Frenchmen with top-hats and pointed beards remind me of Paris.

The Institut Zoologique is the finest I have seen in Europe. It is well built, elegantly furnished, well-equipped. There is an abundance of light, the tables are very convenient and one is provided with every thing one wants. Prof. van Beneden is an older edition of Boveri, or rather, Boveri is a young van Beneden. The resemblance between the two men is most striking. Prof. v. B. has not yet recovered from the severe shock which he received a few months ago when his father died. So far he has given me a great deal of his time, patiently looking over my Myzostome eggs and making many suggestions. I sometimes have difficulty in convincing him that I am right on points. The difficulty is increased by my having to explain everything in French. I usually succeed in making the old boy understand, but the tenses of my verbs are something frightful ("haarsträubend").

I have succeeded in finding rooms in a French family. One of the members of the family—for many years head-cook in a big hotel here in Liège—gets up the most superb dinners. Wish you could enjoy them with me. The fellow is a real artist and I have rarely seen art which I more thoroughly appreciate. If I do not turn up at Woods Hole with the gout it will not be the fault of M. Marius. There are several other gentlemen who take their meals at the same place—a lawyer, an instructor in mathematics, the superintendent of a gun factory. (Liège is great in guns and the teaching of music.) They all speak French and nothing but French and I listen with both ears to all sorts of discussions about the anarchists who are just now parading the streets of Liège with their red flags and keeping the people in constant fear of a bomb. (Everybody expected to hear an explosion last night at nine o'clock but everybody was disappointed. At least I was; a dynamite explosion is the only form of European entertainment which I have not yet studied.)"

Two nights later Wheeler's education was complete, as a dynamite bomb exploded in Liège.

As for his biological education, Wheeler continued to take every advantage of the opportunities offered him. Under van Beneden, he continued to study *Myzostoma,* but he changed his emphasis, as is noted in the title of his paper summarizing his work, "The Behavior of the Centrosomes in the Fertilized Egg of *Myzostoma glabrum* Leuckart." Edouard van Beneden (1845–1910), the son of a highly reputed Belgian zoologist, Prof. P. J. van Beneden of the Catholic University of Louvain, who had done some pioneering work in parasitology, was, as Wheeler noted, in many ways like Theodor Boveri. Both men were working in the same areas of biology, namely embryology and cytology, though their researches were not restricted to these. Edouard van Beneden had first demonstrated the importance of the centrosome in cell division, and had then gone on to explain some even more fundamental processes of fertilization. Using the intestinal round worm of the horse, *Ascaria megalocephala,* whose chromosomes are few and large and thus easier to study, van Beneden showed that the number of chromosomes is the same in each cell and characteristic of a particular species. He further showed that the number is reduced during the formation of the egg and sperm and restored during the sexual process. Prof. van Beneden is also remembered for the journal he founded and edited, *Archives de Biologie*.[11]

Another important cytologist of these days was a young American Thomas Hunt Morgan, of whom we shall have more to say later. In 1894 he had been teaching at Bryn Mawr, but in June he began a year's leave to study in Europe. Wheeler had been asked to take Morgan's place for a year (at a salary of $2000—$500 more than he was getting at Chicago), but he declined saying that he did not feel that he should be away from his work at Chicago for another year. Also, he did not want to be so far away from Milwaukee! Meanwhile, Wheeler greeted Morgan as he passed through Liège in mid-June.

Dr. Wheeler's studies abroad ended when he left Liège on June 29 for Brussels and then London. He spent ten days in and around London, going to Oxford with another former student from Würzburg, Mr. Hill, and visiting Kensington and the British Museum to look at Myzostomids and crinoids. Finally, he sailed for New York on July 11 on the Teutonic—arriving on the eighteenth, and going straight to Woods Hole on the nineteenth to see his friends.

5 / Woods Hole and Chicago

In a letter to Jordan, Wheeler gave an interesting if not completely clear picture of what was going on in mid-season at Woods Hole, or Wood's Holl, as it was still called in 1894. On August 2, he wrote:

Have been here two weeks now, accomplishing nothing in the line of investigation but getting fat and boring people. There is the usual amount of slander and rows, etc., etc. You were down here early in the season and know how it is. Since you left some things have happened which I hope to have a chance to talk over with you in Chicago. Harper is expected down here soon . . . Ryder has lectured to us for the past four days and we are now familiar with the most recondite processes of life. Watase averages 5 cigars a day; Mead ditto. Bumpus smokes like a fiend . . . The old wharves where we used to bathe have been removed and we have to crawl over the barnacles in the beaches near Fay's woods and flounder about among the rocks. Mall has just left. He is engaged to be married and spent a large share of his time down here advising all the young men to get married. I hope to be in Chicago about two weeks from today. Will you bother so late in the season? Wish you were down here to talk over teaching at Chicago. It is a subject in which I can get no satisfactory information from Prof. W. On many matters he appears to be perfectly unreasonable—as usual. Please pardon this absurd letter. Kind regards to Mrs. Jordan and yourself."[1]

Professor Whitman was being almost overwhelmed by his success. His Marine Biological Laboratory was so popular that facilities could not be found to accommodate all who wanted to work there, and expansion appeared impossible because of lack of funds. Whitman had more immediate worries than the fall schedule at Chicago, and he knew his young instructor! For as his diary shows, Wheeler was capable of organizing his own time.

After spending a few weeks at Woods Holl and about the same time at my home in Milwaukee returned to Chicago Sept. 1st, 1894, and began preparing for my year's work. Decided to give a course in Comp. Anat. of Vertebrates and accordingly spent the whole month of Sept. in writing up lectures on that subject beginning with Amphioxus. Taught Comp. Anat. during the Quarter from Oct. 1st '94 to Jan 1st '95 and from Jan. 1st to April 1st '95. Took up Fishes, Amphibia, Reptiles and Birds. Then gave Embryology of Vertebrates from April 1st to July 1st and repeated the same course from July 1st to Aug. 14th when I went on my vacation to Wyoming. During the winter wrote some on my Myzostoma paper for the *Mittheilungen aus d. Zool. Staz. zu Neapol* but had not finished the paper when I left for Wyoming. A complete diary of my six weeks trip in Wyoming to the National Park on horse-back in the company of Norman Wylde and Welcher will be found in a notebook. [This has been lost and only one letter remains telling briefly that horses and provisions were picked up at Casper for the rugged trip into the Park.] On my return Sept. 1st '95 felt very well and at once finished my Myzostoma paper and sent the MS on to Paul Mayer. From Oct. 1st to Jan. 1st 1896 corrected the proof of the article, gave my course in Comp. Anat. completing the Ichthyopsida [lower vertebrates] with my class. Also began work on the Empidae [dance flies] during the Thanksgiving recess paying a visit to the University of Kansas to see Prof. S. W. Williston and W. A. Snow. Both of these gentlemen gave me a good deal of Dipterous material (Empidae, Dolichopodidae). From Jan. 1st to March 1st 1896 completed five papers . . . About the middle of Feb. began preparing lectures on Mammals, the portion of the course in Comp. Anat. which I did not have time to give last year.

Dr. Wheeler may have had good intentions in keeping a diary of those years, for he started one again in March of 1896, after the entry just cited, but too many other activities apparently diverted him. Nor is there any correspondence remaining which might give us a clearer picture of his work. But the *University Register* and some miscellaneous pamphlets and catalogues give us a general idea of what was being taught, or was at least scheduled to be taught. The University calendar was divided into four quarters, and Wheeler taught one or two courses in each. Comparative anatomy was started in the fall and continued through the winter quarter. Embryology then replaced it in the spring and carried over into the summer. In the winter quarter of 1894–95, Wheeler added a course in entomology, which he apparently taught each quarter until the fall of 1896. A summer catalogue for 1896 indicates that the course included field work, and that Wheeler gave practical methods for collecting and preparing insects as an introduction to systematics. Wheeler contracted typhoid fever in September 1896 and missed the fall and winter quarters of teaching. When he returned for the spring quarter of 1897, he again taught comparative anatomy and embryology, and with the assistance of a new instructor also gave elementary zoology. This schedule continued through the spring quarter of 1899, but entomology was never reinstated. Dr. Whitman and most of the rest of the faculty spent their summers at Woods Hole, although President Harper protested, saying the University's summer quarter was being neglected by the zoologists. President Harper and Dr. Whitman must have reached a compromise with the understanding that a few general courses in biology would be given on the campus in the summer quarter, and Dr. Wheeler was one of the men chosen to do this.

Dr. H. H. Newman, who was one of the students in the department, and later a member of its staff, wrote a brief history of the zoology department and said this of Wheeler: "William Morton Wheeler was a member of the department from 1892–1899. While at Chicago he is best remembered by the writer for his excellent course on the Biology of Vertebrates. In this course he did not limit himself to comparative anatomy, but dealt with phylogeny, breeding habits, embryology, behavior, distribution and other subjects. It was one of the best undergraduate courses we have ever taken. Wheeler's research during his seven years at Chicago covered a wide range of subjects and resulted in twenty-two publications, including researches in cytology, morphology, taxonomy, embryology. His chief interest, however, then as later, was in entomology. He had not yet begun his work on ants and other social insects for which he subsequently gained world-wide renown."[2]

In one of Dr. Wheeler's notebooks is a list, neither identified nor dated, containing the following names: Agnes Claypole, Emily Gregory, C. E. McClung, Mr. Turner, Mr. Pickel, Mr. McCarty, Mr. Newman, and Minnie Maria Entemann. Some of these names have been lost in obscurity, but apparently they were graduate students working with Dr. Wheeler during at least one, probably more, of the last years he was at the University of Chicago. As far as can be determined, most were doing problems in embryology, and at least four of them were working with insects. Miss Claypole (later Mrs. Robert O. Moody) received her doctorate from Chicago in 1896 with a thesis on a primitive insect, a springtail, entitled "The Embryology and Oögenesis of *Anurida maritima* (Guer.)"; she later became a teacher in California. Miss Gregory received her doctorate in 1899, having done a thesis on the development of the excretory systems of turtles; she later became professor of biology at Wells College, Aurora, New York. Miss Entemann (later Mrs. Francis B. Key) received her doctorate in 1901 on her work on coloration in *Polistes,* the common paper wasp (published in 1904 by the Carnegie Institute). She became associated with the Race Betterment Foundation in Battle Creek, Michigan and seems to have done nothing more in biology. Mr. Newman was presumably the Dr. H. H. Newman previously mentioned. There is no certainty about the Mr. Turner mentioned, but he may have been Charles H. Turner, who received a Ph.D. *magna cum laude* from Chicago in 1907. If this was the same Mr. Turner, it is interesting to note that his research, as indicated by his publications, was neither entomological nor ethological in the 1890s, but his doctoral thesis and most of the rest of his life's work (he died in 1923 at the age of fifty-six) was on the behavior of insects, and especially of ants, wasps, and bees—the same groups which came to interest Dr. Wheeler at about the same time. That Dr. Turner did not make the mark on science that Wheeler did may have been due not so much to his early death as to the fact that he was a Negro living in an era when opportunities for Negroes, especially for highly educated ones, were almost nonexistent.[3]

Dr. C. E. McClung, who went on to renown in the field of cytology, did not receive a degree from Chicago but apparently was sent to study briefly under Wheeler by Professors Williston and Snow at the University of Kansas. Professor Wheeler started him on some studies of reproduction in a grasshopper. This is described in an obituary for Dr. McClung, written by D. H. Weinrich in *Science:* "At the suggestion of Prof. Wheeler he undertook a study of spermatogenesis of *Ziphidium fasciatum,* a 'long-horned' grasshopper. In this material 'a peculiar nuclear element,' which others working on different insects had considered to be a nucleolus, was shown by McClung to be a chromosome. Although not the first to discover that this element was distributed to one-half the spermatozoa, apparently he was the first to see the significance of this fact in relation to sex determination. This interpretation, first announced in 1901, brought him world-wide recognition and initiated a life-time of researches on the chromosomes and their relation to taxonomy and evolution."[4]

Dr. Wheeler's own research during the 1890s is difficult to describe briefly, and can only be summarized as varied. Except for his year in Europe, he left us no informal comments on his work, and what we know is based mainly on a study of his publications. As mentioned earlier, Wheeler made some investigations on flat worms during his summer at Woods Hole in 1892. These studies were mainly of a morphological nature but included information on life histories, more or less incidentally to his embryological observations. While at Naples and Liège, as discussed earlier, Wheeler concentrated on the cytological and embryological development of marine annelid worms of the genus *Myzostoma.* These worms are hermaphrodites, being functionally male when young and functionally female in old age. There was considerable interest in this at the time because of the argument over the evolution of hermaphroditism. Some workers considered it to be a primitive condition in Metazoa and presumed that dioecious forms were derived by the suppression of one set of reproductive organs; others felt that hermaphroditism was a secondary acquisition. Sometimes the discussion was heated. In 1899 Wheeler must have hoped he had the last word with one of his critics in this display of the sharp wit which was to characterize many of his essays later in life:

The sexual phases of *Myzostoma* are hardly of sufficient general importance to justify more extended comment on Beard's paper, abounding as it does in misunderstanding, misrepresentation and futile speculation. Beard's unwarrantable inversion of the natural sequence of the degrees of parasitism within the genus *Myzostoma;* his gratuitous and confused hypothesis of a choice on the part of a dioecious animal between the Scylla of hermaphroditism and the Charybdis of parthenogenesis; his one-sided interpretation of the observations of Nansen, v. Graff and myself, who have devoted far more attention to these parasites than he has been able to give; his depreciation of the theories of others while demanding belief in his own speculations;—all these matters might be considered at great length, but Beard is not alone in having "more congenial and more important work in hand." I trust that enough has been said in this and my previous papers to convince any fair-minded zoologist that the "complemental male" of *M. glabrum* is one of those tenuous and fanciful creations for which one could have wished that euthanasia, that silent death so becoming to pet speculations when they have ceased to afford either amusement to their originators or edification to their readers.

No mention is made in any of Wheeler's notebooks or letters of how or why he came to give for the eighth Biological Lecture at Woods Hole in the summer of 1894, a translation of Wilhelm Roux's introduction to *Archiv für Entwickelungsmechanik der Organismen*. Apparently Wheeler had not met Professor Roux, nor are there any editorial comments included with the translation, and certainly Wheeler was never an experimental embryologist. Possibly Edmund B. Wilson or Jacques Loeb, who knew Wheeler well at Woods Hole and were aware of his fluency in German, persuaded him to translate this essay which was of such interest to the American cytologists and embryologists, most of whom were working summers at the Marine Biological Laboratory.

Of much more interest, at least as revealing of its author, was the fifteenth Biological Lecture given at Woods Hole in the summer of 1898, namely, Wheeler's paper entitled "Casper Friedrich Wolff and the Theoria Generationis." This paper, written in faultless prose, is a clear and informative presentation "of the two great views of embryonic development that have been and still are held by thinking students of nature—*preformation* and *epigenesis*." Before Wolff's time (1738–1794) philosophy rather than science had attempted to explain early development, and especially differentiation in an embryo. Wolff, using instruments that even in Wheeler's day were considered poor, showed that there was no expanding of a preexisting organism, and that preformation, therefore, could not exist. Rather, said Wolff, all the substance of the embryo is organized. He then offered a theory on how organization might occur, which Wheeler described briefly; and then Wheeler said: "We know that Wolff's main error lay in grossly underestimating the complexity of the problem he attempted to solve. This has always been a great pitfall in attempting an explanation of life. Perhaps it is well that it is so, for Wolff would hardly have had the heart to attempt it if he could have seen the problem with our eyes. And may not we, too, daily commit the same blunder when we lend a willing ear to those who regard living protoplasm as nothing more than a 'complex chemical compound?' " Nor did Wheeler miss a chance to defend the observational sciences: "Remaining within the province of observation which he staked out for himself, and pursuing his excellent method, Wolff was not only able to undermine the theoretical edifice of the predelineationists, but also to lay the foundations for future structures of great promise. Thus all conscientious investigation with good methods leads to subordinate facts of value besides the main line of facts accumulated in support of the theory in hand. Wolff was a biologist in the true sense of the word. He regarded plant and animal life as but slightly different aspects of a single set of phenomena. It can be shown that he anticipated to some extent the modern theories of protoplasm and the cell."

Wheeler's final paragraphs (of which only a part are given here) comparing Wolff and Charles Darwin are probably the most unusual part of this essay:

> Wolff's position in the history of thought on the subject of organic development becomes somewhat clearer when we compare him with Darwin, for whose coming he helped prepare men's minds. Wolff's *Theoria* was published in 1759; Darwin's *Origin of Species* in 1859. Wolff had been preceded by Harvey in much the same way as Darwin was preceded by Lamarck. Both Wolff and Darwin were ideal investigators, patterns for all time. Darwin's love of truth, his perfect fairness and modesty withal, seem to have been Wolffs' possession also . . . Both Wolff and Darwin devoted their lives to the investigation of the same great problem—the development of life on our planet. Both found answers to their respective parts of this problem. Wolff published his answer when he was very young; Darwin waited till he was well along in years. Each was confronted by a formidable, clearly formulated theory of special creation. The theory that confronted Wolff was special creation of all *individual* organisms by a preadamite fiat. Darwin was confronted by a theory of the special creation of all the *species* at the same inscrutable time . . . Both Wolff and Darwin collided with prevailing theological views. Darwin's experience in this matter is well known, and perhaps the less said about it the better. It is not so generally known that Wolff's failure to establish himself as a professor in Germany, and departure in 1769 for St. Petersburg, where he spent the remainder of his life, was probably due not only to professional jealousy, but also to a certain antagonism on the part of religious contemporaries . . . Both Wolff and Darwin left their theories unfinished. They maintained a transformation of simpler into more complex matter, but they did not succeed in demonstrating how this transformation is accomplished. I have already quoted Wolff's confession of ignorance of the way in which

epigenetic development is brought about. The doubts entertained by Darwin and his successors concerning the adequacy of natural selection as a complete explanation of descent are familiar to us all. The absolute completeness of the old *emboîtement* and special creation hypotheses doomed them to a speedy death. Wolff's and Darwin's hypotheses have lived because they represented only parts of a great truth. On this account, also, they have supplied and will continue to supply powerful incentives to investigation . . .

Just as Wolff's followers have split into two schools—one believing in little, the other in much preformation in the germ—so Darwin's followers have split into two schools, the Neo-Lamarckians and the Neo-Darwinians, in obedience to the two psychological tendencies to which I have called your attention. The Neo-Darwinians, in laying great stress on the segregation, stable and complex *intrinsic* structure of the germ plasma and its importance as a vehicle of hereditary characters, and in attributing less value to the *extrinsic* factors, like food and environment, are allying themselves with theorists of the type of Parmenides and Plato. On the other hand, the Neo-Lamarckians who believe in the permanent change-producing effects of the *extrinsic* factors (environment, etc.) on structure, and attribute less value to the architecture of an *intrinsic* vehicle of heredity (germ plasma), range themselves with Heraclitus and Aristotle.

The amount of differentiation displayed during the ontogeny of an organism or during its phylogenetic history will be differently estimated by different workers, till we are in possession of some means of mathematically measuring differentiation and variation. The demand for mathematical measurement is already being made in certain quarters, and this demand progressing science will undoubtedly supply. At present we are quite adrift in our discussions, so long as we ignore the more general and philosophical aspect of the question and thereby overestimate its simplicity. Even if we accept differentiation and the interaction of differentiated products as the root ideas of the evolution of the individual and of the race, we are still at a loss to understand how the initial differentiation arose—how the homogeneous first became the heterogeneous . . .

The pronounced "epigenesist" of today who postulates little or no predetermination in the germs must gird himself to perform Herculean labors in explaining how the complex heterogeneity of the adult organism can arise from chemical enzymes, while the pronounced "preformationist" of today is bound to elucidate the elaborate morphological structure which he insists must be present in the germ. Both tendencies will find their correctives in investigation.

Descriptive embryology did continue to interest Wheeler in the 1890's, and even when he was still recovering from his nearly fatal bout with typhoid fever, he continued to observe and write. In December 1896, he went to southern California to recuperate for three months, and upon reaching San Diego, he rented a room in an abandoned hotel on Coronado Island in the harbor. This was to serve as a laboratory where he could study the various marine organisms he fished out of the Bay. One of these was a parasitic animal found in the octopus, and whose affinities today are still obscure. It is called *Dicyema* and is classed with flatworms by some zoologists, although others relegate it to an entirely separate subkingdom (Mesozoa) between acellular and multicellular animals (Metazoa). Both C. O. Whitman and Edouard van Beneden had worked on European species, and through them one imagines that Wheeler became interested. His single paper on the life history of an American species presented some interesting new facts, but his interpretations added little to the meager literature of the group, and consequently the paper is seldom cited today. Nevertheless, it is interesting to note the breadth of Wheeler's zoological interests, even to including this little parasitic organism which is probably completely unknown to a majority of zoologists.

Another enticing problem was presented to Dr. Wheeler on that same California trip by Dr. G. Eisen of San Francisco; namely, eighty-seven specimens of a new *Peripatus* from Mexico. In 1898 Wheeler published a description of what he called *Peripatus eisenii,* and he intended to follow this with a description of the development of its embryo. Wheeler's preoccupation with ancient, annectent forms was to continue throughout his life; he later described a "singular arachnid" from Texas, and still later was much concerned with the progenitors of various groups of social insects.

If there was a connecting link through all of these varied papers, it may have been Wheeler's interest in tracing phylogenetic relationships through the animal kingdom. In 1899 his foray into vertebrates was published—that is, the summary of his research under Professor Boveri on "The Development of the Urinogenital Organs of the Lamprey." This long paper, with its detailed descriptions, thorough covering of the literature, and seven plates of sixty-seven beautifully executed drawings (mostly from microscopic slides), was and is still a model for scientific research and reporting. Dr. Wheeler was proud of it, we know, for in 1930, when he was asked what he considered to be his fifteen most important papers, he listed this one, along with his doctoral dissertation and his two *Myzostoma* papers from Naples and Liège.

During these last few years of the nineteenth century, Wheeler's writings also included a few book reviews, the previously mentioned memoir on George Baur, and a major paper on "The Free-swimming Copepods of the Woods Hole Region," complete with keys and illustrated descriptions, which is still a standard reference on the subject. A short paper entitled "Anemotropism and Other Tropisms in Insects" is of interest as an indication both of Wheeler's early interest in attempting to understand instinctive behavior and of the influence of Jacques Loeb and his theories of tropisms which at this time were being widely discussed in biology. We shall return to this subject in a later chapter.

Although Wheeler and Loeb were not associated again professionally after 1899, except from time to time at Woods Hole, their lives were interwoven in various ways throughout the ensuing years. Dr. Loeb had married (in 1890) an American, Anne Leonard, whom he had met on a mountaineering trip in Switzerland. She received her doctorate at Zürich, but she had done her undergraduate work at Wellesley College in Massachusetts. There, one of her classmates had been Harriett Emerson, from Rockford, Illinois, whose youngest sister, Dora, would become Mrs. William Morton Wheeler on June 28, 1898. Family tradition holds that Will and Dora met at the Loebs, but there seems to be no record of their first meeting.

As we have seen, Wheeler was at one time very friendly with Lisa Grob, and at a later date engaged to her younger sister, Amalie. Lisa was now married and living in New Braunfels, Texas. Amalie remained a spinster and later lived with her sister and family. We know nothing of Will's break with Amalie, but he was to remain friendly with the Grob sisters for many years. Will's transfer of affections from Amalie Grob to Dora Emerson may have reflected the fact that he had grown apart from his Milwaukee background and was now an established man of science. Dora was not only well educated in science and the humanities, but she came from a distinguished and remarkable family. Perhaps we can make up for our ignorance of the courtship of Will and Dora by telling a bit about the Emerson family and how they came to mean so much to Professor Wheeler.

Ralph Emerson, Dora's father, was a large man both physically and mentally, and his success as a businessman had come about through his own efforts and abilities. He had sent his five daughters to Wellesley College and his son to Harvard. A number of his ancestors had been New England clergymen, and one of these, his great, great, great grandfather, Joseph Emerson, was also the great, great, great grandfather of Ralph Waldo Emerson, the Concord philosopher. Young Ralph Emerson at first tried his hand at teaching (he especially liked mathematics), but his health broke down. When he was barely twenty years of age, he left Andover, Massachusetts, where his father was a professor at the Theological Seminary, and went out to see his older brother, Joseph, at Beloit College in southern Wisconsin. (Joseph would become professor of Latin and Greek at Beloit and have a building named in his honor there.) Then, after some small business trips for his father, Ralph arranged in 1851 to study law in the office of Kersey H. Fell of Bloomington, Illinois. Periodically, in his circuit traveling from Springfield, a young lawyer named Abraham Lincoln would visit the Fell office, and it was there that young Emerson and young Lincoln met. In his *A Short Autobiography,* Mr. Emerson said:

> As I saw law practised I did not like it. I drew warmly towards Lincoln. He appeared to take a kindly interest in me and [we often took] a walk or two together. At last my heart was full. I told Lincoln I wanted his advice. How kindly he turned around with "Well." I told him I wanted to be a lawyer and a Christian. He appeared to think that was not difficult, talking in an evasive way. I pressed the question saying, "Mr. Lincoln, I see all the big money is made by lawyers who aid scamps in doing what is wrong. Now is it possible for a man to make a fair living as a lawyer and yet never be caught aiding any bad man or keeping a bad man from being punished? Is it possible for a young man to start out here at the West and get a lawyer's living on that rule?" He almost stopped—a long sigh came; then suddenly his chin projected forward and we walked on in silence. No word came but I had got my answer. I fancy from that time forward, he appeared to value my acquaintance and to talk with me as he did no other citizen of Bloomington.[5]

Mr. Emerson did not continue reading law, but went back to Beloit and accepted a clerkship obtained for him by his brother, Joseph. The firm which employed Ralph Emerson had two stores—one selling dry goods and groceries, the other hardware. Emerson was bookkeeper for both and did so well that he was soon transferred to Rockford, Illinois to help manage a branch store. In those years, 1852–53, Rockford was at the end of a railroad line, and Mr. Emerson (who had become co-owner of the store) sold goods fifty to sixty miles westward. One of his customers was John H. Manny, who was building a few reapers. Emerson offered to loan Manny materials to build more reapers if he would come to Rockford to do it. Eventually he negotiated a partnership between Manny and Wait Talcott, who was a wealthy local businessman. Emerson soon became a member of the Manny Company, and later he married Wait Talcott's daughter, Adaline. But no successful venture is without its problems: the J. H. Manny and Company found itself being sued by a competitor, Cyrus H. McCormick, to sustain a patent. At this point, Emerson remembered his lawyer friend and recommended to Manny that he hire Abraham Lincoln. Lincoln, to prepare for the case, came to Rockford and went through the Manny factory, but when the trial opened (in Chicago, to suit the judge), Manny and Company was represented by George Harding, a well-known patent lawyer of Philadelphia, and by Edwin M. Stanton, then in his prime as a lawyer. Stanton had "looked down his nose" at the rumpled midwestern lawyer and dismissed him scornfully as "that long-armed baboon." But when Lincoln became President, he appointed Stanton Secretary of War, because, he said, he was the best man he could find for the job. And, incidentally, although the suit dragged on for many months and proved to be costly, it was decided in favor of Manny.

Ralph Emerson continued to prosper in his various business adventures, but one of them brought him great tragedy: the death of his only son. A fire had broken out in a nearby factory, and Ralph, Jr., in trying to help put out the fire, was killed accidentally—just before his twenty-third birthday in 1889. *Ralph Emerson, Jr., Life and Letters,* edited by his mother and published privately in 1891, attests to this great family loss, as does the fact that for many years following the tragedy, the Emersons paid the cost of insurance policies for all the Rockford firemen.

If Mr. Emerson himself did not follow the intellectual life, he enjoyed and appreciated it in others, and especially in his wife. Adaline Elizabeth Talcott had graduated from Rutgers College in 1856 and after her marriage had shown great cultural and executive abilities—often going to national and international conventions for philanthropic and other organizations. Her book *Love Bound and Other Poems,* about her love for her husband and her children, was first published in 1879 and went through three editions—something of a best seller for a book of personal poems.[6]

Mr. and Mrs. Emerson's youngest daughter, Dora Bay, was born on March 7, 1869. She graduated from Wellesley College in 1892, with a major in chemistry, and is listed in the catalogue for the University of Chicago as a graduate student in 1892–93 (the year that William Morton was studying abroad). We do not know if she attended Chicago. She did go to New York City to attend Columbia's Teachers College, and received her Master's Degree there in 1898, just before she was married. During the summer of 1895 she took a course in vertebrate morphology at Woods Hole. Wheeler was not there at the time, but Will and Dora apparently met that summer or soon after, for they were corresponding in mid-1896. While Will was in California recuperating from typhoid fever in the winter of 1896–97, he was visited several times by two of Dora's sisters, Belle and Mary, and their mother, who were traveling in the West. Wheeler noted in his diary that on February 17, 1897 Mrs. Emerson visited his laboratory in San Diego to see his specimens; they then went together in her carriage to Point Loma at low tide, when the collecting was especially good.

That there was a great deal of understanding and appreciation between Will and Dora, and between Will and the Emersons—a special feeling between a remarkable family and their talented future son-in-law—we know from a touching letter he wrote on July 13, 1897:

My dear Mrs. Emerson:—

I beg your forgiveness for not anticipating your kind letter. Had I received it a few hours earlier I should have detained Dora in Chicago—I have grown so selfish—but we did not find it at the laboratory till after she had made arrangements to leave on the three o'clock train yesterday. We are both unspeakably happy. When you saw me last in California life had ceased to have any meaning for me and I was only a machine for turning out zoological papers. Thank God that stage has been passed, and I have the courage to face life and do my duty cheerfully. And this doing one's duty *cheerfully* makes all the difference in the world. It is a matter of continual wonder to Dora and myself that we did not understand the meaning of life before we reached our present age. Perhaps we understand it more thoroughly now. At least we both appreciate our mothers more. When I saw how you had retained your bright intelligence and charming character through an active life—and how Dora resembled you in so many ways, I could not help admiring her—and loving her when I came to know her better. Surely I am the most fortunate of men. I feel how unworthy I am of such a woman—I could draw the ideal man whom she should have as a husband—and the contrast between him and myself would be appalling. How sore I feel when I make the comparison I cannot tell you. I can only hope to make her happy—to improve my character through my love and admiration for her. Perhaps as the years run on, I may approach that ideal man more closely—be a son to you and replace in a slight measure that grand youth of whom Dora has told me something and whom I shall always regret not having known.

I thank you most cordially for your kind invitation to spend a few weeks at the sea-shore. Nothing would please me more than to continue our collecting work on the Atlantic coast. I will let Dora decide this matter for me and do whatever you and she think wisest. There is a great deal of work which I should be doing here—but perhaps it had best be left undone for the present. My health is very good now—much better than it was when you saw me in San Diego. My muscles are harder and my brain in a healthier condition. The repugnance to all forms of bodily exercise, which seems to follow some cases of typhoid, is wearing away, and when the anniversary of my sickness comes around I shall be in better condition than I ever was before in my life. Hoping that I may soon have the pleasure of seeing you again either in Rockford or in the east, I remain

Sincerely & Gratefully Yours
W. M. Wheeler[7]

Dora and Will were married quietly in Rockford almost a year later—with only relatives and a few close friends attending. The event was described in a local newspaper, *The Morning Star,* on June 29, 1898:

The palatial home of Mr. and Mrs. Ralph Emerson, in North Church street, was the scene last evening of the marriage of their youngest daughter, Miss Dora Bay Emerson to Professor William Morton Wheeler of the University of Chicago.

The ceremony which linked the destinies of these excellent young people was performed in the front parlors before the large window which had been decorated for the purpose of presenting an effective background of green and white. From the chandelier in the center of the room were festoons of asparagus fern and mingled with this was the pure white of flowers. The room was prettily decorated and the softness of the delicate fern made an effect of fairy-like beauty.

At the hour of half past eight to the strains of the beautiful Mendelssohn's wedding march, played by the Benedict orchestra, the couple took their places before the officiating clergyman. Their path was marked out for them by two dainty little ribbon bearers, Miss Harriet Hinchliff and Master Ralph E. Thompson, niece and nephew of the bride.

The words of the service which joined the hearts and lives of this couple were spoken by Professor Joseph Emerson of Beloit College, who is an uncle of the bride. During the progress of the ceremony the sweet and subdued strains of music by the orchestra were heard blending with the voice of the minister as he read the service.

The bride was daintily gowned in white satin with trimmings of duchess lace and wore a bouquet of orange blossoms . . . After the ceremony there was a period of informal congratulations from the assembled friends and then the company was conducted to the dining room which was appropriately decorated in the green and white and refreshments were served.

The article then gave a brief biographical sketch of both Dora and Will, mentioned that their wedding trip would be delayed until after the summer term at Chicago, and listed the out-of-town guests. Included in the list were Will's mother, his sister Olive, and his aunt and uncle, Mr. and Mrs. William E. Anderson.

6 / Texas: Enter the Ant

Among those receiving the Ph.D. degree from the University of Chicago in the spring of 1899 was Wesley W. Norman, head of the School of Biology at the University of Texas in Austin. That same spring, the School of Biology at Texas had been divided, and Norman had become head of the zoological division. Then, because of increased enrollment, both the zoological and botanical schools were moved from the basement of the university building (Main Building) to the entire third floor of the new east wing. After this successful spring, Dr. Norman left for Woods Hole, where he planned to spend the summer in his physiological research, but he contracted typhoid fever and died suddenly on July first. The position was offered to William Morton Wheeler, who three years earlier had barely survived an attack of the same disease. A few weeks later Wheeler resigned his position as assistant professor at Chicago and accepted an appointment as professor of zoology at the University of Texas (for a salary of $3000, which was only a few hundred less than the president of the University received). He took the train to Austin early in September 1899, after what must have been an intense few weeks of discussion and negotiation.

It is needless to express my feelings on the loss of Wheeler [C. O. Whitman wrote to President Harper on September 4, 1899]. You have known how highly I valued him, and if you have properly estimated him, you certainly know that the University can not afford to lose such men. Why you have let him go is more than I am able to understand, unless you have decided to let my department suffer. Can it be that you think Wheeler is not as well entitled to a professorship as men who have received such recognition in other departments of science? Within a few months Zoology loses two men [Dr. Watase had gone to the University of Tokyo] that are known throughout America and the whole scientific world as biologists of the first order. When such men go, everybody knows why they go, and they do not ask what is the matter with the men, but what is the matter with the University. It is the reputation of the University that suffers, and suffers more in proportion to what is expected of it.

You seem willing to hold me responsible, at least, for Wheelers' case. I am willing to be held to account for what I do, but not for what I do not do. If you really think so poorly of me as to imagine I could advise Wheeler to leave us, I hope you will take pains to disabuse yourself. I recommended Holmes to Winston [G. T. Winston, first president of the University of Texas] and do not know how he came to think of Wheeler. Wheeler came to me with his mind made up, and I did all in my power both with him and with you to save him to the University.

Will you kindly tell me what you have in mind to do under the circumstances?[1]

The University of Texas had begun in 1883 with approximately two hundred students, but by 1899 the enrollment had increased to almost six hundred. In spite of the lack of good students at the beginning, the university had encouraged scholarly work and had

The Main Building of the University of Texas, Austin, taken about 1900. (University of Texas Library)

gradually added to its equipment and its library, which included 35,000 volumes in 1899. An effort had been made to build a distinguished faculty, and among Wheeler's colleagues were such men as F. W. Simonds (geology), W. L. Bray (botany), H. Y. Benedict (mathematics), and Sidney E. Mezes (philosophy). In 1899 there were only three buildings: Main Building, housing all of the classrooms as well as the library and auditorium, a boy's dormitory, and a small chemistry laboratory. Wheeler settled quickly in the new wing of Main Building and was soon teaching classes, as he wrote to E. O. Jordan in Chicago on October 13, 1899:

> I am very much pleased with my new position. Although I have 160+60+10+8+2 students in my five classes I really have much less work to do than I had in Chicago. There is plenty of assistance and I have $1100 appropriation + equipment fund + the lab. fees of all these students + a library fund of $100. The students are mostly farmer boys with good heads & better muscles, good manners and thoroughly in earnest in their work.
>
> The Univ. will have at least 200 students more than last year (1000 in all). In 3 years we shall have more than Chicago.
>
> I am delighted with Texas and its people and not at all of the opinion of the man who said that if he owned Texas and hell he would sell Texas and live in hell. (The man would be right in saying this of Chicago!) The fauna is everything I expected and more. It is far richer and more tropical than that of California. The climate is very healthful. The sun is hot to be sure, but there is always a breeze and the air is very dry. I have a larger house on a three acre lot with peach and fig trees. We are keeping a horse and I shall soon add a cow, dog and possibly a burro, or donkey (price $1.50!) to my establishment. The house is very near the country—five minutes walk from the great woods of live oaks draped with Spanish moss and mingled with huge cacti now in full fruit. When you and Mrs. Jordan grow tired of Chicago you must bring the boy and stay with us. I forgot to say that the roses are now in full bloom—all the varieties of the northern hothouses and more besides.
>
> Mrs. Wheeler and baby Wheeler [Adaline] will be here Oct. 25th if all goes well.[2]

In the *University Record* for October 1899, Wheeler pointed out that the University of Texas, because of its location in an extremely rich faunal belt, had great potentialities for excelling in zoological work. He said that zoology should be taught first as an essential to a liberal education, second, as a basis for medical training, and third, as a subject with economic applications. He continued:

It is, moreover, obvious that much of the energy of every school in the University must necessarily be directed to the training of students in the elements of a given subject, and that few students are able to devote much of their time to research. These conditions must retard the study of the vast zoological resources of Texas. The instructor who wishes not only to impart the basic principles of his subject, but also to induce his students to contribute, in some slight measure, to the actual growth of science, is bound to organize the work of his school in such a way that research problems may be solved gradually by the combined or successive labors of several students. Thus the structure, development, habits and geographical distribution of some interesting animal or group of animals may be worked out piecemeal by several students during successive years. Even beginning students can aid greatly in such work by collecting and preparing the materials to be used by themselves or other students in subsequent years of the University course.

For the sake of calling attention to the enormous field of study which is, so to speak, the natural heritage of the zoological student in Texas, a few of the innumerable subjects suitable for investigation in the laboratory may be indicated. These subjects are suggested by a few weeks' acquaintance with the Texas fauna—a year's acquaintance would increase the list a hundred fold. The conscientious study of any one of these subjects could not fail to reveal new facts which would alter or complete our general conceptions of certain zoological phenomena.

After a page and a half of discussion of some of the local vertebrates and some of the possible research problems relating to them, Professor Wheeler turned his attention to the invertebrates:

The work to be done in Texas on invertebrate forms cannot be estimated, as it seems to have no limits. The crustacea and mollusca of the State have received some attention from systematic workers, but the whole subject of their habits and distribution, not to mention numberless problems in their development and anatomy, has not yet been touched. The leeches and earth-worms of the State, together with the fresh water Oligochaeta are almost or quite unknown.

When we come to the insects the opportunities for work are indeed magnificent. To the zoologist acquainted only with the more scanty insect fauna of the Northern States, Texas presents an embarrassment of riches. A short ramble among the woods and hills about Austin during a season not especially rich in insect life suggests these problems for investigation: The habits and development of the ant lion (*Myrmeleon*), the singular larvae of which dig their funnel-shaped pitfalls even in the streets of the city; the embryonic development of the "dobson" (*Corydalis*), which is a very primitive and typical insect well adapted for class work; the habits of *Pronuba yuccasella* and its allies, the remarkable moths which fertilize the blossoms of the yuccas; the embryonic development and spermatogenesis of numerous large bugs (*Hemiptera*), of the "walking sticks" (*Phasmidae*), "devils' coach horses" (*Mantidae*), mole crickets (*Gryllotalpa*), and other conspicuous Orthoptera; the habits of the leaf-cutting and agricultural ants; the habits of the social and solitary wasps and bees; the development and metamorphoses of many of the undescribed caterpillars and beetle larvae; and systematic studies without end of many groups of insects, some of which, like the Diptera, Neuroptera, Odonata and Orthoptera, have scarcely received much attention even in the Northern States.

In the same publication for the following June (1900), Professor Wheeler had a longer article on "The Study of Zoology," which was a more detailed consideration of what he thought zoology was then and what it could be ideally. He said that man was interested in zoology mostly because of what it could tell him about himself. "As usually understood, Zoology includes only enough of the study of man to do justice to the comparative aspects of the science. In a sense Zoology is *the* study of all others which leads most directly to an understanding of the human species. Man is the final result of a long and intricate development extending through the unmeasured past. The science of history is necessarily limited to a tiny fragment of this enormous period—the vastly greater and in many respects the more significant portion of man's development, anti-dating all history, can be dimly traced only in the structures and activities of the animal organisms more or less closely related to man through the bonds of hereditary ascent."

Wheeler then explained that for convenience the science of zoology had been divided up into several subordinate sciences, and that often these had been developed by different methods with somewhat different aims. He described comparative anatomy ("the basic study in Zoology"), microscopical anatomy ("it furnishes the necessary preparation for the study of growth and development, including the problems of heredity and sex"), development or embryology ("the science which attempts to explain through what processes animals have become what they are—must illumine the very depths of all other zoological disciplines"), and physiology and its important application to hygiene problems. He included natural history, too, which he said had long been part of zoology, but that now it had passed through the anecdotal stage and was being placed on a scientific basis. "It deals with the relations of animals to one another and to their environment, and comprises such subjects as symbiosis and parasitism, the sexual relations of animals, the various forms of family and social life among animals, their nest-building, breeding and feeding habits, together with the comparative study of instinct." Dr. Wheeler felt that zoologists should study geographical distribution, which "will enable us wisely to exploit the resources of our country without the destructive waste at present in vogue," and paleontology for a better understanding of spatial relationships of animals. Classification, or taxonomy, however, said Dr. Wheeler, has no place in the classroom because it "is in a state of perpetual flux, since it is being continually altered and improved from day to day as our knowledge of the animal world expands." But, he added, because it is important in explaining evolutionary affinities, it does have a place in the laboratory and museum.

During the four years of Wheeler's stay in Texas, the classical courses in zoology already mentioned were included in the departmental offerings. The introductory course, first biology and later zoology, was taught by Wheeler, assisted by his departmental instructor, Augusta Rucker. Wheeler also taught comparative anatomy (fall term), histology (winter term) and embryology (spring term). Miss Rucker taught physiology. All freshmen in 1899 and 1900 were required to take a course in physiology and hygiene, and Wheeler and Miss Rucker taught this jointly with the department of chemistry. Wheeler added to the curriculum entomology, advanced zoology (whose special problem studies would be published as "Contributions from the Zoological Laboratory of the University of Texas"), a weekly seminar on zoological literature, and in 1901, a course in zoological drawing.[3]

As previously noted, Wheeler had five courses to teach in the fall of 1899, and some of these had large enrollments, but in spite of the heavy teaching load, he seemed to enjoy his work. His greatest pleasure, however, seems to have been his association with his teach-

William Morton Wheeler as professor of zoology at the University of Texas, Austin, about 1900.

ing assistants and graduate students (masters' candidates, as no doctorate was yet offered by the University of Texas). In 1937 one of these students, C. T. Brues, wrote in retrospect:

Members of the University of Texas faculty and old-timers among the ex-student body who were on the campus just about the beginning of the present century will have personal recollections of Professor Wheeler. He came to Austin in 1899 as professor of zoology, and although he remained for the brief space of only four years, his influence marked the beginning of a new era for zoological science in the University . . .

At the time Dr. Wheeler served the University the Department of Zoology included only a single instructor and a couple of young student laboratory assistants in addition to the professor in charge. Consequently he had to lecture and conduct all the zoological courses. This entailed an enormous burden of routine work, and his lectures were always meticulously prepared and delivered with faultless attention to detail in content and sequence . . . In addition [to department offerings] he conducted a weekly seminar and during two winters gave a course in scientific German!

There were very few graduate students, but the breadth of interest and high standard that he maintained is clearly shown by the fact that the little group of students trained in the old two-room laboratory produced a large proportion of professional biologists. These included university professors and others in widely separated fields—general biology, physiology, medicine, vertebrate embryology, and entomology. In those days facilities for investigations were hopelessly meagre, but scientific contributions immediately began to flow from the laboratory to the number of some fifty publications during the four years from 1899 to 1903. Of these fully one-half were the product of Professor Wheeler's own pen. Such sudden and sustained activity at once attracted the attention of zoologists throughout the world and served to add greatly to the prestige of the University of Texas among scientific men . . .

Personally, Professor Wheeler was a most charming person when one came to know him, and like all really great men never held himself aloof from anyone with whom he had dealing of any kind. He was extremely fond of his students although at first they quite frequently did not realize this and clothed him in a coat of dignity that quite camouflaged his real delightful personality. Consequently most of the casual students in his formal courses never came to know him well enough to regard him with the same admiration and affection that his own special students and colleagues invariably cherished . . .[4]

Augusta Rucker was one of those cherished students and colleagues of Professor Wheeler's, and she contributed a great deal to making his stay in Texas both pleasant and memorable. She was born in Paris, Texas, went to a private school, and under the influence of an excellent teacher went on to the University of Texas, where she received her B.A. degree in 1896 and her M.A. degree in 1900. She relieved Wheeler of a fair portion of the teaching load, and she shared his enthusiasm for research. Between 1900 and 1903,

Miss Rucker published five papers which were mainly anatomical descriptions of previously undescribed Texas invertebrates (three on a unique arachnid). It is interesting to speculate on how Miss Rucker's life might have been different if she had continued to be associated with Wheeler, but she stayed behind in Austin to continue teaching when he left in 1903. After twelve years as a teacher in the zoology department, however, she resigned to enter medical school at Johns Hopkins, where she received her M.D. in 1911. As a doctor, she spent most of her life in New York City, working primarily among women and children, trying to aid the poor and campaigning for better public health. Dr. Rucker died in Hyannis, Massachusetts, on December 26, 1963, when she was reported to be ninety.[5]

Another one of Dr. Wheeler's students was Carl G. Hartman (1879–1968), who received his B.A. degree in 1902, his M.A. degree in 1904, and his Ph.D. in 1916—all from the University of Texas. Dr. Hartman, who is noted particularly for his embryological studies of the opossum, wrote an essay in 1959 entitled "The Making of a Scientist." In it, he said:

> I suppose that the making of a scientist depends more on the influence of some one teacher than on any other factor. In my case it certainly was William Morton Wheeler, Professor of Zoology at the University of Texas during 1899–1902. He was ably assisted by Augusta Rucker . . . Perhaps it was Dr. Wheeler's reputation among the students which induced me to register for Zoology immediately upon my arrival. The move was a lucky break for me.
>
> Dr. Wheeler was considered by many to be the most brilliant biologist in the country. He was also a great linguist, a voluminous reader in *belles lettres* as well as science, particularly the philosophical aspects of the subject . . .
>
> Wheeler's lectures were fascinating. I took his entomology course and was permitted to accompany him on tramps in the field in those years when he was writing his famous *"Ants."* I like to claim that I helped him write the book by carrying his vasculum and eating his milk chocolate! I did, however, earn the chocolate by making some photographs for the book and helping with the digging into ant burrows, notably the extensive nests of the leaf-cutting [ants] . . . Dr. Wheeler was also somewhat of an ecologist, and the rich fauna of the Austin region was spread out before me with considerable system and order. The distribution of plants and animals was interpreted for me by a master with all of nature as a laboratory. For example, it was apparent that the Carolinian fauna

of the Atlantic and Gulf region followed up the flood plains of rivers, interdigitating with the Sonoran or Western fauna of the uplands between river valleys. A stone's throw would make a difference in the habitat of the agricultural ant . . . and the fungus-growing [ant].

Dr. Wheeler left Texas to accept the post of Curator of [Invertebrates] at the American Museum of Natural History in New York, inviting me to go with him. When he soon thereafter went to head the Bussey Institute of Harvard he again offered me an assistantship, which I also refused. My brilliant fellow student, Charles Brues, accepted and later succeeded his mentor as Professor of Economic Entomology at Harvard. Brues' boon companion, A. L. Melander, became a prominent entomologist out on the west coast. Other graduate students who came under Wheeler's influence in Texas at the turn of the century were: Walter Hunter, the psychologist; the gynecologist Harvey Matthews; and Jesse McClendon, the biochemist. As for myself, I applied at Columbia University for a fellowship in Germanic Languages and was, fortunately I think, turned down.

Dr. Wheeler was at his best with graduate students; he was bored with undergraduates. Witness the following three instructions I received in the summer of 1901 from his summer haunts at the Woods Hole Marine Biological Station. The diction is mine.

June 15. "For the fall class in entomology lay in a supply of well-fixed hellgramites and cockroaches." Both were easy to secure; the former abounded under rocks in the river and were easy to get at low water; the haunts of the latter I knew very well for I had previously worked in a local bakery.

July 15. "As I shall be very busy, I want you to take over the laboratory in Entomology."

August 15. "I will be too busy to fuss with the Entomology course at all; you are to take full charge of the course." That was my initiation in the teaching of science! The experience in entomology brought me an appointment for the summer of 1904 as Field Entomologist to study the cotton boll worm. Unfortunately, a bout with malarial fever prevented me from getting this desirable experience.

Dr. Hartman's first embryological studies were made on bats because, as he himself pointed out, his office provided him continually with plenty of specimens. Young Hartman did not, however, forsake his field studies. "As I had made little progress in one year on the subject I had selected for my master's thesis, the embryology of the bat, I decided to spend the summer of 1903 studying solitary wasps which I had noticed frequently on my tramps through the woods with Professor Wheeler. In two months of roaming along the sandy, open post-oak woods on the banks of the Colorado River, I filled a dozen notebooks with my observations and experiments and made many photographs. Result: my Master's Thesis on *The Habits of Some Solitary Wasps of Texas;* it was later published as a Bulletin of the University of Texas, and I occasionally see it listed by dealers in old books even today, 50 years after publication. I still have solitary wasps as my hobby, making snapshots and motion pictures and publishing papers for technical and popular journals."[6]

Dr. Hartman mentioned a number of the graduate students at Texas at the turn of the century, but the two who were then and continued to be most closely associated with Dr. Wheeler were Axel L. Melander (1878–1962) and especially Charles T. Brues (1879–1955). The two young men first met in 1893, when the Brues family moved from Wheeling, West Virginia, to Chicago, Illinois. When the boys entered high school, they had expected to specialize in chemistry, but were soon converted to entomology by a gifted biology teacher, Dr. Herbert E. Walter (1857–1945). Brues, in writing of Walter after his death, said:

> His graduate work began at Brown University in 1892–93 under the tutelage of H. C. Bumpus, at that time professor of comparative anatomy. Their early association, already begun during several summers at Woods Hole, led him to Brown and proved to be the beginning of a lifelong friendship. The next year was spent in Germany, following the habit that then prevailed among young aspirants to a zoological career . . .
>
> On his return from Europe he took a position as teacher of biology in the North Division High School in Chicago, where the writer and his boyhood friend, A. L. Melander, had the great good fortune to receive their first instruction in the mysteries of biological science at his hands. Then, as later, Walter was a marvelously fine and enthusiastic teacher whose equal I have seldom known. He took a great interest in secondary education during this period of ten years, but wanted to return east and complete the graduate study he had begun in Germany.[7]

Walter did return east, taking his doctorate at Harvard University in 1906 and going on to a career as a distinguished teacher at Brown University. His well-written text, *Biology of the Vertebrates,* was a favorite with beginning biology students for years, and frequently reprinted and revised—a nice tribute to an able teacher and scholar.

But back in the days that Melander and Brues were high school students, they benefited not only from Walter's teaching ability, but also from his contacts, for Walter and Wheeler had many friends in common (especially in Woods Hole), not the least of whom was H. C. Bumpus. What was more logical than that young Brues and Melander should enroll at the nearby University of Chicago to study with Wheeler, a top-notch zoologist who was also interested in entomology! "But," the story goes, "on the way to matriculate they met Dr. W. M. Wheeler who advised them to go with him to the University of Texas where he had just accepted the headship of the zoology department." Brues and Melander received their A.B. degrees in 1901 and their M. S. degrees in 1902 from the University of Texas. Their early papers, including their master's theses, were published in various journals as "Contributions from the Zoological Laboratory," and were all on insects, representing various orders (the majority on flies—Wheeler's major entomological interest up to then).

Dr. Wheeler's enthusiasm for teaching and field work seemed boundless. One of the activities he organized was a Marine Zoological Laboratory on the Texas coast at Galveston. In a letter to his old friend Jordan dated January 27, 1900, Wheeler described the prospects:

> Last week I spent 3 days in Galveston settling on a site for the M.B.L. of Texas Univ. Galveston is a pretty town on a flat sandy island with streets lined with oleanders—unfortunately not in bloom when I was there—with long rock jetties running out into the Gulf. We hope to secure a fine old state quarantine station out over the water—an ideal building for a lab.—needing only a coat of paint and some new piles under it. I suppose its woodwork contains a few stale yellow-fever germs with here and there a bubonic plague bacillus, but there will be no mosquitoes in the lab. at any rate. If we fail to secure this station we have the laboratories of our medical school which is on the shore near one of the jetties. I can't say much about the fauna. Corals are beginning to grow on the stones of the jetties and the pelagic fauna is very rich. There are plenty of salt-water terrapin and endless luscious oysters growing everywhere.

The Marine Laboratory was actually held in the medical school building from June 4 to 16, and Wheeler kept a brief diary filled mostly with lists of materials collected, both marine and terrestrial. Miss Rucker, Melander, Brues, and two or three other students made up the complement of this venture. Melander and Brues, as might have been expected, concentrated on collecting insects, but a number of excursions were made by all to various interesting areas in the bay to look for crabs, sea anemones, shrimp, and other invertebrates. This twelve-day session seems to have been successful, but was not held again by Professor Wheeler—for reasons we can only surmise. One, perhaps the important one, was that Wheeler had found a new research subject for which he did not need a marine setting. He had found *ants*.

In a letter dated April 5, 1900, Wheeler wrote to Jordan: "The spring here is already far advanced. The fields and the Univ. campus are a sea of blue lupines with hosts of other spring flowers—pink, yellow and crimson intermingled. The profusion of flowers is not as great as that in California but vastly greater than in the northern states. The trees are in full foliage and the days are like those of early June in Illinois. The whole winter has been mild and sunny with scarcely a frost. Since January the mockingbirds have become more and more vocal. Now they [are] veritable paragons of song."

A little bit further along in his letter, Wheeler went on to say: "There is no end of interesting work here for the zoologist. I have left morphology and have gone into the study of the habits of ants, of which we have no less than 40 species within a few miles walk of the University. I am filling my house with artificial ants' nests after the Janet pattern. We have made some interesting discoveries in some of our species during the course of the year. The other day we found the hitherto unknown ♀ of the Eciton ants, a tropical group which extends into Texas from Mexico and Central America. We are going to settle the *Eciton-Labidus* controversy right here." In this instance, one generic name had been applied by earlier workers to the worker caste, another to the very different-looking males. This was only one of several problems Wheeler and his students were able to solve during these years of exposure to the exciting ant fauna of Texas.

In a letter written seven months later to Jordan, Wheeler had some more interesting comments: "With us every thing goes quietly. There is the regular University work and the silent *vie de Provence*—a few intimate acquaintances—plenty of leisure after the manner of southern living and not even the excitement of a mediocre theatrical performance to interrupt the quiet of our days.

I have taken to ants heart and soul and find them more interesting than anything I ever studied before. Our fauna is most interesting, full of surprises. It is a mixture of the subtropical and the northern fauna."

As to how Wheeler became so suddenly attracted to ants, which he had never studied before, we have his own words, written almost twenty years after his move to Texas:

When I took up my work at the University of Texas in the fall of 1899 as a morphologist accustomed to well-furnished northern and European embryological and anatomical laboratories and libraries, I found so little apparatus for the work in which I had been trained, that I fell into a peculiar listlessness and was for some weeks unable to concentrate my attention on any subject that seemed worthy of investigation. One day, while I sat on the bank of Barton Creek, near Austin, in the very spot where, as I later learned, McCook had worked on the famous agricultural ant . . . I happened to see a file of [ants], each with its piece of leaf poised in its mandibles. I vividly remember the thrill of delightful fascination with which I watched the red-brown creatures trudging along under their green loads, and it seemed to me that I had at last found a group of organisms that would repay no end of study. At that time there was no active myrmecologist in the country. McCook had completed his work and Pergande was no longer deeply interested in ants. Prof. Emery, however, and later Prof. Forel extended helping hands to me and forthwith sent me their numerous and important publications, and several of my students, notably C. T. Brues, A. L. Melander, C. G. Hartman and W. A. Long, never wearied of accompanying me on long excursions into the dry, sunny woods and canyons about Austin.[8]

While the above paragraph no doubt contains part of the truth, one must take other facts into consideration as well, for as a matter of fact Wheeler was not the first person to be intrigued by the ants of Texas. In 1899 Texas was still in many ways a frontier, much of it still undisturbed from a biological point of view, and, as Wheeler pointed out, with an extremely rich flora and fauna which was still mainly unstudied. Yet Texas' colorful past had included several naturalists. In the early 1820s some of the German intellectuals fleeing persecution had settled in Texas (as well as Milwaukee), and a number of these men had university training in science. "But," says S. W. Geiser in his *Naturalists of the Frontier,* "the career of the scientific explorer, if it has its glories and its powerful stimulus to intellectual development, has also its dangers, psychological as well as physical. The frontier has broken scientists as well as made them. Isolation from the libraries and museums in the centers of scientific activity, and from both the appreciation and the criticism of fellow naturalists, in many cases has dampened the zeal of explorers who had earlier shown great promise." These naturalists, many of whom are well described by Geiser, worked mainly from the 1830s until the Civil War, and again in the 1870s. But, instead of describing and classifying the specimens they had collected, often with great difficulty, these pioneer workers sent them back to the Smithsonian or to the Philadelphia Academy or to various European museums. And although the names of some of the frontiersmen were commemorated in the names of animals and plants they had collected, most were forgotten even in the centers of scientific learning. Nor were they or their work remembered in Texas itself.[9]

These frontier naturalists were all individualists, but perhaps the most eccentric one of all was Gideon Lincecum (1793–1874). He was an intelligent man, self-taught in many fields, including various sciences, and he made extensive collections in many fields of natural history, including insects. Lincecum, however, had the unfortunate habit of embellishing his observational notes with interpretations of his own which tended to make the scientific world question the validity of his studies. He had some peculiar notions about society and religion, and he expressed his opinions in his scientific notes, often in unpleasant phrases or objectionable words. As if his own idiosyncrasies were not enough, Lincecum interested a Mr. S. B. Buckley, state geologist, in ants and their behavior. Buckley then published several incredibly poor taxonomic descriptions of new species of ants. Between the two men, any understanding of Texas ants was, to say the least, greatly obscured. However, a minister in Philadelphia, H. C. McCook (1837–1911), who was studying ants when he could spare the time from his pastoral duties, became interested in these questionable reports from Texas. He decided to check them out for himself. To use his own words:

In the summer of 1877 I was able to visit Texas for the purpose of settling, if possible, the questions which had been raised as to the accuracy of the reports of Buckley and Lincecum. As will be seen, the observations of Dr. Lincecum were, in many important points, confirmed during that visit, and thus a strong degree of authenticity given to other facts recorded by him which I was not so fortunate as to note. In addition to this, a number of new facts were discovered, and many, which were but partially known, completed or enlarged. The point chosen for the field of my studies was Austin, the State capital, not only because of its high and healthful locality, but because it is in the general range of Lincecum's observations, and was known to be a populous habitat of the insects. I arrived in the city on the 4th of July, and on the 6th instant was established under canvas at "Camp Kneass," three miles southwest from Austin, beyond the Colorado River, on one of the hills that roll upward from Barton Creek, and about half a mile from that stream . . . My observations were continuously conducted at this point for two weeks (July 19), and afterwards for several days in the immediate vicinity of Austin. The sole object of my visit was to study the habits of these insects and the Cutting Ant . . . which is also an inhabitant of that region. My whole time after arriving on the field was, therefore, given for the three weeks of my stay to work which I had proposed to do.

McCook's description of the scenery, though made over twenty years before Wheeler saw it, can perhaps help us to understand how ants caught Wheeler's scientific fancy so quickly:

Camp Kneass is located upon the highlands of the Colorado River of Texas. The flat table-land stretches away toward the east and south, dropping here and there into depressions which form rich valleys. The soil is unequal in depth, varying from three feet or more to several inches. It is black, unctuous, very sticky in wet weather. The bed-rock is a limestone which crops out frequently, and lies in punctured and striated masses over the surface. The formicaries of ants are scattered over the soil in vast numbers, within a few paces of one another. My tent-door was not a half-dozen steps from several large communities, and the tent itself was a regular gangway for the busy creatures. These formicaries are, for the most part, flat, circular clearings or disks, as described hereafter, communicating by well-worn roads with the surrounding herbage. They are made in the light soil of the hill-slopes, in the deep, black earth of the highlands and vales, among the rocks, everywhere, indeed. They abound along the roadsides; they are met in all parts of the city of Austin; in the very streets; on the trodden sidewalks; in the gardens and yards. Even in the open court of the hotel where I lodged there was a community of ants in full activity. The court is paved solidly with stone, but through the cement which joined the slabs the ants had but a gateway, and into and out of this they were passing all day long.

The result of the Reverend McCook's three weeks of observations in Texas, plus his astute use of Lincecum and Buckley's notes, was a widely known treatise of 311 pages, published by the Academy of

Natural Sciences of Philadelphia in 1879 and called *The Natural History of the Agricultural Ant of Texas.* This was the state of affairs when William Morton Wheeler happened into Austin, Texas to take up his duties in a new environment, and there can be little doubt that the preliminary (and sometimes questionable) studies on the ants of Texas by Lincecum, Buckley, and especially McCook played a role in directing his attention to these insects.

Another impression of Texas and the Wheelers was left by Professor John Henry Comstock (1849–1930) of Cornell University, who was one of the best known entomologists of the time. That Professor Comstock made a visit to Austin was perhaps indicative of the rising fame of young Wheeler, as well as a sign that the great teacher and adviser of so many entomology students knew a promising young man even when he was not a product of his own classroom. In *The Comstocks of Cornell,* Mrs. Comstock wrote of her husband's trip in March of 1903.

At Austin he was the guest of Professor and Mrs. William Morton Wheeler and his enjoyment of the stay with them has been a happy memory. Certainly they are two of the most delightful people that have graced university circles in America. Harry [John Henry] wrote me at various times during his trip:

"I reached Austin in time to take breakfast with the Wheelers and received a very cordial welcome.

"This forenoon Professor Wheeler took me into the field where we found many things new to me. The collecting is excellent here, and the fauna is very different from that of Miami. Our collecting was done almost entirely by looking under stones. We found two kinds of tarantulas, several scorpions, and a considerable number of spiders new to me.

"It has been a delightful day. Fortunately for me the Wheelers are not Sunday people. So Dr. W. and I have had two collecting trips, a long one this morning, and a shorter one this afternoon, both of which were successful.

"My stay at the Wheelers has been a delight and an exceedingly profitable one for the spider work. I had splendid weather the entire week till this morning, but left Austin in a pouring rain. I wish you knew Mrs. Wheeler—she is a charming woman.

"I am riding through the land of the agricultural ant. It is an exceedingly abundant species here. One day Professor Wheeler pointed out the place where McCook camped when he made the observations for his book on these ants.

The Wheeler family in 1900: Dora, William Morton, Ralph, and Adaline.

"During our walks, Wheeler has told me enough wonderful things about ants to make a large volume. He is making a very exhaustive study of the group.

"I had a whiff of Texas atmosphere this morning, on my way to the train. I was about 300 feet from a man when he shot and killed another man. I had heard the shots and saw the crowd gather about the fallen man. I then went on to the station. Before our train left Austin I heard that the man that was shot lived only a few minutes."[10]

Wheeler's avid field work sometimes took him to New Braunfels, a town settled largely by German immigrants, where Lisa Grob Dittlinger and her family, and now Amalie Grob, were living. One of the Dittlinger children still remembers his enthusiasm when he discovered colonies of leaf-cutting ants along the Guadalupe River. She recalls that when Wheeler would arrive at their home, her father would turn over his horse and buggy to him and "Uncle Willie" would be off combing the territory up and down the river.

All in all during his stay in Texas, Wheeler wrote approximately thirty papers of varying length and content on ants. The majority of these were concerned with the behavior of the many ants he found around him, at first in Texas, then south into Mexico, then north and east. He spent his summers in Illinois and New England (at Woods Hole briefly now and then but mainly at his summer home in Colebrook, Connecticut), and he collected ants and observed their colonies wherever he happened to be. Mrs. Wheeler helped him with the photography—an important technique for his new-found field studies. He encouraged his students to do extensive field work, too, and as a result of his inspiration and guidance, the first fifty-one "Contributions from the Zoological Laboratory of the University of Texas" were mainly the products of field studies, often supplemented with laboratory observations. The majority were on insects (except for the little arachnid previously mentioned)—many on ants, although there were several on flies and other insects. Embryology had become a thing of the past for Wheeler. He did, however, review one book on embryology during these years, as well as books on general zoology and evolution. His earlier papers on animal behavior also date from this period; these will be reviewed in a later chapter.

How Wheeler obtained the necessary literature and type specimens to enable him to make any sense of his field observations and to fit what he saw into what was already known is something of a puzzle. Modern scientists consider themselves hampered if the library of the institution where they are working does not contain every book and journal, no matter how obscure, pertaining to their field of interest. The University of Texas, which had no tradition in the field study of arthropods, could not have had the necessary aids for Wheeler and his students. Wheeler had been exchanging reprints with contemporaries since his days at the Milwaukee Museum. Now he apparently wrote to the Americans and Europeans who were working on ants, and within a short time he had provided for his laboratory a relatively complete set of the current literature. He helped further by translating into English and publishing in this country two important papers on ants—one from French and one from German.

Wheeler's diaries and personal notebooks were sparse in his Texas years, but in one note dated 1900, he had listed his expenses at Woods Hole (Wheeler, along with many other biologists, had helped

Dr. Whitman give a course in nature study for teachers) and then written a reminder to himself to send his ant papers to Adele M. Fielde at a specified address in New York City. He had met Miss Fielde at Woods Hole, where she had been for several preceding summers studying ants, and where she would be until 1907. Wheeler must have had great admiration for her, as apparently did everyone who knew her or knew of her life and work. Wheeler not only made extensive use of her ant publications, he adopted the artificial ant nest she designed, and he helped her in her work by identifying the ants she used in her experiments.

In the *Memorial Biography of Adele M. Fielde, Humanitarian,* written by Helen N. Stevens and published in 1918, there are many details of her most unusual life, though little about her scientific interests.[11] She was born, brought up, and educated and then taught school in New York state. When she was twenty-five, she became engaged to a missionary, the brother of her college roommate. Shortly thereafter he left for Bangkok, Siam, where he wanted to work with the Chinese. Miss Fielde, after a year of preparation, was to join him both in marriage and in mission work. However, upon arrival, she discovered that he had died soon after her ship had left New York. Miss Fielde was not one to turn back, and she devoted the next twenty to twenty-five years of her life trying to help the Chinese, especially the women, first in Bangkok and later in Swatow, China. Her leaves of absence, which were few and generally short, were spent mainly in studying. In 1883, however, she came back to the states for two years to learn obstetrics (Chinese women could not go to a male doctor) and to "investigate organic evolution." She matriculated at the Woman's College of Philadelphia and also studied at the Philadelphia Academy of Natural Sciences. She spent the summer of 1883 at the seashore laboratory at Annisquam, Massachusetts—the laboratory considered by many to have been the link between Penikese and the Marine Biological Laboratory at Woods Hole.

In 1892, Miss Fielde, no longer sympathetic to the missionary dogma, came back to the states and settled in New York City. There, for the next fifteen years she lectured and wrote (to sustain herself financially) and worked to promote political and social reforms. All this was during the winters, for she spent four months each summer at Woods Hole carrying on her scientific studies. Then in 1907, when she was sixty-eight, she moved to Seattle, Washington, where she devoted all her time to civic activities, including child welfare, prohibition, and women's suffrage. And she abandoned ants! Adele Fielde (1839–1916), unlike so many of us, spent her life not only doing what she wanted to do but doing it exceptionally well, and in diverse fields. But, to quote an entomological friend of Wheeler's, William T. Davis: "Just think what she might have accomplished if she had only concentrated on ants!"[12]

Wheeler's association with Adele Fielde serves to remind us that despite his relative isolation in Austin from the scientific world, his summer visits to Woods Hole and other points in the northeast helped him to maintain contact with a wide circle of biologists. After three years at Texas he began to grow restless with the limitations of the University. This is reflected in his comments in *The University Record* of 1902, where he notes the cramped quarters, insufficient number of microscopes, and inadequate library facilities. "Investigation which might otherwise extend over a wide field is necessarily limited to a rather narrow channel," Wheeler remarked. "Even the identification and description of our common Texas animals is accomplished only with difficulty or not at all."

Whether Wheeler actively looked for another position we do not know, but in the fall or winter of 1902 he corresponded with his old friend, H. C. Bumpus, who was then the director of the American Museum of Natural History in New York. In a letter dated January 9, 1903, Dr. Bumpus offered Wheeler a position as curator of invertebrates at the museum at a salary of $3500 per year. Wheeler accepted, but he said he preferred not to start until October first, as he wanted to complete some work during the summer in west Texas and Mexico. He also mentioned, in a letter dated January 16, 1903, that he wanted to publish two books in the near future—a monograph on North American Formicidae and a popular, illustrated book on ants. He had, or would have, he said, nearly all the materials by the end of the summer. What did Bumpus think of those ideas?

Wheeler also said in that late January letter that he was sending Bumpus the publications "of my laboratory," and that he realized that "much of this work is crude and inchoate, but it has been done without a single word of encouragement from anyone here in Texas." He further lamented the poor prospects for financial help from the state legislature for the University, but he added that he would truly regret "the probably lifelong separation from some of the very good friends I have made since coming here."[13]

7 / In and Out of New York

> Only a few rods from the house [Rockwell Hall] and at the foot of a rock-strewn orchard, a broad expanse of meadow spreads out between the wooded hills. Before the memory of the oldest men now living in Colebrook this meadow was the bottom of a lake which the early residents had created by damming up a stream that still winds slowly through the tall sedges. This stream soon leaves the meadow to hurry over a rocky bed as the charming little torrent which lends its name to the hamlet. The meadow is so damp with the water from the surrounding hills that to explore it in comfort one must wear rubber boots. It yields an abundant crop of hay during the summer, but grasses and sedges are not by any means its only vegetation. It is bright in places with glowing cardinal flowers and purple fringed orchids, not to mention humbler plants like the dwarf cornel, the cinquefoil and the partridge berry. The expanse is dotted over with a few large boulders covered with moss and other vegetation, and almost submerged in the peaty soil. There are a few stumps in the last stages of ligneous decay, some fragments of drift-wood and a very few flat stones lying more loosely on the surface. Much of the meadow soil is thrown into hummocks formed by tufts of coarse grass or great clumps of moss.

Thus did William Morton Wheeler describe one of his favorite areas for studying ants. This description was published in 1903, but the meadows and woodlands of Colebrook continued to interest Professor Wheeler throughout his life.

Colebrook is a small village in the northwestern part of Connecticut, in the Litchfield hills, not far from Winsted. For many years, Colebrook was the location of the summer home of Ralph Emerson and his family, and when William Morton married Mr. Emerson's youngest daughter, Dora, he joined the summer migration. For the first few years the Wheelers stayed at Rockwell Hall with the Emersons. This lovely old house, now filled with antiques and surrounded with beautiful gardens and trees, is still standing at the town center and is a fitting monument to the early settlers, the Rockwells, who built it.

According to the histories of the area, three Rockwell brothers were pioneer settlers in Colebrook in the middle of the eighteenth century. One brother, named Samuel, had six sons: Samuel, Jr., Timothy, Solomon, Reuben, Alpha (reported to be the first child born in Colebrook in 1767), and Martin. These young men, with their father, built an iron foundry (part of the ruins are still present on the bank of the local stream) and, beginning about 1801, were for many years "engaged in the iron business." Timothy, the second son, married Mary Burrall from Canaan, Connecticut, and proceeded to build his new wife a house just south of the present town center. However, he died suddenly in 1794 before it was finished. The youngest son, Martin, soon thereafter married his sister-in-law, Mary, and finished the house his brother had started. The house, of course, has had numerous additions to the original structure, which was completed in 1795. Meanwhile, Reuben, the fourth son of Samuel, built a home north of Rockwell Hall, and the two brothers "lived side by side for forty-five years with the most brotherly affection." Also, Martin represented Colebrook in the state legislature for six sessions.[1]

Martin Rockwell had a number of daughters, and the oldest, Eliza, married the Reverend Ralph Emerson, who was pastor of the Congregational church in a neighboring town. Two other daughters, Mary and Charlotte, inherited Rockwell Hall after their father's death. However, they were spinsters, and the estate passed, at the death of the last sister in 1894, to Eliza's son, Ralph Emerson, of Rockford, Illinois. Mr. and Mrs. Emerson loved this old house and its setting, as did their daughters and their families, and all were welcome. As a tribute to his father-in-law, his hospitality and his encouraging interest in scientific pursuits, William Morton Wheeler named an ant, *Leptothorax emersoni,* after him, saying: "I take pleasure in dedicating this species to Mr. Ralph Emerson in memory of the many happy hours we spent together at Rockwell Hall in Colebrook." That the respect and admiration was mutual is shown, for example, in a letter dated October 9, 1911, written by Ralph

Emerson to his wife: "Prof. Wheeler left here [Colebrook] yesterday. You know how fond I am of him and was very much interested in conversing with him."[2]

Several years later, shortly after Professor Wheeler had moved to New England, Mr. Emerson bought and presented to Dora and her husband the "Sackett House"—a roomy, white frame farm house down the road a scant half mile south of Rockwell Hall. This was renamed Emerson Farm. There the Wheelers could be found for at least part of almost every summer for the remainder of their lives—Dr. Wheeler in his study or out roaming the countryside with his collecting gear, Mrs. Wheeler entertaining relatives and friends, and the two Wheeler children, Adaline and Ralph, helping at home or playing with various Emerson cousins. In 1921, the Wheelers acquired some land on a nearby lake, too, where the children could swim or go boating while their father collected in the surrounding woods and their mother prepared a picnic lunch or supper.

It was an idyllic existence that was welcome to a busy suburban family. For in September of 1903, the Wheeler family—now four, after the birth of Ralph three years earlier—took up residence in the New York suburb of Bronxville, and Wheeler commuted daily to the American Museum of Natural History, located at Seventy-sixth Street and Central Park West in Manhattan.

The Wheeler summer home in Colebrook, Connecticut. Wheeler's study occupied the ground floor of the wing at the left.

Because Dr. Wheeler kept no diary during these years and little of his correspondence from this period has been preserved, it is difficult to know his daily routine as curator of invertebrates, but from the publications of the museum, we can make several generalizations. Curating the invertebrate collections took a considerable amount of his time, and he made a number of changes in the public exhibits so that they were more attractive. He also gave several popular lectures at the museum, at least during the first year or two, both to adults and children—much as he had done in the Milwaukee Public Museum. Some of his talks were entitled "Mimicry and Protective Coloring in Animals," "Shore and Island Life of the Bahamas," and "The Habits of Ants." "Ants, Bees and Wasps" was given to children of museum members one Saturday morning, as was "Egypt and Her Neighbors." Why the latter, we do not know, unless he filled in for someone else. In the early spring of 1905, he gave eight lectures on ants at Columbia University, and this series was the basis for his ant book, to be published in 1910 by the Columbia University Press. Of this we shall have more to say in the next chapter.[3]

During his four years at the American Museum, Wheeler spent a considerable amount of time in the field, both in the museum's interest as well as his own. In May and June 1904, he joined Dr. Frank Chapman, curator of birds, on an expedition to the Florida keys and the Bahamas, and among the specimens he sent back to the museum were corals, sea fans, a variety of insects, myriapods, and various mollusks. In May of the following year, he went on an expedition to New Mexico, Arizona, and California, ostensibly to study desert flora and invertebrate fauna, but mainly to learn as much as possible about ants and their behavior. In March of 1906, he went with an expedition organized by the New York Botanical Garden to the islands of Puerto Rico and Culebra. During four weeks he collected vast numbers of ants—reportedly over 5,000 specimens. In a letter from Culebra, dated March 4, 1906, he wrote back to Bumpus saying that the insect fauna was lean, but that there were interesting isopods, myriapods, thysanurans, and so forth. He then added some local color by saying that Culebra was the "Treasure Island" of Stevenson, and people were still digging. He also said that every night they could see lights out over the water as the smugglers ran their boats from the Danish St. Thomas Island on the east to Puerto Rico on the west. Then the cryptic remark: "The island is run by Capt.

Walling and a species of stinging ant (Solenopsis geminata)." Captain B. F. Walling was commandant of the Naval Station on Culebra and provided the expedition members with shelter on board his ship as well as landing boats and horses for shore explorations.

Dr. Wheeler's summer trip for 1906 proved to be an especially long and fascinating excursion for him. The major part of the summer he spent with Professor T. D. A. Cockerell of the University of Colorado in Florissant, Colorado, digging for plant and insect fossils in seemingly never-ending beds of Miocene shale. In a letter to Dr. Bumpus dated July 16, 1906, he described some of the scenery: "I wish I could send you the weather we are having here. I sleep under four blankets and this morning the little gullies about Florissant were filled with ice—the remains of a lively storm we had yesterday afternoon. Today, Sunday, I climbed Topaz Butte (9200 ft.) walking five miles each way—quite a stunt for an old decrepit individual like myself. I wish you could see the wild flowers here—wonderfully beautiful species closely allied to those of Siberia and Greenland. I saw one whole meadow purple with the curious *Elephantella groenlandica,* and a remarkable lousewort (*Pedicularis*). So far I have collected thirty-five species of living ants—among them two or three North European forms. This is really a remarkable list for a single small locality at such an altitude (8150–9200 ft.)." Wheeler also reported that his fossil collecting varied considerably in amount from day to day; for example, he had "collected seventy-five fossil insects day before yesterday, and twenty yesterday." On July 20, just before leaving Florissant, he said that he had about a thousand specimens of fossil ants alone. "In my old age I can find a job breaking stone on the road, I am sure, as I have split about ten tons of hardened volcanic mud."[4]

Many years later, Cockerell reminisced about this trip as follows: "I think of the summer of 1906, when Wheeler came out to Colorado to take part in an expedition to Florissant. We were hunting fossil insects, and digging a trench in the volcanic shales, uncovered many remarkable species, which have since been described. But Wheeler also looked for living ants, and we sprawled on the ground while he showed us the red *Polyergus* and described to us its slave-making operations. He had an almost uncanny knowledge of the ants, and could recognize most of the North American kinds at a glance. Thus we spent the days hunting and observing, and in the evenings discussed many matters far into the night."[5]

Elsewhere Cockerell described the Florissant beds and subsequent field trips in these words:

> The shales, considered to be of Miocene age, resulted from showers of fine volcanic ash, which fell into a lake, and covered the remains of plants and animals. In the course of time numerous layers accumulated, and these may now be dug out and split apart, revealing specimens preserved by delicate impressions, or as thin films of carbon. Even such minute and fragile objects as aphids are found in quite recognisable form. In 1877 Dr. S. H. Scudder, then the greatest authority on fossil insects, spent a summer at Florissant, and obtained a very large collection. This material, along with other specimens derived from other sources, enabled him to produce a great volume on Tertiary Insects, published by the U.S. Geological Survey.
>
> It was currently supposed that the beds had been mainly dug out, so that little would remain to be discovered, but on visiting the place we found that they were practically inexhaustible, and would continue to yield fossils indefinitely to those industrious enough to hunt for them. Even in Scudder's treatment some important groups, such as the ants, bees and wasps, were omitted. The opportunity for discoveries was too good to be missed, so for several years we had Florissant Expeditions in the field, always productive of good results. My wife said that hunting fossil insects had all the fascination of gambling, with none of the attendant evils. Slab after slab would be turned up, usually with nothing to show, but once in a while a good specimen would appear, and the new species certainly averaged more than one a day. On one occasion my wife turned up a piece of shale in a new locality, and there was a butterfly, the spots still showing on the wings.[6]

Wheeler and Cockerell never found time to share another expedition, but their friendship was lifelong, mutually stimulating, and sometimes exasperating for both. Of all of Wheeler's professional acquaintances, Theodore Dru Alison Cockerell was perhaps the most unusual—a man of varied talents, and with the boundless energy to pursue the many facets of his interests. He was born near London on August 22, 1866, which made him a little more than a year younger than Wheeler, and he died in January 1948, more than ten years after Wheeler. In an obituary for the *Journal* of the New York Entomological Society, C. D. Michener wrote: "The loss will be felt by all who knew him for his quiet wit, charming whimsicality, kindliness, and his personal interest in fellow biologists endeared him to all. Although for many years especially interested in the taxonomy of wild bees, he was interested in and wrote about so many other fields that there is scarcely a taxonomic biologist who has not examined some of his papers. He wrote extensively on scale insects, land snails, slugs, fossil insects, fish scales, sunflower taxonomy and genetics, and paleobotany. Although he regarded himself as an amateur in botany, he described 32 new plants from New Mexico, in addition to others from other areas."[7]

Professor Cockerell seems to have had little formal schooling as a child, due apparently to his frailness, but he was encouraged by his family in his interest in the outdoors—his "scientific pursuits." He said that this interest was crystallized in 1879 when, as a boy of twelve, he was taken to the Island of Madeira. "It was in Madeira that I made what I think of as my first scientific discovery. I was apparently the first to find and report on the caterpillar of the finest of Madeiran butterflies, *Pyrameis indica occidentalis*. The find was duly reported in Lang's book on European Butterflies."

In 1887 young Cockerell developed tuberculosis and had to leave his job in London and go to a healthier climate. He came to live in the mountains of Colorado, and he later attributed his long life to the excellent climate of his adopted state (he became a United States citizen in 1898). For three years he lived in Wet Mountain Valley, Colorado, worked on various jobs in order to earn enough money to keep himself alive, but spent as much time as possible on his ambitious plan to catalogue "the entire Fauna and Flora of Colorado, recent and fossil." He had still to reach his twenty-fifth birthday.

In 1890 Cockerell returned to London seemingly cured of tuberculosis, and worked at the British Museum (Natural History) for about a year. During that time, he helped prepare the second edition of Alfred Russel Wallace's *Island Life*. He felt particularly fortunate in this association, and throughout his life he enjoyed telling of this profitable and pleasant relationship. Cockerell was already an eager naturalist when he met Wallace, but Wallace fired his imagination with evolution and its mechanisms, especially with studies of tropical islands and endemic species (one of Cockerell's favorite subjects), geographical distribution in the broad sense, and the right of a man to study any and all aspects of the natural world which were of interest to him. Cockerell said they discussed all of the debatable biological and sociological questions of the day.

After a year at the British Museum, Cockerell was appointed

curator of the Public Museum in Kingston, Jamaica (due at least in part, he thought, to the influence of Dr. Wallace). The lush tropical flora and fauna were a revelation to young Cockerell, but he had to leave after two years because of a recurrence of tuberculosis. This time he came to the United States permanently, first working in New Mexico for a decade or so, and then finally moving back to Colorado, where he spent most of the remaining years of his long life—mainly associated with the University of Colorado.

Cockerell's bibliography, which, perhaps understandably, was not published until 1963, contains 3,904 items.[8] Many of the papers cited are extremely short, though length is not necessarily an indication of importance. Dr. William A. Weber of the University of Colorado, who assembled the bibliography, remarks that Cockerell did often publish prematurely, but not for self-aggrandizement or notoriety. "The key to Cockerell's productivity was, I believe, the fact that he believed in publication as a means of pronouncement, and that he was obsessed with the drive to communicate to others the thoughts of his lively intellect, the facts he had learned, and to carry on dialogues on the great issues confronting the world." Furthermore, when Cockerell began to publish, most scientific journals accepted and encouraged short communications.

In 1950, a book called *Rocky Mountain Naturalists* was published in Denver and dedicated "to Theodore Dru Alison Cockerell, Naturalist, Teacher, Poet, Friend." The author of this book was Joseph Ewan, and the last chapter of his book discusses briefly, though with sensitivity, the life of his friend Cockerell. Mr. Ewan regretted, he said, that he had only come to know Professor Cockerell after his retirement from university teaching, but that to be around him even then "was to be carried forward as though caught in the stirrup of Pegasus." He described the elderly Cockerell in these words:

> In his seventy-fourth year Professor Cockerell was slender, as he had always been, of average height, soft-voiced and even a little sibilant of speech yet steady of finger and with little dimming of the wide blue eyes that his students remember so well. He usually dressed in a light grey suit and wore his coat and waistcoat in even the warmest weather invariably closed at the throat by a more unconventional ample dark green cravat of the sort commonly associated in this country with artists. He was crisp but unhurried in conversation; in movement purposeful.
>
> To one entering the room the most obvious impression was the general abundance of everything. Reprints stacked in neat piles ready for

mailing; boxes of foreign bees from Tanganyika and Australia sent by the British Museum of Natural History for the critical determinations that Prof. Cockerell alone could give; his own larger reference series of bees and other insects maintained for the identification of unknowns received from correspondents here and abroad; and that unclassifiable miscellany of set-aside mail, books, photographs, twine for parcels, paper-clip dishes, withered plantings, an old pepo or two with those bizarre chimaeras, a turkey vulture feather brought in from a walk on the mesa —all this was present in no disarray, but rather in the tradition of the "naturalists' cabinets" of another period. Of course there were books, many of them acquired by gift from the authors and duly inscribed, and others purchased by the professor, many on his trips to foreign countries. Some books were still brightly jacketed and lent color to the many indispensable reference books in bindings of sombre buckram. The weight of a long file of the *Zoological Record* in fading orange bindings sagged several shelves. But in the midst of all the books and insect boxes, maps, and prints, and giving the whole room a touch of timeless dignity, was a copy of the well-known three-fourths portrait of Darwin. It was an impressive copy, perhaps three feet wide by five feet long, its dark tones leading up to the shaggy browed face of Charles Darwin. Obviously the portrait gave a mood to the room, much as similar portraits lend moods to the meeting places of learned societies and in the same spirit trivia could hardly be contemplated there.[9]

No two scientists could have been more unlike than T. D. A. Cockerell and William Morton Wheeler. Yet each seemed to understand and appreciate the other, although Wheeler did once, perhaps oftener, say that Cockerell was a terrible nuisance with "his abominable way of writing little notes . . . and scattering his publications about the world in footnotes and in other inaccessible papers with endless changes in nomenclature."[10] Yet Wheeler respected Cockerell and was glad whenever Cockerell reviewed one of his papers or books favorably. And the publishers usually were grateful for Cockerell's manuscripts, even though they arrived in his cramped and sometimes almost illegible handwriting (he never had a typewriter). Wheeler's correspondence files contain many letters from Cockerell, and all are in his minuscule script. He was a prodigious letter-writer, and no subject seems to have been sancrosanct—even criticisms of Wheeler's ant studies.

In 1907 Wheeler decided to extend his horizons with a summer in Europe. Also he wanted to meet the European myrmecologists whom he admired and who had been so helpful to him in his initial studies of ants. In a paper entitled "Comparative Ethology of the European and North American Ants," which was published in 1908 in a German journal honoring Dr. Auguste Forel, Dr. Wheeler described his trip in the following words:

> Eight years ago Professor Forel published a series of observations which he made on the North American ants while on a journey from Toronto, Canada, through Massachusetts and the District of Columbia to the Black Mountains of North Carolina. These observations are of unusual interest to the student of the North American fauna because they were recorded by one thoroughly acquainted with the European species, and one whose unusual powers of observation have rarely missed the essential in the objects of his investigation. For the territory which he covered the comparison he draws between the European and American species is remarkably accurate. But North America comprises an enormous and diversified area, and many of the species observed by the eminent myrmecologist extend their range into territory very unlike that in which he sojourned. As my study of the ants for several years had been confined to North America, I was glad to have an opportunity during the summer of 1907 of visiting Europe and of forming my own impressions, especially as I am under lasting obligations to Professor Forel for leading me at once to an intimate acquaintance with the Swiss species. The circumstances under which I developed this acquaintance were full of the delight of companionship with a genial and inspiring personality, and the emotional appeal which Europe, the mother country of our race, always makes to the American. And if the American happens to be a naturalist, this appeal is wonderfully enhanced. No one can remain unaffected by the monuments of a great civilization, but it is certain that none of these can give the kind of pleasure which the naturalist feels on first hearing the skylark or the nightingale, or on first beholding an alpine meadow flooded with gentians or the great colonies of the fallow ants among the pines.

Dr. Wheeler then went on to describe his journey briefly. He left New York on May 9, and his steamer stopped for a few hours in the Azores, Gibraltar, and Genoa, where he was able to make at each place a hurried collection of ants.

> From Genoa I proceeded at once to Professor Forel who was then residing at Chigny on the shores of Lake Leman. This portion of Switzerland is classic ground for the myrmecologist, for it was at Geneva that J. P. Huber pursued his famous "Recherches sur les Moeurs des Fourmis

Indigènes" and at Vaud that Forel, as a mere lad, made the observations embodied in his splendid "Fourmis de la Suisse." Here, too, another eminent myrmecologist, Professor Carlo Emery, took up the study of Formicidae. Certainly I could not have selected a better spot in which to become acquainted with the European species. Although Professor Forel was busy with preparation for moving his household to Yvorne and was weighted down with much other work, he nevertheless welcomed me into the bosom of his charming family and found time to conduct me to the favorite collecting grounds of his youth, the meadows of Vaud and the Petit Salève near Geneva. He also directed me to the most favorable localities in the Jura, Canton Vallais (Fully, Sierre, and Sion), Canton Ticino (Monte Generoso and Monte Ceneri), and the Grisons (Upper Engadin). These and several other localities I visited during June. During July I stopped to see my former teacher Professor Boveri in Würzburg, where I had an opportunity to make a couple of myrmecological excursions with Professors Hans Spemann and K. B. Lehmann. From Würzburg I proceeded to Dresden to see my university chum Professor Escherich, now at the Royal Academy of Forestry at Tharandt. He and Mr. H. Viehmeyer conducted me to the fine collecting grounds in the Dresden heath, where the latter gentleman has been making many valuable observations on ants. At this point my collecting trips ended as I had to devote some time to the study of the natural history museums of Berlin, Hamburg, Bremen, and Altona, before returning to the United States.

This was the formal report of his trip, but some of Wheeler's informal comments are even more interesting. He neglected to say in the above paragraphs that he had revisited, after an absence of thirteen years, the Naples Zoological Station, but in a letter dated May 24, 1907, he wrote to Dr. Bumpus that there was a new wing on the station, that Anton Dohrn's son was about to succeed his father as director, that Eisig and Mayer were both very gray, and that Mayer had told him all about the fine things Bumpus was doing in the museum. He continued:

I shall never recover, I fear, from the shock which Naples has given me. Almost the last traces of local color have gone, with the costumes of the people. The cab men no longer bother me and I walked from the Mola San Vincente to the aquarium without seeing a beggar, with signs of prosperity and cleanliness on every side, fine stores, new buildings, good German restaurants, fresh Munich beer, etc. Santa Lucia has been completely cleaned up. The population seems much less dense probably because thousands have gone to the United States during the past fourteen years. The boys in the streets have learned some English and "23 skidoo" is heard in every side. Vesuvius has changed its outline materially and is no longer smoking. It is a beautiful, rounded, purple cone with white dikes on its surface, built to prevent the lava from flooding the towns at its base. The life in the harbor has quieted down. A tender carries you ashore and on board again and you are not "held up" by the boat-men at the Mole. It is a warm day and the whole beautiful town is swimming in a magnificent haze of purple and the air is full of the cries of the people in a minor key as of old. How I wish you and Mrs. Bumpus and Carey were here again to appreciate all this!

Dr. Bumpus replied to Wheeler that he was delighted to have his letter from Naples "bringing up recollections of the pleasant months that we spent there." Bumpus also mentioned that he had attended a museums' meeting in Pittsburgh and had a long talk with Carl Akeley and others. And then he added: "I cannot help feeling rather lonely around twelve o'clock when there is no one with whom to go to lunch."

On July 21, Wheeler wrote to Bumpus from Dresden:

My short stay in Würzburg was very pleasant. Boveri has grown much broader and stouter and is in superb health. He devotes much of his spare time to painting. It seems that he was at one time doubtful whether to become a zoologist or an artist. The versatility of the German professors is something wonderful. The Professor of Hygiene, Prof. Lehmann, postponed his lecture in order to go ant-hunting with me in Steinberg. Prof. Spemann, who is Boveri's replacement, also accompanied me with four students and the assistant, Dr. Zanic. I felt very proud, I can assure you. They were all very much impressed with my acquaintance with the European ants and took copious notes and specimens! Mrs. Boveri is well and goes to Boston in a few days with her little daughter for the summer, while Boveri goes to the Engadin. Würzburg has changed very little since we were there. The Germans, however, seem to me to have changed greatly. They are better behaved, dress better, eat better things, and are very self-conscious without being as conceited and self-sufficient as formerly. Boveri and all his men expressed themselves in terms of the greatest admiration and appreciation of the work now being done by the American zoologists. Harrison and Jennings, especially, are favorites. Wilson and Morgan are admired but not altogether approved of.

Wheeler commented that Dresden pleased him so much that he intended to stay a week before going to Berlin. He also noted that

"Escherich, our old friend from Würzburg, is now director of the forestry academy at Tharandt a few miles from here, and I shall go out to see him this afternoon." Ten days later he gave further reports of several German museums, none of which he liked as much as the one in Dresden. Apparently Bumpus had delegated Wheeler to look over the natural history museums and report back to him on his impressions.

Although at the time Wheeler did not say much about his visit with Forel, in a review of Forel's *The Social World of the Ants Compared with That of Man,* published in 1930, Dr. Wheeler wrote a sympathetic review not only of the book but also of Forel's career. Some of his remarks are as follows:

The significance of the "Social World of Ants" can be fully appreciated only by the myrmecologist who has enjoyed intimate personal acquaintance with its author. Auguste Forel was born in 1848 and began to study ants and other insects before his eighth year. The observations which he made during his childhood and youth were brought together and published in 1874 as a quarto volume, the "Fourmis de la Suisse," which is one of the landmarks of myrmecology, and has lost none of its value during all the years that have since elapsed. This was only the beginning of his penetrating studies of the ants, which were carried on without interruption till 1922, when illness and failing eyesight compelled him to write his last myrmecological paper and to dispose of his very valuable collections to the Museum of Geneva. Since 1874 he has published hundreds of important monographs and notes, containing descriptions of several thousand species, subspecies and varieties from all parts of the world and treating of every aspect of ant structure, distribution, psychology, and behavior. But all this was only a delightful avocation! He was at the same time one of the foremost European neurologists, practicing psychiatrists, and social reformers. His publications in these capacities, as listed by some of his Viennese friends and admirers on his 60th birthday, comprise 10 important works on brain anatomy, 31 on normal psychology, sleep, hypnotism, and suggestion, 23 on psychiatrical and criminal psychology, 34 on the deleterious effects of alcohol, 11 on sex and social ethics. His long experience as a psychiatrist, at one time as the director of the large hospital for the insane in Zürich, and since in private practice, his untiring, sympathetic, and unselfish devotion to his patients of all social classes, his extraordinary intuition, coupled with a singularly ardent and straight-forward personality, yielded him a rare knowledge of his fellow men. So comprehensive and penetrating was his insight into human behavior that there is scarcely one of its protean aspects which he has not touched upon with illuminating comment in his publications.

"Wheeler once described to me [Cockerell once wrote] some of the incidents of his visit to Forel and told with relish the parting comment of the Swiss, 'Wheeler, you are outwardly calm, but inwardly perturbed. I am outwardly perturbed, but inwardly calm.' Wheeler declared that this was essentially true; as we think of it now, he combined in one individual a high development of the emotional and intellectual faculties. One can imagine that he might have been a great religious or political leader, had he not adopted the principles and practices of the scientific worker."[11]

In 1914 one of Wheeler's students, William Mann, visited Dr. Forel and related amusingly of how some of the local villagers looked at him when he asked directions to the Forels. It seemed that it was not unusual for men with mental problems to come and stay at the Forels for a certain length of time while the doctor treated them. Dr. Mann described his stay: "Next morning, carrying my bag, I walked up the road to Yvorne, where Dr. Forel welcomed me, introduced me to his family, and showed me the guest room. It was a thrilling week for me, revelling in his study with his ant collection. He would leave me alone for hours, while with a hand lens and a notebook I went over one species after another. Toward the end of the week he brought out some empty insect boxes and turned me loose in his duplicate collection. The result was between five and six hundred different species of ant, all identified by the master and all for my own collection." Further on Mann ended his description of his visit with Dr. Forel: "I shall always remember him as a stooping, bearded, heavily eyebrowed, kindly man, one of the truly great scientists of his time and one who cared as much for the welfare of his fellow man as he did for his beloved ants."[12]

Like Mann, Wheeler also received a contribution from the Forel collection. The final beneficiary was, of course, the American Museum, for it received some 3,500 specimens, representing about 1,400 species of exotic ants and about 800 type specimens, which were invaluable to future students.

It was never Wheeler's good fortune to meet the second great European myrmecologist of the time, Carlo Emery. Emery and Forel had been born in the same year, and both in their latter years were semi-paralytic invalids. As youths, both had begun their studies of ants, unknown to each other, on opposite "sides of the same mountain" near Vaud, and then, as Wheeler said in the obituary he wrote for Emery in 1925, both men "cooperated in enormously extending and deepening our knowledge of the ant faunas of all parts of the world and in encouraging younger men to take up the study of these fascinating insects." We shall say more of Wheeler's indebtedness to both Emery and Forel in the next chapter.

During the summer of 1908 there was not much correspondence between Bumpus and Wheeler because Wheeler was away from the museum only during July, when he supposedly spent most of the time resting and relaxing with a couple of friends in Maine. One was his old friend, Franklin P. Mall, from Clark and Chicago days, and Wheeler wrote to Bumpus that he was "helping Mall to loaf." Also, "we have lots of fun generalizing together on universities, their policies and we both wish you could be with us."

Since both Wheeler and Bumpus were soon to leave the American Museum, we should pause a moment in our narrative of the career of the former and say a few words about that of H. C. Bumpus. As previously stated, Bumpus and Wheeler had first met in 1889 at Woods Hole, and they remained lifelong friends, though there were long intervals when they did not see each other. In the interval since the two were together in Germany and Italy in 1894, Bumpus had spent the years making the department of biology at Brown University one of the best in the country, though somewhat unorthodox. One of his innovations was the provision for immediate and continued work in the biological laboratories for students from the freshman year on up. Like C. O. Whitman, Dr. Bumpus saw to it that all students had a chance to study many kinds of animals from many points of view. From this came an amusing story about his first attempt to provide a cadaver for his premedical students. As this had not been done before and Bumpus was uncertain of the reaction of the nonbiological community, he quietly obtained an unclaimed body from the local medical examiner. Several days later, while he and Mrs. Bumpus were eating breakfast, Dr. Bumpus happened to glance at the local newspaper. Suddenly a peculiar expression crossed his face, and he dashed from the dining room without a word of explanation to his wife. Later, he told how he had noticed a story on the front page regarding a missing person, and he recognized the face as that of the cadaver. The body was "patched up" and returned to the city morgue, to be later "found" and identified.

Bumpus, like all of the students who had come under the influence of C. O. Whitman, had a thorough training in biology and all its broad philosophical aspects. But though Dr. Bumpus was an enthusiastic and successful teacher (as indicated by the growth of his departments first at Olivet and then at Brown), his professional interests matured along different lines than pure scientific research and university teaching. He knew and appreciated these things, but he felt that the general public should also be allowed to share in the knowledge and excitement of the natural world. Because Bumpus was a capable and energetic administrator, he had unusual chances to carry out many of his ideas.

Hermon Carey Bumpus (1862–1943) was born and brought up in New England, and as a child had shown great interest in plants and animals.[13] Even during his boyhood in the Boston suburbs, he kept pet snakes, skunks, and other interesting creatures. Bumpus did his undergraduate work at Brown University, attracted there mainly by Professor A. S. Packard, who had been one of Louis Agassiz's students. Like so many of Agassiz's students, Packard had absorbed Agassiz's methods so well that he accepted Darwin's theories in spite of his teacher and encouraged his students to do likewise. Thus Bumpus was doubly fortunate, for his undergraduate training and his graduate studies were done under two of Agassiz's most distinguished students, Packard and Whitman.

Bumpus' first work with the public was in 1897, as organizer and first president of the Audubon Society of Rhode Island. There was a growing concern along the New England coast over the increased killing of wild birds for commercial use, especially gulls and terns, whose white wings were fashionable on ladies' hats. As president of the society, Bumpus immediately began an educational campaign to stop this, and he worked with legislators toward the eventual passage of suitable protective legislation for waterfowl. His ever-growing concern for wildlife was evident in 1900 when he was made chairman of a national committee for bird protection within the American Ornithological Union, and in 1905 when he became director of the National Association of Audubon Societies of America.

At the American Museum of Natural History in New York City, Bumpus had his first major opportunity for public education. He was hired in 1900 by the museum's president, who said about his institution: "I believe it to be today one of the most effective agencies which exist in the City of New York for furnishing education, innocent amusement and instruction to the public." The American Museum had been founded some thirty years before by Albert S. Bickmore, another former student of Agassiz and friend of A. S. Packard. He was still associated with the museum as a trustee and as curator emeritus in the department of public instruction during Bumpus' regime. Bumpus was just the man to execute the aims of the president, and after his first year he was promoted from presidential assistant to director of the museum. Bumpus traveled around the United States and Europe visiting other museums, good and bad, and often had his curators do likewise (as Wheeler did). From these visits he formed a clearer idea of the best ways to present natural history subjects to the public. He then brought in a young sculptor, James L. Clark, who was a student at the Rhode Island School of Design, to improve the taxidermy. He even arranged for Clark, who was later to become head of the department of preparations and installations, to go to Chicago and study with Carl Akeley, who was creating such effective, realistic exhibits at the Field Museum, but who, up until then, had kept his methods more or less secret.

Bumpus not only was in charge of the museum administration (such as making up the yearly budgets), of the scientific work (hiring staff, supervising expeditions and publications, and so forth), but also of the educational aspects, including exhibits as well as public lectures and classes. Bumpus' son, in a biography of his father, tells of one incident: "Not long after assuming his new duties Dr. Bumpus issued a special invitation to the children of the New York schools to visit the museum for a nature talk to be given on a certain day. As the hour approached he sensed that something unusual was happening in the street, and looking from the window he saw that the approaches to the museum were blocked with thousands of children. Realizing that only a fraction of the throng could be taken care of in the lecture room, he hastily marshaled the entire museum force to handle the emergency. As the children entered, they were allocated to various exhibition halls of the vast building, and there were entertained by the curators and the scientific staff, summarily pressed into service. The talks were no less instructive because impromptu."

A somewhat different view of the American Museum was given in the following letter, written by an outsider in December 1907. It was addressed to Samuel Henshaw, director of the Museum of Comparative Zoology at Harvard, by a young man who would himself one day be director of that museum, but who was then interested mainly in learning more about reptiles. The writer was Thomas Barbour.

Dear Old SH. I went first thing this A.M. to Am. Mus. N.H. after running the gauntlet of various . . . door keepers I got to Bumpus' office. He was out did not know when he would be in. Then I went to Wheeler he was extremely pleasant asked kindly for you and took me to see Allen and Chapman. I said howdy and where are your *Solenodons*. They said down in the taxidermist shop. Chap. looked daggers at me for asking about them and would not shake hands. I went down to the tax. shop and saw two [mounted] only one is any good: the other is [mounted with] head sticking out of a hole in the rocks. They are all over ruddy brown with a yellowish spot on top of head! The tax[idermist] was a German and I jollied him well, I could soon see that Wheeler was not interested walked away. Then the tax. volunteered that they represented a new species. The third specimen was perfectly rotten and Verrill put it in sulphuric acid [which] decalcified it and the skeleton is not good. Then I walked around awhile—there were no labels on the critters—and went into Bumpus' place again and told him how fine they (the S's) were. He delivered me a long lecture on museum management and then I got back to *Solenodon* again. I don't really think anyone up there knows where they really came from . . . [I] wondered if he got them from Haiti or San Domingo. B. said S.D. not Haiti and then said way up in the interior and you can't get any more, etc.—the usual formula. I asked if he thought those the last on the island; he said no—but its very hard country to get about in. I [chatted] some more but got no more information! Wheeler has some fine well named series (Vaughn) of Hawaiian corals. [Would] like to [exchange] for Echinoderms. I told him to write you. Wheeler runs the whole of invertebrate zoology reptiles and fishes. I guess there are 40 in the dept of taxidermy and God only knows how many paleontologists and grafters. Oh, Lord. Such a place I was never in. Men yelling orders to one another and washing the floor at the same time. Had a pkg. of photos. under my arm and I walked in a door where a "buttons" ran up to me and said Hey! Where are you going with that bundle? I did not answer and he waltzed up and picked at my sleeve. I whirled on him and told him to go to hell—damn quick. And I was allowed to go in peace.[14]

In 1906 Dr. Bumpus was instrumental in forming the American Association of Museums, and he became its first president. Through this society, Bumpus attempted to raise the goals of all American museums by emphasizing that their functions were twofold: first, they should encourage good scientific research; second and equally important, they were obligated to keep the public interested and informed as to scientific knowledge and its advances. As evidence of his belief, Bumpus had the bookplate of the American Museum redesigned to read: "For the people. For education. For science." Many of the museum trustees and curators did not understand this concern for education, and Dr. Bumpus eventually felt restrained by the more conservative elements in the museum and so resigned as director in 1910.

The next two positions that Dr. Bumpus filled were not in science, but because of his ingenuity as an administrator, his accomplishments were indeed remarkable. For three years he was business manager of the fast-growing University of Wisconsin—an office which had never before been established in an American university. Not only did he reduce the financial chaos he found, but he organized an efficient handling of the three million dollar annual budget. Bumpus then left Wisconsin to become president of Tufts University in Medford, Massachusetts. He was there from 1915 to 1919, and because of the war led anything but a quiet life. There he not only removed Tufts from its precarious financial standing, but left it fiscally sound and operating smoothly, with increased alumni support and with the largest enrollment of students in its history.

In the archives of Tufts University there are many letters of recommendation for Dr. Bumpus when he was being considered for the presidency, and the recurring theme throughout these is his amazing originality. While this trait in his personality was obvious to his friends and associates even in 1915, it was not until after he left Tufts that Bumpus made his greatest contributions to society. Dr. Wheeler, in his letter, had written almost as a prophecy:

Professor Bumpus combines, in a very extraordinary manner, abilities which taken singly would make any one man a great success in life. He has not only a thorough university training as an investigator, so that he is able to appreciate the value of work done by teachers and students in any single branch of knowledge, but has also a very penetrating insight into human knowledge and experience of men, rarely found among university men, combined with an extraordinary capacity for executive and purely routine business work. All of these abilities have been manifested to an unusual degree throughout his whole career. He not only has the ability to select men able to cooperate with him, but is also able to induce them to work with one another in the most enthusiastic and serviceable manner. In other words, he is a master not only in stimulating cooperation in his human environment, but in getting good "teamwork" out of all the men he may select to carry out a particular program. He is possessed of unusual originality and ingenuity in overcoming difficulties and in suggesting lines of endeavor serviceable to science and to the community in which he lives. He has a magnetic personality and of such a character that all students and teachers who come in contact with him not only admire him but acquire a great affection for him.

In the years following his resignation from the American Museum of Natural History, Bumpus had continued his association with the American Association of Museums, and, in fact, was chairman of the Committee on Outdoor Education. Bumpus had long been interested in conservation, but he felt that any progress would be impossible without the cooperation and interest of an informed public. The national parks seemed to Bumpus to be a logical place to begin on this problem, so through his committee, he set to work to organize a nationwide educational program in the national parks. Its purpose was to interpret the natural, archeological and ethnological features of the parks to the throngs of visitors. With the approval of the National Parks Service and with a liberal grant from the Laura Spelman Rockefeller Memorial, Bumpus submitted a highly original plan for placing at strategic points throughout the national parks what he called "Trailside Museums." The natural features of the parks were to be the exhibits, while the buildings were to contain readily available sources of information and interpretation about them. With the enthusiastic cooperation of the park executives, Bumpus created a model museum in Yosemite to serve as a demonstration. The success of this experiment was so complete that a succession of "trailsides" was established in other national parks, and eventually, as hoped, the United States National Parks Advisory Service took over the program. Bumpus, as National Parks Advisory Board chairman, continued to be its guiding spirit.

Dr. Bumpus is also credited with originating the natural history shrines, which are roadside or trailside signs giving information about some nearby object or natural phenomenon. He also helped the visitors in the parks by designing self-guiding pamphlets, first in Yellowstone, and then in a number of other parks. Today they are an accepted part of the educational program of every national park, as well as of many state and local parks. Over the years, as Bumpus worked in the national parks and observed their increasing use, he felt that efforts should be made to preserve more wild areas. He became chairman of an advisory committee to the Secretary of the Interior for the consideration of further acquisitions of parkland. Two areas, in which he was especially interested, later became Everglades National Park and Big Bend National Park. Today, thirty years later, when the problem of properly maintaining the parks we have, to say nothing of acquiring new ones, has become critical, we can only say that Dr. Bumpus was a true visionary—a man far ahead of his time.[15]

Dr. Bumpus' ideas on public education seemed to have interested Dr. Wheeler very little, at least in the days he was at the American Museum. Wheeler's research turned more and more to pure science, as his publications show (he averaged about ten ant papers a year). But his writings also show that his mind was not completely engrossed with learning more and more about ants. Ants were only a part, granted a large and interesting part, of the whole picture of animal behavior.

In 1904 in *The Auk,* a journal primarily of interest to ornithologists, Wheeler published a short article entitled "The Obligations of the Student of Animal Behavior." He said that most of us, because we love animals, tend to endow them with powers they do not possess. A student of the science of animal behavior should feel "compelled to submit traditions concerning animals to searching and depurative criticism," and then "rebuild our knowledge of animal behavior on the securer foundations of careful observation and experiment." There are many joys as well as pitfalls for the student of animal behavior, and Wheeler interestingly described some of these. He concluded that this study was no longer in the anecdotal stage, as many of the authors of nature books in this country seemed to think. "At the present time the animal anecdote is admissible only in works of art, like the fable, the animal epic, or the animal idyll, or for the purposes of destructive criticism. In other words, its chief scientific use is negatively didactic, or for the purpose of illustrating how not to study and describe animal behavior." Wheeler then added one of his delightful asides in a footnote: "Those who cannot repress a feeling of disappointment on learning that there is no evidence to show that animals can reason like themselves, may find consolation in the fact that the very naivete of animals—their limitations and stupidity, humanly speaking—is a fact of great interest and beauty. Who will deny that the very absence of the reasoning and reflective powers enters very largely into our aesthetic appreciation of the actions of our domestic animals and of our own children?"

In the popular press of the early 1900s there were many stories purporting to describe "the ways of the wood folk." These articles and books were intended mainly for children, one guesses, but adults read many of them, too. Two of the better known writers involved in what became the "nature-fake furor" were Ernest Thompson Seton and John Burroughs. In an article in *The Atlantic Monthly* for March of 1903, John Burroughs called attention to certain "abominations" in some of the then current nature books. The authors were evidently writing for money and not for the sake of truth. They were humanizing animals, and the worst offender was a minister turned writer, William J. Long. A number of scientists, including several curators at the American Museum, joined in the condemnation of such books in the pages of *Science.* One of these was Wheeler. Long, who was a man of about Wheeler's age, had described in one of his stories a woodcock, seen sixteen years before, which supposedly had put a cast of mud on one of its broken legs. In "Woodcock Surgery," Wheeler took exception to this saying that sixteen-year-old recollections tended to be hazy at best, and stories from hunter friends were notoriously inaccurate. This same volume of *Science* carried several other comments by various men, ending with a rebuttal by Long, and the editor then indicated that he hoped the discussion was ended.

Wheeler's horizons were enlarging professionally, too. For some time there had been agitation among American entomologists to form a national society whose purpose would be "to promote the science of entomology in all its branches, to secure co-operation in all measures tending to that end, and to facilitate personal intercourse between entomologists." On the evening of December 28, 1906, during the meetings of the American Association for the Advancement of Science in New York City, between two and three hundred men interested in insects and their relatives met at the American Museum of Natural History and formally organized the Entomological Society of America. Professor J. H. Comstock from

Cornell was chosen to be the first president, and the speaker of the evening was William Morton Wheeler. The topic for this first annual address was "Polymorphism in Ants," which "was illustrated by an extensive and very beautiful series of lantern slides."[16] A year later, at the Christmas meetings in Chicago, Wheeler was elected to succeed Professor Comstock and to become the second president of the society.

At about the same time, late 1907, another interesting offer was being made to Dr. Wheeler. A new graduate school of applied science was being formed at Harvard University, in the old Bussey Institution in the Jamaica Plain-Forest Hills section of West Roxbury, within the present city limits of Boston. A new faculty was being assembled there, and Dr. Wheeler was asked to come and be the professor of economic entomology. How Wheeler happened to have been considered for this position or when he was first approached, we do not know. But Wheeler's rising reputation as one of the country's leading entomologists, his broad background in zoology, and his Woods Hole associations, were all well known in scientific circles at Harvard. Furthermore, he had first visited with the entomologists at the Museum of Comparative Zoology in 1889 (during his first trip from Milwaukee to the Marine Biological Laboratory at Woods Hole), and had published in *Psyche,* the journal of the Cambridge Entomological Club, for many years. His first paper there, in fact, had been in 1890, and over the years seventeen of his papers had appeared in it. So it was that on March 5, 1908, Professor Wheeler wrote a letter to President Charles W. Eliot of Harvard accepting his offer to join the faculty. And in the following September, he and his family left New York City and its suburbs and moved north to become part of Boston's academic community.

Wheeler retained the title of honorary curator at the American Museum for many years. He wrote Bumpus that his greatest regret on leaving the museum was "that I shall no longer be so intimately associated with you and your plans for the advancement of science and education in New York City and the country at large."

8 / "Ants Are to Be Found Everywhere"

Wheeler was forty-three when he moved to Harvard University in the fall of 1908. He was to remain there for the rest of his life. He was already a scientist of international repute, but it was the publication of his 663-page volume *Ants, Their Structure, Development, and Behavior* by the Columbia University Press early in 1910 that placed him in the forefront of biology. This seems an appropriate place to pause in our story of Wheeler's life and take stock of his contributions to the study of ants. There can be no question that despite the fact that he had moved four times since receiving his degree from Clark University, Wheeler had had eighteen exceedingly productive years. That he should have produced after only a decade of study a book on ants that remains a standard reference more than half a century later is a tribute to his tremendous energy and erudition. *Ants,* affectionately dedicated "to my wife Dora Bay Emerson," was to be reprinted without change in 1926 and again, long after his death, in 1960.

"It is probably not too much to say that Dr. Wheeler's *Ants* is the best book on entomology ever published in this country," wrote T. D. A. Cockerell in *Science* on June 3, 1910. "Here we have morphology, anatomy, embryology, psychology, physiology, sociology, paleontology, zoogeography, taxonomy, and even philosophy dealt with in an illuminating manner! The ant is presented to us as the hub of the universe, and if there is any biological subject which may not be suggested by the study of myrmecology, it is probably of small consequence."[1]

Wheeler's lucid and forceful style and his enthusiasm for ants are apparent from the opening pages:

> It is certain that the ants occupy a unique position among all insects on account of their dominance as a group, and this dominance is shown first, in their high degree of variability as exhibited in the great number of their species, subspecies and varieties; second, in their numerical ascendancy in individuals; third, in their wide geographical distribution; fourth, in their remarkable longevity; fifth, in their abandonment of certain over-specialized modes of life from which the other social insects seem not to have been able to emancipate themselves, and sixth, in their manifold relationships with plants and other animals—man included.
>
> Ants are to be found everywhere, from the arctic regions to the tropics, from timberline on the loftiest mountains to the shifting sands of the dunes and seashore, and from the dampest forests to the driest deserts. Not only do they outnumber in individuals all other terrestial animals, but their colonies even in very circumscribed localities often defy enumeration."

After his introductory chapter, Wheeler devoted three chapters to the structure of ants and one to their immature stages and development. The next two chapters, on polymorphism, followed with only slight change his long paper on that subject published by the American Museum in 1907. Two later chapters, one on honey ants and one on fungus-growing ants, had also been published earlier in the *Bulletin* of the American Museum of Natural History, and indeed parts of all of his chapters had in some measure been anticipated in some of Wheeler's many papers. Now for the first time all of this was brought together in a single very readable volume, the author's original research interwoven with that of many other workers, principally European, whose work had previously been largely unavailable in English. There were chapters on the distribution and fossil history of ants, but by far the greater part of the book was taken up with discussions of the life histories and behavior of various groups of ants, their nests, and their relationships to plants and to other animals, especially to those living as parasites or inquilines in ant nests. One chapter treated the classification of ants, and an appendix consisted of a set of keys for the identification of the genera of North American ants. Wheeler expressed the hope that he might someday prepare a monograph of all the North American species of ants so that students might be able to identify them readily in the course of adding further to knowledge of the biology of ants.

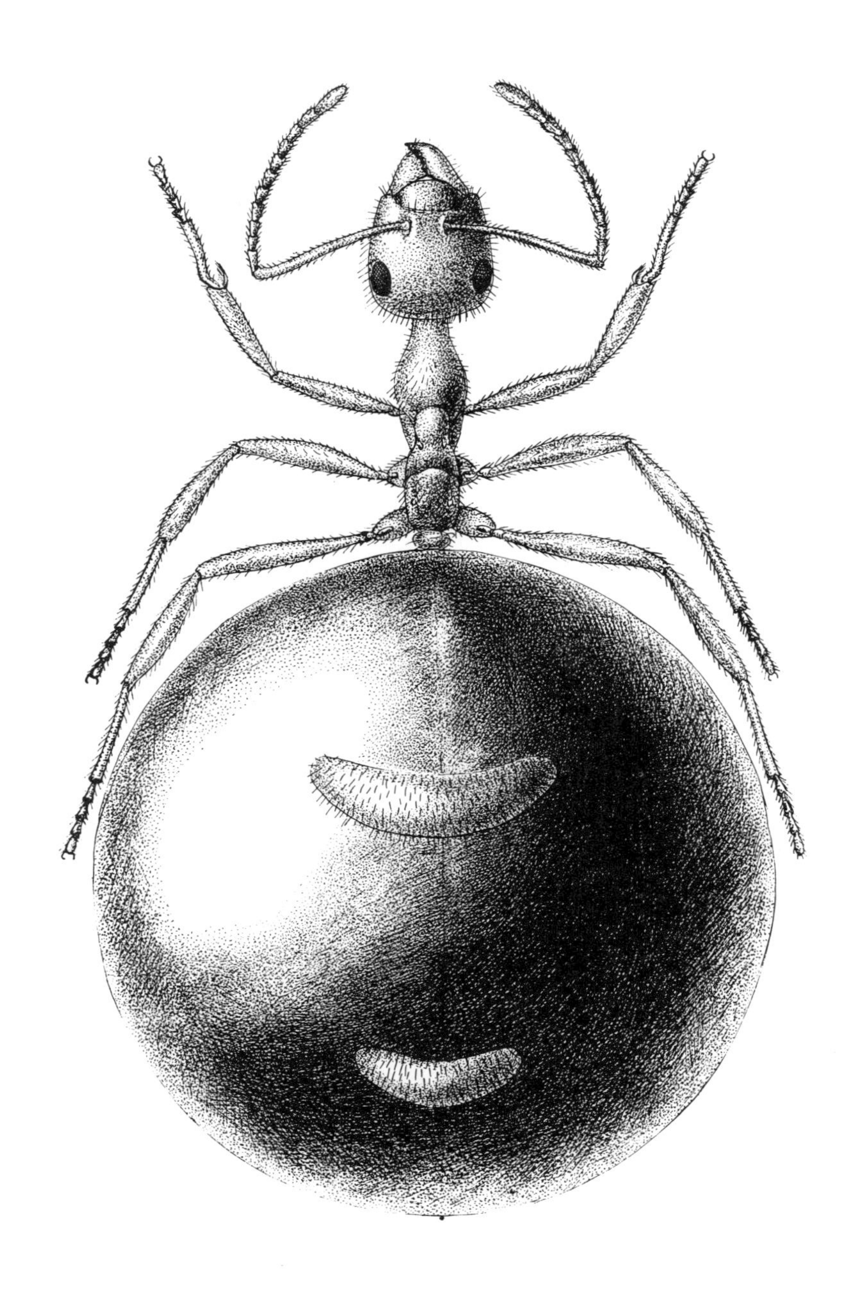

A figure from *Ants:* a replete worker of the honey-ant *Myrmecocystus.* (Columbia University Press)

Ants is still in print, and rather than describing that classic work in further detail, it may be of greater interest to look backward over the decade which culminated in the book and then briefly forward to the remainder of Wheeler's career as a myrmecologist. Wheeler's attraction was clearly to the ant as a living animal, exciting not only in its own right but as a touchstone for the study of ecology and sociality.

We have already described the beginnings of Wheeler's interest in the ants of the vicinity of Austin, Texas. Within a short time he had learned various techniques for keeping colonies of ants in the laboratory (reviewed in an appendix of *Ants*), permitting much closer and more continuous observation than would be possible in the field. His first paper on ants was published in the spring of 1900 in *The American Naturalist,* a journal to which Wheeler contributed many times and of which he was made an associate editor later that same year. In this paper he described his discovery of several nests of army ants along Shoal Creek, near Austin—nests at first found to contain nothing but workers. In fact, army ants were known up to that time chiefly from workers; the queens and males had been described in a general way, but they were so utterly different from the workers that they had been placed in different genera, and in no case had all the castes of any one species been associated. Over a period of days, Wheeler and a student found two queens in nests of a species now called *Neivamyrmex nigrescens,* and these unusual queens were described and figured. Wheeler also described the laying of eggs in artificial nests and the care of these eggs by the workers. A year later, in the same journal, he described the male of this species and several others, and also took occasion to summarize other recent studies of army ants by Forel, Emery, and himself. Brief as these two papers were, they stood as the definitive papers on the army ants of the United States until the taxonomic studies of M. R. Smith, published in 1942, and the still more recent research on behavior by T. C. Schneirla of the American Museum of Natural History.

In his earliest studies, Wheeler was also much attracted to the Ponerinae, a group often believed to represent the most primitive of living ants. In a paper published in *Biological Bulletin* in the fall of 1900, he discussed the life histories and behavior of "three beautiful species" found near Austin, Texas, and in the following issue of the same journal he added two species of more northern distribution which he had encountered during summer sojourns in Rockford, Illinois, Colebrook, Connecticut, and Woods Hole, Massachusetts. As he pointed out, almost nothing was known of the biology of these ants, yet they promised to prove particularly interesting as representing early stages in the evolution of social behavior in ants. He found that in fact their colonies were very small, that the queens differed little in size and structure from the workers, that the larvae were fed with large pieces of prey (rather than regurgitated food), and that there was little or no interchange of food between workers. Yet no animal is without specializations. Wheeler described the unusual manner in which workers of the genus *Odontomachus* "snap" at their prey. The large mandibles are spread widely when the workers are hunting, and several sensory hairs project forward from their bases. When potential prey is detected, the ant "darts forward and suddenly closes its mandibles with a very audible click. The signal for closing the mandibles seems to be given the moment the long sense-hairs touch the object. The mandibles are brought together with such force that if they strike a solid object the ant is thrown backwards—often to a distance of three or four inches—occasionally even to a distance of ten or twelve inches. The ant alights on its feet, like a cat, and again advances to repeat the act." When Wheeler placed a living housefly in an *Odontomachus* nest, several ants "began snapping at it like a pack of angry dogs. With each snap a leg or wing was severed and often thrown to a distance of 2 or 3 inches. In less than a minute all the limbs had been shorn from the trunk." When egg yolk was poured on the floor of the nest, the ants detected it and began "snapping at it as if it were a deadly foe."

Some ponerine ants are specialized in that they hunt only a certain type of prey. For example, in another paper in *Biological Bulletin,* published in 1904, Wheeler showed that *Leptogenys elongata* is a specialist on terrestrial Crustacea ("sowbugs"). The larvae of Ponerinae are also specialized in possessing various spiny or glutinous tubercles which are evidently protective or serve to attach the larvae to the walls of the nest. Wheeler's first papers on Ponerinae include descriptions and drawings of the larvae and represent the first of many studies of ant larvae, studies which were extended after his death by his student G. C. Wheeler.

Another interesting primitive ant studied by Wheeler was *Cerapachys augustae,* which he described in 1902 and named for its discoverer, Augusta Rucker. Prior to that time, the genus *Cerapachys* had been known only from Africa, Asia, and the East Indies. Various workers had noted a resemblance of these ants to army ants, but Wheeler's studies showed many similarities to the ponerine ants; for example, the colonies were small and the queens and workers were similar. It seemed possible that *Cerapachys* might be close to the common ancestor of the army ants and the ponerine ants. Later studies have shown that in fact *Cerapachys* is a "robber ant," preying in groups on colonies of other ants; its resemblances to army ants may be the result of evolutionary convergence rather than close relationship.

An amusing by-product of Wheeler's early interest in these primitive groups of hunting ants was his series of exchanges in the pages of *Science* with O. F. Cook of the U.S. Department of Agriculture, a series that reveals the type of withering sarcasm that Wheeler was capable of leveling at work he considered shoddy or unreasonable. Cook was concerned with the introduction of natural enemies of the cotton boll weevil, a serious pest that had entered the United States from Mexico in 1892. He had settled on the "kelep" (the Indian name for a species of ponerine ant) for introduction from Central America into Texas for the control of this insect. Dr. Cook published a lengthy report on the biology of what he called "the cotton-protecting kelep of Guatemala," characterizing that insect as "not a kind of ant" but a wholly new and unstudied type of insect.

Dr. Cook's paper [Wheeler wrote in *Science* on December 1, 1905] displays so many mis-statements of fact, such inadequate knowledge of the work that has been done on other species of ants, and such a wilderness of unkempt argument and speculation as to entitle it to a high rank as an example of what a scientific essay should not be . . . He evidently wishes to make us believe that the kelep . . . is really a creature *sui generis* which the advanced systematist would do well to regard as the sole representative of a distinct family, the Kelepidae. Here Dr. Cook shows admirable self-restraint, for it might just as well be made the type of a new phylum (Kelepata) or subkingdom (Kelepozoa) . . . Dr. Cook's amazing estimate is attributable to a confusion of ideas concerning certain well-known phenomena among social insects in general and to a lot of inconclusive, not to say slovenly, observations on the kelep in particular.

Wheeler went on to point out that there was nothing in Cook's report to suggest that the kelep was not a fairly typical ponerine ant.

These ants form small colonies and are rarely dominant insects; furthermore, Texas already had a rich ant fauna, and it seemed unlikely that such archaic and unadaptible insects could become established. When Cook replied that "adequate ignorance of literature is a necessary qualification for learning the habits of a new insect like the kelep," Wheeler replied with the following very quotable remarks:

> There can be little doubt that scientific investigation is often impeded rather than furthered by too much attention to the "literature of the subject." Many a piece of zoological research may be perverted from the outset by an incessant appeal to what has been written . . . Investigation and publication are, however, two very different matters. One may investigate a thousand things, experience all the thrills of first discovery . . . But whenever one does decide to publish, it is necessary to reckon with the great "paper memory of mankind," the conserved experience of other workers who have loved and investigated the same things. It then becomes a duty to study the "literature of the subject," if only for the purpose of bringing the new work into intelligible, organic relation with the old. Failure to do this may be justly interpreted as carelessness, sloth, ignorance or conceit.

This was the fourth time Wheeler had held forth on this subject in the pages of *Science,* and readers of that journal may well have been growing slightly bored with the kelep. At about the same time, as a matter of fact, Wheeler was carrying on a similar exchange in *Science* with W. E. Castle, who would later become Wheeler's colleague at the Bussey Institution at Harvard. This exchange related to the Dzierzon theory, which maintained that male honeybees developed from unfertilized eggs, females (queens and workers) from fertilized eggs. This concept had been extended to other social Hymenoptera, and Wheeler brought together evidence that worker ants, although unmated, in some cases laid eggs which gave rise to other workers. Mrs. John Henry Comstock, of Cornell, for example, had reared workers of the common ant *Lasius alienus* from eggs clearly laid by other workers. Castle defended the Dzeirzon theory vigorously and suggested that Mrs. Comstock's ants may have "previously been with males." Wheeler retorted that the mating of workers with males was unknown, and that worker ants do not even have receptacles for semen. "Academic convictions like those advanced by Castle," Wheeler remarked, "can be of service only in prejudging a field of inquiry; they can be of no imaginable use in stimulating or furthering research except indirectly through the spirit of contradiction aroused by their dogmatic character."

To a certain extent Wheeler was venting the antagonism toward genetics which he felt throughout his life.* But in this case time has shown him to be correct. Although Dzierzon's rule is generally applicable to Hymenoptera, as Wheeler recognized, there are fairly numerous instances of true parthenogenesis, that is, of females being produced from unfertilized eggs. The appropriate genetic mechanisms have evidently arisen many times independently, and among the ants only in some of the more specialized genera, including *Lasius*.

Such tiltings with other biologists in the pages of *Science* fortunately took little of Wheeler's time. His research soon extended from the more primitive groups of ants to various more specialized genera. For example, in 1902 two papers were published on agricultural ants (*Pogonomyrmex*), and in the first he addressed himself to Gideon Lincecum's claim that these ants actually sow grass seeds around their nests and even weed their gardens prior to harvesting the grain. Charles Darwin had communicated Lincecum's paper to the Linnaean Society, and McCook had admitted its plausibility in his widely known book on the agricultural ant of Texas. Wheeler noted that seeds stored in the nests of these ants sometimes germinate and are no longer suitable as food, and when this happens the ants may be seen removing these seeds and carrying them to a refuse heap around the nest, where they sometimes produce a ring of grasses. To say that these ants sow and tend these grasses, Wheeler remarked, "is as absurd as to say that the family cook is planting

* This is nowhere better demonstrated than in a footnote in his 1904 paper *Dr. Castle and the Dzierzon Theory*. Castle had just published the pioneering paper *The Heredity of Sex,* only three years after the rediscovery of Mendel's laws and two years before Castle himself introduced Drosophila as an experimental animal. Wheeler refers to Castle's paper as one "in which the *disjecta membra* of certain Darwinian, Weismannian and Mendelian theories concerning three such intricate subjects as heredity, sex and parthenogenesis, are stirred to the point of turbidity, garnished with a few accessory hypotheses, and served up in a pamphlet of thirty pages. As if such messes could be either palatable or digestible! . . . It turns out to be merely another case of the old fallacy of juggling the phenomenon to be explained—in this case, sex—back into the germ-cells and then pulling it out again *à la* Little Jack Horner, with the naïve assurance of having contributed something 'new' to science."

and maintaining an orchard when some of the peach stones which she has carelessly thrown into the back yard with the other kitchen refuse chance to grow into peach trees."

Wheeler's move to the American Museum in 1903 removed him from the immediate presence of the ants that had stimulated his first interests in myrmecology. But at the same time he was freed of teaching responsibilities and given ample opportunity to travel. The museum also provided an excellent outlet for his publications, the *Bulletin,* which he proceeded to fill with a barrage of papers on ants. These included major papers on the fungus-growing ants and on the honey ants, which in modified form appeared in his book, and also a 93-page paper on polymorphism, which was presented in abbreviated form at the first annual meeting of the Entomological Society of America before becoming, again with various modifications, Chapters VI and VII of *Ants.* Wheeler developed an elaborate terminology, based on Greek stems, for no less than twenty-seven different "castes," many of which occurred only in certain genera of ants, others only as a result of certain types of parasitism. His terms included such tongue-twisters as "phthisogyne" and "ergantandromorph"—names which have fortunately disappeared from the literature since Wheeler's day. In a masterful treatment of polymorphism of ants published in 1953, Professor E. O. Wilson of Harvard University reduced the number of distinct castes to ten and suggested generally simpler terms (for example, "worker minor" instead of "micrergate").[2] Wilson may be regarded as a "second generation" student of Wheeler's, since he did his graduate work under F. M. Carpenter, a student of Wheeler's at the Bussey Institution who stayed to carry on the field of entomology at Harvard.

Polymorphism and caste determination interested Wheeler for the remainder of his life, and he returned to the subject many times, his final views being summarized in his posthumous book *Mosaics and Other Anomalies Among Ants* (1937). We shall reserve further discussion for a later chapter.

Wheeler was also very much interested in various phenomena relating to parasitism, and he recognized the universality and complexity of parasitism among ants. In a long paper published in *The American Naturalist* in 1901, "The Compound and Mixed Nests of American Ants," Wheeler first described his studies of *Leptothorax emersoni* alluded to in the preceding chapter. This ant (now called *Leptothorax provancheri,* an earlier name provided by Emery) lives in special chambers inside the nests of a larger ant, *Myrmica brevinodis,* but maintains connecting galleries with the *Myrmica* colony. The small *Leptothorax* workers climb over the *Myrmica* workers and clean them extensively; in return the *Myrmica* workers feed them with droplets of regurgitated food. Wheeler believed that the study of "guest ants" such as these might provide clues to the origin of more advanced types of social parasitism. In subsequent sections of this paper he reviewed other known cases of symbiotic or parasitic relationships between different species of ants, making much use of the earlier work of Auguste Forel and Father Erich Wasmann on this subject. This provided the basis for his extensive treatment of the subject in his book *Ants.*

Another aspect of parasitism is presented by the many insects other than ants that have become adapted for life in ant's nests, the so-called myrmecophiles. Wheeler's second and third papers on ants, published in 1900, dealt with myrmecophiles which he and his students had discovered during his first months of field work on ants. The first was a small cricket, *Myrmecophila,* which was found in the nests of five different species of ants near Austin. These crickets apparently subsist on the oily substances on the bodies of the ants, and are able to avoid being attacked by the ants because of their small size and their speed. The second was a remarkable cockroach which he and his students A. L. Melander and C. T. Brues found while digging up a nest of leaf-cutting ants. These tiny cockroaches live in the fungus-garden of the ants and apparently feed on the fungus, a case of what Wheeler called "myrmecolepsy," that is, stealing from ants. Curiously, of the more than seventy specimens of this cockroach found by Wheeler, all had incomplete antennae, and it seemed possible that the worker ants, in the process of trimming the fungus hyphae, also inadvertently clipped off the antennae of their household guests. Both the cricket and the cockroach were classed as "synoeketes," or indifferently tolerated guests, in the extensive treatment of myrmecophiles in Chapters XXI and XXII of *Ants.* Of even greater interest are the true guests, or "symphiles," which are actually licked, fed, and even reared by the ants; these include principally various beetles, many of which have special glands which produce substances which are fed upon by their hosts. Finally there are many parasites which attack the ant larvae but provide nothing in return. Wheeler studied one of these in 1901, a phorid fly whose larva wraps itself around the neck of certain ant larva and partakes of the food provided for the latter by the worker ants. The 1907 paper on polymorphism included original observations on several symphiles and true parasites. As he pointed out, these often result in the production of ants of abnormal form, and thus "the inferences from these cases have a value approaching those derived from experiment."

Myrmecophiles (and termitophiles, which are similarly adapted for living in termite colonies) were perennially interesting to Wheeler, and he was to review the subject again in *The Social Insects* (1928) and other later publications. His students C. T. Brues and W. M. Mann made a number of contributions to this subject, and Wheeler kept up an active correspondence with persons throughout the world who were working in this field. Both myrmecophily and social parasitism were topics of special interest to Father Erich Wasmann, a Jesuit scholar who spent the greater part of his life in Holland. Wheeler and Wasmann were contemporaries (Wasmann was six years older than Wheeler and died six years before him, within a few months of the death of Auguste Forel), but it would be hard to find two persons passionately interested in the same subject who were philosophically more dissimilar. Wasmann was an outspoken anti-Darwinist, one who, according to Wheeler, was continually clothing his excellent original observations with Jesuit philosophy and thus removing the subject "from the field of legitimate biological inquiry." Wheeler often acknowledged his indebtedness to Wasmann for facts regarding the guests and parasites of ants, but when it came to interpretation the two were continually embroiled in argument. The first exchange occurred in 1905. Early in that year, Wasmann published a long paper on the origin of slavery in ants in *Biologisches Centralblatt,* a journal published in Leipzig. Wheeler responded with a paper in the October issue of the same journal, pointing out that Wasmann had given the impression that he, Wasmann, was responsible for the theory that slavery had evolved from temporary social parasitism. But the fact is that Wheeler had published several papers relating to this subject in 1903 and 1904, and then a formal proposal of the theory in a paper which appeared on February 14, 1905—one day before the appearance of the initial section of Wasmann's paper. All of this now seems slightly absurd in light of the fact that both admitted that Forel and others had played important roles in this field and of the fact that Wheeler later rejected the idea as oversimplified. Only a year later he remarked in his paper on colony founding that he believed that slavery "has no direct ontogenetic or phylogenetic connection with the condition I have called temporary social parasitism."

A more serious disagreement developed between Wasmann and many of his contemporaries over the origin of the relationship between ants and myrmecophiles. Wasmann claimed these relationships could not be explained by natural selection, but that special "symphilic instincts" had arisen as modifications of the ants' original brood-rearing instincts and that their guests had been developed by "amical selection," somewhat akin to man's selection of bizarre varieties among his domestic plants and animals. Wheeler, Karl Escherich, and others believed that no special instincts were required and that the relationships had been shaped by natural selection, the myrmecophiles being comparable to disease organisms or to the parasitic relation of the cuckoo to its hosts. Wasmann also tended to misuse the word "mimicry" as applied to ant guests and to operate with "other dubious, or at any rate highly speculative, conceptions [including] the relations of his personal Deity to the ants and myrmecophiles." As for Wasmann's symphilic instincts, Wheeler wrote in 1918: "If three of my maiden aunts are fond of pets and prefer cats, parrots and monkeys, respectively, I am not greatly enlightened when the family physician takes me aside and informs me sententiously that my aunt Eliza undoubtedly has an aelurophilous, my aunt Mary a psittacophilous and my aunt Jane a pithecophilous instinct, and that the possession of these instincts satisfactorily explains their behavior. It is only too apparent that the physician has merely called the stimuli that severally affect my aunts by Greek names, plus a suffix denoting 'fondness,' assumed their existence as entities in my aunts' minds and naively drawn them forth as 'explanations.' It is high time that such scholastic methods of conducting biological inquiries were abandoned."

These remarks are particularly amusing in the light of Wheeler's own tendency to create new terms from Greek stems. His verbal exchanges with Father Wasmann continued for many years, and did much to enliven his books and essays. But one should not obtain the impression that Wheeler's life was all controversy or that his barbed pen was the major source of his stature. A much more important factor was the sheer bulk of his original work. Wheeler was the first thoroughly scientific student of American ants, and this provided him with a limitless field for his great energies. As he pointed out in his 1905 reply to Wasmann: "Compared with the ant-fauna of North America, that of Europe is in many respects very meagre, not to say monotonous. Hence it is easy to see how during the nearly forty years of diligent observation on the part of Forel and the more than twenty years devoted by Wasmann to similar studies, temporary social parasitism . . . should have passed unobserved. I am convinced that had these savants been able to study our much richer . . . fauna they would have long since detected not only the regularity of the parasitism I described . . . but also many other interesting facts which have hitherto escaped observation."

Wheeler was not long in extending his interests into the American tropics. His first trip to Mexico was taken while he was still in Texas, and there were many later trips to the West Indies, Central America, and British Guiana. On his two trips to Australia, in 1914 and in 1931, he studied the exciting ant fauna of that continent, many of the results of his several trips being published in his book *Colony-Founding Among Ants* (1933). He eventually published on the systematics of ants from all over the world. Yet the ant—least of all the dead ant, impaled on a museum pin—was never an end in itself, but always a means to an understanding of some facet of nature, be it parasitism, sociality, instinctive behavior, or evolution.

When Professor G. H. Parker wrote to Wheeler to congratulate him on the appearance of *Ants* in 1910, Wheeler replied that he had "now reached the stage of Doctor [A. S.] Packard since having written one book I can now cut it up into pieces and paste these together in a different combination in order to make other books."[3] Wheeler took his own advice, for his scientific and popular essays often appeared later in his books, and some of the topics covered in *Ants,* for example, were once again treated in different form in *Social Life Among the Insects* (1923) and *The Social Insects* (1928). But his horizons had broadened in these later books, and we shall reserve discussion of them for another chapter.

There remains one aspect of Wheeler's myrmecological studies that we have not yet discussed, and that is his contribution to ant systematics. At first Wheeler depended upon European workers, particularly Carlo Emery, for the identification of the ants he was working on. In 1902 he published in *The American Naturalist* his own translation of Emery's keys to the subfamilies and genera of ants of the world, and these formed the basis for the keys to North American ants which appeared as an appendix to his 1910 book.

Even much later, in his great tome on the ants of the Congo published in 1922, he relied heavily on Emery's basic work on the classification of the ants of the world.

However, it was soon apparent to Wheeler that in so large and poorly known a group of animals it would not be possible to depend indefinitely on colleagues in Europe. If he wished to continue his field and laboratory studies of the behavior of ants and their guests and parasites, he would have to become involved in their taxonomy, particularly since, as he often pointed out, knowledge of their ecology and behavior often shed much light on their structure and systematics. Wheeler began describing new species of ants in 1901, and he continued to do so throughout his life. From the beginning, his descriptions were detailed and often illustrated with carefully executed line drawings. He solicited ants from many sources, and in fact in the preface to *Ants* he remarked that he would "be glad to identify ants for anyone who is interested in their study." Throughout the last thirty years of his life he spent much time working over material collected by himself and sent to him by others from all over the world. This involved a great deal of correspondence and routine descriptive work, which often left Wheeler somewhat buried in minutiae, but such was his conviction as to the importance of this work that he persisted with it even through the days of his fame as a writer and lecturer. Mrs. Wheeler used to refer to his taxonomic work as his "knitting," and it may well be true that he looked upon this work as a relaxing contrast to more abstract matters.

One of the problems which Wheeler faced initially was the work of S. B. Buckley, who, though untrained in this field, published descriptions of seventy-six supposedly new species. His descriptions were, in Wheeler's words, "fearfully and wonderfully made. With a persistency, which at times seems almost intentional, the author selects for description the worthless, insignificant features of the ant's body, and passes without a word over the important, distinctive characters . . . sometimes he mistakes the sex of the form he is describing."

Emery and Mayr had attempted to identify Buckley's species, but with little success. Wheeler felt in a better position, since he had collected ants in many of Buckley's haunts in Texas; yet even he was able to salvage only a few and suggested consigning the remainder to the "taxonomic rubbish-heap." As to the several of Buckley's names that he was able to preserve, Wheeler remarked that "if I have succeeded in throwing a little light on some of his species, I could wish this to be regarded as a tribute to a pioneer naturalist who long ago searched the woods and hill-slopes of Texas, collecting ants and observing their ways 'with much pleasure and satisfaction.' "

Wheeler's first revision of a genus of ants (*Leptothorax*) appeared only three years after he began his studies. As his student W. S. Creighton remarked many years later, "this paper must have come as a profound surprise to Emery, Forel and Mayr, for in it they could clearly see the end of the supremacy which they had enjoyed in the field of North American ant taxonomy." Other revisions followed, including major papers on the large genera *Camponotus* (1910) and *Formica* (1913). There were also a great many papers of more limited scope, as well as many lists of ants of specific areas, with or without keys for their identification. These included papers on the ants of North Carolina (1904), of New Jersey (1906), of Texas, New Mexico, and Arizona (1908), of Connecticut (1916), and a long paper on the mountain ants of western North America (1917). There were also many papers which were concerned with other geographic areas: Japan (1906), Cuba (1913), Borneo (1919), China (1922), and so forth. An 1139-page *Bulletin of the American Museum of Natural History,* published in 1922, dealt with various aspects of the ants of Belgian Congo, based mainly on the American Museum's Congo Expedition. This included a lengthy account by Wheeler on the systematics and distribution of African ants as well as a key to the genera of ants of the world. Another major section of this volume, dealing with the relationship of African ants to plants, was contributed by Joseph Bequaert, then at the American Museum but later to move to Harvard. Wheeler's student W. M. Mann and his colleague I. W. Bailey contributed shorter sections. Other later papers dealt with the ants of the Canary Islands (1927), Hawaii (1934), Oceania (1935), and many other areas. Like many other taxonomists, Wheeler tended to be lured away from the native fauna in his later years, attracted no doubt by the many exciting and unusual ants from other parts of the world.

Wheeler's plan for a field book of North American ants, conceived in the early days of his ant studies as a means of encouraging interest in these insects, was revived in 1936 as a joint project with W. S. Creighton. Wheeler's sudden death in 1937 left the project entirely in the hands of Creighton, who expanded it into a major monograph on the ants of America north of Mexico which was published in 1950 by the Museum of Comparative Zoology at Harvard.[4] The higher classification of the ants of the world, which Wheeler had not modified greatly from that of Emery (aside from describing a number of new genera), has now passed largely to Professor W. L. Brown, Jr. of Cornell. Brown was a student of F. M. Carpenter's, and thus in a sense a second generation student of Wheeler's, like E. O. Wilson. But that is not to say that any of these men have followed in Wheeler's footsteps unquestioningly. Every scientist is a product of his times and builds upon those elements in the scientific record which he finds relevant.

Mention must also be made of Wheeler's interest in fossil ants. The work which had been done prior to 1910 was summarized in a short chapter in *Ants*. Generally speaking, ants do not fossilize well, but several amber deposits contain excellent fossils. The Austrian myrmecologist Gustav Mayr had monographed the ants of Baltic amber in 1868, but much material had accumulated since that time. In 1914 Wheeler published a major new treatment of the Baltic amber ants, basing it on nearly 10,000 specimens (many of them the property of the Geological Institute of Königsberg and destroyed during World War II). In this paper, Wheeler described twenty-one new genera and forty new species, nearly doubling the numbers previously known. More than half of these genera are still living, and the evidence suggests that during the time when Baltic amber was formed (some 30 million years ago) "the habits and instincts . . . were nearly if not quite as advanced as those of existing forms." "The general impression," Wheeler wrote, "is one of wonder at the great exuberance of the group in the early Tertiary of Europe and the conviction that since this period the family has not only failed to exhibit any considerable taxonomic or ethological progress but instead suffered a great decline in the number of species . . . at least in Europe."

As we have mentioned, Wheeler collected ants with T. D. A. Cockerell in the well known fossil shales of Florissant, Colorado, in 1906. He never worked up this material, and in 1923 Cockerell urged him to do so, remarking in a letter that "you will never get through the recent ones in a lifetime . . . Why can't you publish half a dozen of the more *interesting* Florissant ants? They are of much greater importance than the multitudes of recent species"[5] Although

Wheeler had long planned to study the North American fossil ants, he became involved in so many other projects that he was unable to fulfill his plans. In 1925 he turned the project over to his student F. M. Carpenter, who published a major paper on the subject in the *Bulletin of the Museum of Comparative Zoology* in 1930.[6]

In Wheeler's time, in fact, until quite recently, no fossil ants were known from prior to the Tertiary. In 1967 Carpenter, Wilson, and Brown announced the discovery of a fossil ant from the late Mesozoic, represented by two workers and therefore clearly social. Wheeler would have been greatly excited by this and by many other harvests that have been reaped from seeds planted in his books, lectures, and personal contacts.

Wheeler's major efforts in ant taxonomy were directed toward the description of genera, species, and varieties and toward the production of keys and faunal lists which might serve as guides to the study of ants in specific areas. He was not an innovator and in fact often resisted changes in terminology and procedure. For example, he followed tradition in calling the propodeum (first abdominal segment, fused with the thorax) the "epinotum," the antennal flagellum the "funiculus," and so forth. In a letter to S. A. Rohwer of the U.S. National Museum in 1909, he remarked that it is better "to use nomenclature which has been used for a long time . . . I think it is best to be very conservative in all such matters especially as the description of species becomes more comprehensible when you use the same nomenclature as your predecessors [rather than] adopting the results of a comparative study covering the whole insect group and for that reason becoming necessarily highly hypothetical."[7]

Wheeler's descriptive studies of ants were very much in the tradition of Forel and Emery, even though they involved a cumbersome, pentanomial system of nomenclature not in use in other groups of animals. Forel, in his early and classic studies of the ants of Switzerland, had found it expedient to apply varietal names to local populations which differed in color or in minor structural features. Soon he began extending this to museum specimens from other areas, using two categories below that of the species, "subspecies" being distinguished by less prominent features than species and "varieties" by very minor differences in color or hairiness. Since the genera were often divided into subgenera, many ants came to be distinguished by a series of five names: genus, subgenus, species, subspecies, and variety. In his later years, Forel often described ants hastily and with little attention to the real relationships of the forms he was treating. Carlo Emery also followed "the weird and wonderful pentanomial system," but with more restraint than Forel, and with much more concern for the broader aspects of the classification of ants.[8]

Wheeler accepted the pentanomial system and adhered to it throughout his life. For example, in his revision of *Formica* (1913) he reported the discovery of two females and several workers at the Stony Brook Reservation in Boston, of an ant differing from a known subspecies in having the thorax somewhat more slender and the front of the head smooth and shining. He described this as a new variety, and it became known as *Formica* (*Formica*) *sanguinea rubicunda* var. *sublucida* Wheeler. This interpretation was based on the fact that these specimens differed only slightly from other varieties of *rubicunda,* a form which differed from other subspecies of *sanguinea* by somewhat more prominent features; *sanguinea* in turn belonged to a complex of species making up the subgenus *Formica* of the genus *Formica.*

Wheeler often defended this strictly typological method of naming ants and even suggested that research workers on other groups of animals might some day come to use it. In a letter to Professor W. E. Hocking of Harvard in 1917, Wheeler remarked that "in taxonomy it seems to be important to recognize even the finest divisions of species . . . when practical to do so. The main danger seems to be that in continually splitting, one tends to magnify the importance of the finest categories and to detach them from the whole. If we could only carry on our classifications to minute detail without losing the connections, we should, I believe, be doing useful work."[9]

At the same time Wheeler recognized the awkwardness of such long names. Thus in his 1910 book he suggested that in practice the varietal name be used as if it were specific, by-passing the specific and subspecific names. He felt that this was justifiable "as the variety among ants is very nearly equivalent to the species among many other groups of animals, such as birds and mammals." As late as 1935 he wrote to W. S. Creighton: "Of late I have been trying to get rid of many of the subspecies and varieties of ants by elevating them to species and subspecies, kicking them upstairs so to speak, largely because the nomenclature is becoming too complicated."[10]

It seems odd that Wheeler, who frequently stressed the importance of habitat and behavior, so often insisted that museum specimens differing from others by minor features of structure or color be ranked as varieties, when in fact he himself recognized the prevalence of species which are difficult to identify on the basis of structure but still behave as full species. He seemed well aware of what we now speak of as the biological species concept and yet in practice was content to follow Forel's example of hierarchies based on degree of structural or color difference. To Wheeler, taxonomy's function was simply to "put a name" on an animal so that he or others would "have a handle" for further study. As the years passed and he accumulated more material from odd corners of the earth, he tended to write shorter descriptions, with fewer illustrations, in the effort to "cut through" the mounting complexity of the subject —the tendency of many older taxonomists.

Wheeler was correct that many of his varieties were in fact species. The ant he described from Stony Brook, cited above, is now regarded as a full species, occurring throughout much of eastern North America, and is called simply *Formica sublucida* (with or without a subgeneric name). But many other varieties are now recognized simply as aspects of variation, often within a single nest series, which are sometimes interesting but unworthy of including in a formal system of nomenclature. The term *subspecies* is now used in the sense of a geographical race, that is, a variant occurring consistently throughout a portion of the range of the species, without regard to the prominence of its morphological differences. Thus the many hundreds of ants described by Wheeler have had various fates: some are now regarded as species, others as subspecies, and a rather considerable number are no longer used at all, but simply cited as synonyms of other forms. That many of his names have fallen into synonymy is a consequence of our changing concepts of the nature of species and subspecies and of the further application to ants of the ecological and behavioral studies that Wheeler himself espoused. In the introductory section of *The Ants of North America,* Creighton remarks that Wheeler described more than 270 forms of ants from this continent.

"Yet [he continues] it may be stated that the most valuable contribution which Wheeler made to the taxonomy of our ants was not in the large number of new species which he described. It was rather that Wheeler carried his taxonomy into the field and amplified structural distinctions with others relating to habits, distribution and ecology. For no amount of study of expatriated specimens can furnish these vital details and until the taxonomy of our ants was brought back to this country it perforce remained as lifeless as the preserved specimens on which it was based. It is Wheeler's singular distinction that by his indefatigable efforts he made our ant taxonomy the living thing it is today."

Before concluding this chapter, we should comment briefly on Wheeler's energetic pursuit of ants in the field. Wherever he went, and on whatever business, he took occasion to collect the local ants and whenever possible to observe them in nature or in artificial nests. Late in his life he published in *Scientific Monthly* a short article titled *Some Attractions of the Field Study of Ants,* in which he expressed his sentiments regarding the pleasures of field work. He also listed some of the equipment useful for the study of ants and some of the problems awaiting study. These problems included such matters as intracolony communication, orientation and homing behavior, the determination of castes, interrelations with myrmecophiles, and the population size and life-span of colonies. He concluded with the comment that "we only need more myrmecologists." The problems he listed are the very ones that are now being pursued by several groups of biologists, and he would be pleased and sometimes surprised at the results being obtained. Still, considering the importance of ants in nature and as subjects for research on social phenomena, he would surely reiterate, were he alive today, his plea for more myrmecologists!

9 / Professor at the Bussey

The Bussey Institution, which housed Dr. Wheeler and a number of other eminent scientists for over twenty years, had had a checkered past. In 1908, when Wheeler came to Bussey, it already had a thirty-seven year history as an undergraduate school of husbandry and gardening. Long before the Congressional Morrill Act, which established the state agricultural colleges, such a school was provided for in the will of Benjamin Bussey, and it was eventually established in 1871 as a department within Harvard University. Another gift from Mr. James Arnold in 1872 enabled the University to begin to develop some of the Bussey land into what would become the Arnold Arboretum.

In his history of Bussey, published in 1930, Wheeler discussed its early years, but in fact the Bussey Institution was not really much of a success prior to 1908. This was why, presumably, the president and trustees of Harvard decided to discontinue the old undergraduate school, and reorganize it (as they did the old Lawrence Scientific School) into something more vital and useful. Bussey was to be "an institution for advanced instruction and research in the scientific problems that relate and contribute to practical agriculture and horticulture," and it became part of the Graduate School of Applied Science. Four fields were chosen for immediate development: economic entomology, animal heredity, experimental plant morphology, and comparative pathology; plant anatomy, forestry, and economic botany were added later. The undergraduates who were at Bussey at the time of reorganization were allowed to finish, but after 1910 only graduate degrees were given. A further revamping occurred in 1915, when Bussey became a Graduate School of Ap-

William Morton Wheeler as professor and dean of the Bussey Institution, 1916.

plied Biology, the staff organized as an independent faculty, and Dr. Wheeler appointed dean. This was the *status quo* until 1929, when another major change took place; but more of that later.

The Harvard administration may have intended to set up an institution where problems in applied biology could be solved, but in fact they did much more than that. The outstanding faculty they chose soon proceeded to make Bussey a mecca for basic biological research. As Wheeler himself saw it: "The economic entomologist, like the physician, should strive to make himself unnecessary by prophylactic measures, or perhaps I should say that he should prevent plant diseases rather than spend his time studying symptoms and applying remedies to diseases after they have become established."[1]

Associated with Wheeler at Bussey at first were William E. Castle, working in animal genetics, and Theobald Smith, a comparative pathologist. Charles T. Brues, Wheeler's former student in Texas, joined the Bussey faculty in 1909, after having been curator of invertebrate zoology for four years at the Milwaukee Public Museum. Edward M. East, whose field was plant genetics, came from the Connecticut Agricultural Experiment Station. Then came I. W. Bailey (plant anatomy) and Richard T. Fisher (forestry) in 1914, and Oakes Ames (economic botany) in 1915. Of all these men, it is difficult even now to say which one contributed the most to the advancement of biology. They were all individualists in their own way, and the students at Bussey, regardless of their own particular fields of study, must have found the atmosphere of the Institution continually stimulating, if not sometimes filled with controversy.

The connection between the biology taught on the Cambridge campus and that given at the Bussey in Jamaica Plain was at best nebulous. The old Lawrence Scientific School in Cambridge was incorporated either into Harvard College or Bussey (for example, Castle moved from Lawrence to Bussey), and most courses continued as formerly. Most of the Bussey faculty gave courses in Cambridge, too; and Bussey students took courses in Cambridge and often assisted in the teaching on the main campus, which involved daily commuting by subway. The number of students at Bussey during the early years was never great—a dozen or so perhaps at any one time. Those who had little or no work in Cambridge lived on the Bussey grounds in the old Watson mansion (Benjamin Watson had been a member of the old horticultural faculty and had collected many exotic plants for the arboretum). The house had been converted into a dormitory with a dining room which the students ran themselves, with various members of the faculty often joining them for lunch. Impressions of Wheeler's early years at

Bussey are difficult to put together because Wheeler left almost no personal papers covering those times. There appears to have been a warm relationship between the students and staff. In fact, so it was said, about the only visible difference between them was that the staff offices were slightly larger with room for more books. Furthermore, the members of the small staff appeared to prefer the freedom and simplicity of the Bussey to the more academic atmosphere of the departments in Cambridge.

The research and teaching facilities were located in the main stone building, which had been built in 1871–72 for the instruction of the undergraduates in the agricultural school. It was old and drafty, with ancient and temperamental plumbing, and the heat often did not even get up to the third floor where the library and some of the student offices were located (leading some of the students to keep a pot of hot tea brewing constantly). There was some teaching at Bussey, such as the laboratory work in the beginning entomology courses, although the lectures were given in Cambridge, at least during the winter semesters. Brues assisted Wheeler with these courses (7a, Morphology and Classification of insects, and 7b, Habits and Distribution of Insects), and often gave them when Wheeler

The Bussey Institution in Wheeler's time. This building is still standing in Jamaica Plain, Massachusetts, and is currently being used as a laboratory by the state department of health. (Widener Library, Harvard University)

was away or busy. Brues also gave first one and later two courses in economic entomology (Practical Entomology and Forest Entomology).[2]

The most stimulating interchange between the students and faculty of Bussey seems to have been at the weekly or biweekly seminars. Although guest speakers were sometimes present, these discussions usually were started by one or two students or faculty with a review of a current paper or papers of interest. The paper could be in any language (Professor Wheeler frequently assigned foreign language papers to his students), and the speaker was expected to present its contents and defend it. In this defense, the differences among the faculty, particularly on genetical questions, were often very evident. Dr. Castle was usually the least formidable, saying that if the student was going to hang himself, time would do it and he did not need to help. Dr. Smith left Bussey in 1915, but he seems to have been remembered by colleagues as a quiet, kindly man. East and Wheeler, however, were "devil's advocates," for both had logical minds and strong opinions, and if one did not pick up some flaw in logic or argue a debatable point, the other would. If a student escaped easily at the seminars, he might still meet his nemesis at the oral examination in which he had to defend his thesis. Students never knew ahead of time who might sit in on the defense, and sometimes as many as a dozen people might be there—from the Cambridge faculty, any or all of the Bussey faculty, and sometimes guests who happened to be visiting in the area.

As previously mentioned, four fields were chosen for the initial development of Bussey in 1908, and two of these were economic entomology and comparative pathology. The choice of these two fields may have been influenced by one of the most pressing problems of the day in the northeastern United States: the great increase in defoliation of shade and forest trees by the caterpillars of the gypsy moth, *Porthetria dispar*. The story of the introduction of the gypsy moth into the United States is well known, for it has been told many times as a good example of what can happen when a plant or an animal is brought into an area where it has not previously been known and where it has no natural enemies. In 1869 a French naturalist who was interested in silk production possibilities imported certain caterpillars from Europe to his home in Medford, Massachusetts. Among them was the gypsy moth, which soon escaped from its cages to the surrounding woodlot. The species flourished in its new environment and gradually spread across Massachusetts and to the neighboring states. It took almost twenty years, however, for the seriousness of this pest to be recognized. Then the state and federal governments inaugurated a control program, and since that time over a million dollars have been spent annually in attempts to control the pest.

In 1905 the state of Massachusetts, cooperating with the Federal Bureau of Entomology, imported a large number of parasites and natural enemies of the gypsy moth from Europe and Japan. Any immediate beneficial results apparently were not noticeable, but in 1907 a number of diseased caterpillars were observed. No one knew for sure what the disease was (though it was called "wilt"), or where it had come from, but many had hopes that it would destroy the gypsy moth once and for all. Others, the more cautious workers, felt that while it probably would not eradicate the pest, it might keep its numbers considerably reduced. In the *Annual Reports* for the year 1908–09, the dean of the Harvard Graduate School of Applied Science, Wallace C. Sabine, a physicist, wrote:

> In the early spring an arrangement was made with the State Gypsy Moth Commission and the Bureau of Entomology in Washington whereby Professor Thaxter [botanist] and Professor Smith were to cooperate in the investigation of the browntail and gypsy moths. The investigation of Professor Thaxter was directed along the line of fungus parasites. The investigation of Professor Smith was especially directed to the wilt disease as a possible bacterial parasite of the gypsy moth. Professor Mark [zoologist] held himself in readiness to investigate the wilt disease should it prove to be protozoan in character, and later in the spring took up this investigation. The work could only in part be regarded as Bussey work, but it at least illustrates one of the great fields of usefulness in which the Bussey Institution stands ready to serve. In this general investigation work was also done during the summer by Mr. Reiff at the Bussey Institution on infection by the wilt disease.

The Gypsy Moth Laboratory in Melrose, Massachusetts, was the center for the work on that pest and its close ally, the brown tail moth. When Wheeler arrived at Bussey, he found the above-mentioned Mr. Reiff already working on these moths. In the years 1909 through 1911, William Reiff published a number of papers, several in *Psyche,* on his work. In fact, in the first collection of *Entomological Contributions from Bussey,* nine of the forty papers were by

Reiff. Reiff's background and his fate after he left Bussey are unknown. A summary of his work while at Bussey, however, can be found in the chapter on virus infections in *Principles of Insect Pathology* by E. A. Steinhaus, published in 1949.[3] Reiff, after reading accounts of work done in Europe on caterpillars and disease susceptibility, "concerned himself with the experimental production of 'flacherie' (believed to be bacterial in nature) in gypsy-moth caterpillars. In 1911, Reiff reported favorable results on extensive field tests in which he fostered the 'flacherie' among gypsy moth populations in several localities in Massachusetts. He used the terms 'wilt disease' and 'flacherie' synonymously; thus apparently he did not differentiate between the true wilt diseases, or polyhedroses, and the bacteria-caused 'flacheries' . . . Furthermore, from a scientific standpoint, so much of Reiff's work on this subject is open to criticism (see Escherich, 1913) that it is difficult to evaluate the effect, if any, that his experiments had on the clarification of the nature of the wilt disease of the gypsy-moth caterpillar."

Thus, as can be seen, Wheeler's old friend and specialist in problems of forest entomology had early taken exception to Reiff's work. Wheeler himself was apparently not very happy about Reiff. The thirty-sixth contribution from Bussey was a sixty-page pamphlet published by the Massachusetts state forester: "The Wilt Disease, or Flacherie, of the Gypsy Moth," by Reiff. In the introduction, Wheeler's help was acknowledged, but on March 10, 1910, Wheeler had already notified the state forester that Reiff did not have a Ph.D., should not set himself up as a specialist, and that he, Wheeler, regretted the "unpleasantness in the newspapers." *Mr.* Reiff had apparently made himself *persona non grata* around Bussey, and perhaps elsewhere, too.

The problem of the gypsy moth was not abandoned at Bussey with Reiff's departure, and we shall return to it, but meanwhile another mystery exists regarding the early days of Bussey. When Wheeler accepted President Eliot's offer to come to Harvard, he had asked for two assistants: one to be an instructor, and the other a technician to aid in collecting and mounting material. For an instructor he suggested F. Silvestri, who was also interested in ants. This suggestion did not materialize, for reasons we do not know, and Brues was hired to the position in 1909. However, during Wheeler's first year at Bussey, he was assisted by a Paul Hayhurst, who was interested in aphids. The third through the sixth contributions are papers by Hayhurst which had been published in various entomological journals in 1909, but after that his name vanished from the entomological scene.

During the many years that Wheeler and Brues were associated at Harvard, first at Bussey and later at the Biological Laboratories, they had about forty doctoral students, many of whom first received master's degrees, and about ten students who received only master's. Many of those students went on to make major contributions in their fields of special interest. The degrees granted by the Bussey, and throughout the Graduate School of Applied Sciences, as long as it existed, were Master of Science (S.M.) and Doctor of Science (S. D.) The first doctorate in entomology in the new school was given in 1911 to Edward G. Titus for his work on a group of economically important beetles, the clover leaf weevils and the alfalfa weevil. Dr. Titus was on leave from Utah College, and most of his work had been done in that state, where these introduced insects were just being recognized as major pests. Why he came to Harvard to finish his study we do not know, but in his thesis entitled "Monograph of the Species of Hypera and Phytonomus in America," he acknowledged Dr. Wheeler's help and thanked him "for his sincere kindness, his encouragement and advice throughout the work." This study presumably was a suitable subject for a school of applied science "ready to serve" its public.

The gypsy moth problem continued to confront New Englanders, and Bussey continued to be one of the centers of scientific investigation on the matter. To quote again from Steinhaus: "It was not long after this [the discovery of the wilt disease] before the wilt disease was generally recognized by foresters and entomologists alike as being a widespread infection in gypsy-moth larvae throughout the entire gypsy-moth infested area in New England. The nature of the disease and its primary cause then became the subject of investigation by several men, chief of whom were Glaser and Chapman (1912–1916)."

These two men, James W. Chapman (1880–1964) and Rudolph W. Glaser (1888–1947), were graduate students at Bussey when they did their initial studies in insect pathology—important work contributing much to our general understanding of viruses, and in particular to insect virus infections. Chapman received his doctorate in 1913 with a thesis entitled "The Leopard Moth and Other Insects Injurious to Shade Trees in the Vicinity of Boston," and

Glaser's degree was bestowed a year later, his thesis entitled "Caterpillar Diseases." Both men continued in this cooperative investigation of insect viruses and other infectious microbes until 1916, when Dr. Chapman went to the Philippines to teach zoology at an institution now called Silliman University. Glaser, later at Rockefeller Institute and then Princeton University, continued throughout his life working in the field of insect pathology. Although Chapman had not been one of Wheeler's "ant men" while at Bussey, once he had settled in the Philippines he took up the study of ants with notable success, something which no doubt pleased his mentor tremendously. Chapman remained in the Philippines for most of his life, surviving the Japanese occupation of 1942–1945 in a series of harrowing adventures recounted by Mrs. Chapman and himself in their book *Escape to the Hills* (1947).[4] Chapman was later made a research associate of the Museum of Comparative Zoology at Harvard, where his ant collection now resides.

Wheeler continued to publish voluminously on ants during his early years at Bussey. That he took his duties as "economic" entomologist seriously, however, is demonstrated by a Lowell Lecture delivered under the auspices of the Boston Society of Natural History in the spring of 1912. It was titled "The Influence of Insects on Human Welfare" and is especially revealing of Wheeler's belief that basic research and preventive measures are of far more value than "control" as such—an idea that many contemporary economic entomologists have yet to grasp. Parts of Wheeler's (unpublished) essay were prescient indeed. For example, he pointed out the need for increasing the yield of agricultural products to meet the "stress arising from the recent great increase in human population [and] the certain prospect of an even greater increase in the future." Wheeler estimated that insects caused an annual injury to resources in the United States amounting to $700 million (the figure is now several billion, despite a great increase in the number of economic entomologists). He also pointed out, in 1912, the importance of improved transportation and increased commerce in carrying pest species around the globe. The ants were, of course, mentioned—as examples of the complex relationships of insects, their environment, and their parasites. Then follow these revealing comments: "Since injurious insects as well as other parasites are living organisms reacting to a living environment, which we call the host, entomology as an applied science is primarily concerned with the problems of animal behavior and especially with the problems of instinct . . . Of course, it is impossible to study instinct exhaustively without recourse to psychology and metaphysics and without a preliminary knowledge of the insect's special identity, i.e. its taxonomy or classification, of its geographic distribution, or chorology, of its structure and development, or morphology, and of the function of its organs, or physiology, but it is evident that all these subjects are really ancillary or contributory to the study of the central problem, that of behavior."

Wheeler noted that there may be upwards of two million species of insects, "and our knowledge of the structure, development, distribution, and behavior of any one species, even the commonest, is more or less superficial and inadequate. Any one of these 2,000,000 species, moreover, may, under easily conceivable conditions, assume economic importance."[5] Clothing as he did such reasonable considerations in such eloquent language, it is not surprising that Wheeler soon began to attract students of high calibre from a wide variety of sources.

Two familiar names appeared on the list of students at Bussey in the fall of 1913: C. T. Brues and A. L. Melander, both candidates for the S.D. degree. Brues, for reasons not apparent, never received his doctorate, but, in June 1914, Melander did receive his S.D. His thesis was "A Taxonomic Study of the Empididae, a Family of Dipterous Flies," a study begun in 1900 in Texas. While they were students at Bussey, Brues and Melander produced a *Key to the Families of North American Insects,* profusely illustrated by Mrs. Brues and published privately in 1915. This was later expanded into a book called *Classification of Insects,* covering the world fauna, and still later (1954) expanded by F. M. Carpenter to cover fossil insects as well. Brues, Melander, and Carpenter's *Classification,* published by the Museum of Comparative Zoology, has been something of a best seller for a technical book.

Melander had been on leave from his teaching position at Washington State College in Pullman, and he returned to take up his duties again after his year at Harvard. A year later, in a letter dated May 17, 1915, he wrote to Wheeler saying that the president of the College was leaving, and he thought it would be "great" if Dr. Wheeler would consider filing an application for the vacancy. The present president was receiving $6,000 plus a house, and the College needed someone who would place more emphasis on basic research.

In his answer dated June 18, 1915, Wheeler thanked Melander, but declined, saying he was too old, and, anyway, he preferred to continue in research. Wheeler was just fifty and still had many active years ahead of him.[6]

In spite of his letter to Melander, Wheeler did take on administrative work during the academic year 1914–15. As previously mentioned, Bussey Institution was separated from the other graduate schools, and retitled the Graduate School of Applied Biology. Its teaching staff was enlarged, called an independent faculty, and Dr. Wheeler became the dean. Wheeler's old friend, T. D. A. Cockerell, had written to him on another matter, namely, his opinion of the American Association of University Professors, and Wheeler's reply includes an amusing commentary on his role as dean:

> I quite agree with your general standpoint with regard to the situation. Of course I joined the Association as I also believe in trade unionism to a certain extent. I am afraid that they would not have permitted me to join the Association if they knew that I had become Dean, as deans are, of course, supposed to be handmaids, so to speak, of the university presidents. I must say that with the somewhat more intimate knowledge of the trials of the university president, acquired since I have become Dean, that my former antagonism to university presidents in general has somewhat softened. You see how far I have degenerated, but this is all due to the fact that I, like yourself, have come to realize that the academic mind, as developed by most faculty members in the university, is pretty nearly as bad as the ecclesiastical mind.[7]

Though Dr. Wheeler may have tried not to take himself and academic formalities too seriously, the weight of his responsibilities did seem to affect his role at Bussey. Many of the students, particularly those who were not majoring with him, found him distant and difficult to approach. Professor Brues, or perhaps I. W. Bailey, were much easier to "bother" for help in meeting the little, daily problems of independent research.

Two who were not overawed by the dean, however, were Francis X. Williams (1883–1967) and William M. Mann (1886–1960), who came to Bussey in 1913 (as had Melander) and who received their S.D. degrees in June of 1915. Williams had received an A.B. from St. Ignathius College in California in 1903, a B.A. from Stanford University in 1908, and an A.M. from Kansas University in 1912.

His experiences and interests were wide: for he was an inspector for the California Horticultural Commission after he left Stanford; he curated the Snow Entomological Collections while working for his master's degree; he taught economic entomology one year at a small school near Boston; and, after he received his doctorate, he followed the crowd to the U.S.D.A. Gypsy Moth Laboratory for a year's work there. Dr. Williams' papers reflect his wide interests. His early ones are on Lepidoptera, and include one on "The Butterflies and Hawk-moths of the Galapagos Islands," a result of the California Academy expedition to those islands in 1905–06. Dr. Williams accompanied this seventeen-month expedition as field entomologist—a marvelous opportunity for a young biologist. Williams' research at Kansas was on a large and difficult group of digger wasps, "The Larridae of Kansas," and this monograph has become something of a classic, mainly because the behavioral observations were done so well and with so much insight. At the Bussey, Williams elected to do a research problem on a very different topic: the habits, structure, and development of two common fireflies. Why he took up this study is unknown, but the resulting thesis was very reminiscent of Wheeler's earlier embryological papers. Wheeler and Williams also published one paper jointly, a study of the structure of the light organ of the New Zealand glow-worm. A few years before his death, Dr. Williams wrote to us of his times at the Bussey:

Dr. Wheeler took a keen interest in his graduate students, and almost regularly each morning (except Sundays) would come to "my room" on the third floor [and] look at my slides and other work. From him I learned all my insect embryology, and much more. As a class of few students—Dr. W. M. Mann and I once, we went on day trips [to the Blue Hills, south of Boston]. Dr. Wheeler for a time conducted a very interesting entomological seminar at his home in Jamaica Plain, a mile or so from the Bussey. Quite informal.

I always liked and admired Dr. Wheeler . . . Finally, on Dr. Wheeler's recommendation, I secured a position as one of the entomologists on the H.S.P.A. [Hawaiian Sugar Planters' Association] Experiment Station staff . . . I stayed there 31 years and from my viewpoint, these were my best years.

I had the temerity, being very short of cash, to borrow part of my train fare to San Francisco, where my people lived, from Dr. Wheeler—thirty dollars I think."[8]

Williams' years with the H.S.P.A. were indeed productive ones. He traveled widely in the tropics in the search for parasites of pest insects, and wherever he went he made extensive collections and detailed life history studies of diverse insects. His studies of Philippine (1919) and tropical American wasps (1928) remain unrivalled in their field. His paper "The Natural History of a Philippine Nipa House" is a fascinating account of the animals that shared his home for two and a half years. In Hawaii he compiled a book, *The Insects and Other Invertebrates of Hawaiian Sugar Cane Fields,* and authored a remarkable series of papers on the aquatic insects of Hawaii. This last series of papers records his discovery of the first known damselfly with terrestial larvae, a remarkable insect that has invaded a wholly new habitat for its group, foraging for prey in leaf litter at the base of ferns.

After his retirement, Williams returned to California, where he continued to work on the systematics and biology of wasps almost until the day of his death. He married late in life, and he and his wife occupied an attractive home in the suburbs of San Diego when we visited them in 1954. Williams was always a modest and unassuming person, and there are few persons who recognize the great importance of his contributions and the superb quality of his publications, all of them lavishly illustrated with his own carefully executed and artistically pleasing sketches.

William Mann came to Wheeler through Melander, and the details of this are recounted in his *Ant Hill Odyssey,* published in 1948. Of all of Wheeler's students, Bill Mann was probably the most colorful, and his book is written in a racy and amusing style and should be read in the original; it is difficult to paraphrase. Briefly, however, Mann was born in the northwestern part of the country, loved natural history from childhood, and decided that a college education would be the best way to get into the field professionally. He sent for college catalogues, "some of which were confusing, but I did learn that Harvard University was not in New Haven, Connecticut, where I had written." His choice was the State College of Washington at Pullman, and there he met A. L. Melander, professor and head of the department of entomology. In fact, except for a laboratory assistant, he was the whole department. Through Melander, Mann began to know not only entomology, but many entomologists, including Brues (then in Milwaukee), and many of the zoologists up and down the west coast. This was especially true during the summer that he spent, with Melander, at the Puget Sound Marine Biological Laboratory as an assistant in zoology. Representatives from many institutions were there, including a "bunch" of ichthyologists from Stanford. "Then came a big day: Dr. David Starr Jordan, President of Stanford University, dropped in and talked to us about the biological laboratory at Penikese, where he had studied under Agassiz and been diverted from red algae to become the world's greatest authority on fishes. That settled it." Mann transferred to Stanford to continue his college training.

Following his graduation from Stanford, Mann planned to join the Stanford Expedition to Brazil as their entomologist. But, before he left, he met Dr. Wheeler, who was on a trip to the west coast. Mann said that Wheeler accompanied him "on collecting trips to all my favorite spots. At one point he mentioned, 'Heath has told me that you want to come to Harvard and study with us.' Of course I did." But Mann had heard about the expense of a Harvard education and felt that he should work for a year or two before attempting it. According to Mann, Wheeler concluded the discussion abruptly with: "You had better come direct to Boston from Brazil. Harvard will always take care of a student who will work."

When Mann arrived at Harvard, he was taken into the Wheeler home in Jamaica Plain, which was a short walk from Bussey. He was then given a laboratory of his own in Bussey, and he settled down to working on the material he had collected in Brazil. This consisted mainly of distributing materials to specialists, many of whom had supplied funds for collecting. Mann continued: "The ants I was to work up myself, and I began mounting and labeling them. As I worked day after day, my profound ignorance of the ant kingdom became evident. Bottle after bottle that I had marked X for rarity contained common ants of the tropics. Large and showy ones were usually species that had been well known for years, but here and there was a rarity or a new species. Of the two hundred and twenty-three species determined, about forty were new to science."

Dr. Mann's recollections of his days at Bussey are brief, but interesting:

In addition to our own laboratory work, most of us took courses at the University. I attended a series of lectures on the central nervous system, given by Dr. George Parker, who, as Wheeler said, had his lectures so letter-perfect and well organized that they passed through the students just like a dose of castor oil. An enjoyable course in botany was with Professor Fernald—we did some herbarium studies but also made many field trips . . .

Near the Bussey were big colonies of an eastern mound-building ant (*Formica exsectoides*), whose nests teemed with different kinds of inquilines. The little, oblong, reddish-brown Hetaerius beetle, which is covered with huge scales and bears little clusters of golden hair, had been rare in the Western states but occurred here by the dozens. We built artificial ant nests of glass and stocked them with small colonies so we could observe the habits of the ants and the beetles. When a beetle wanted food it would wave its two front legs to attract the ant which would come to it and feed it a drop from its own mouth. Once when the ants were shifting from one compartment in their nest, one which I had purposely let dry out, to another newly opened and dampened area, I saw one of the beetles crawl onto the rear of a worker ant and ride along, not unlike the clown (for which Hetaerius is named) riding on the south end of a mule.

The Cambridge Entomological Club met one evening a month, usually in the library on the third floor of the Bussey building. In addition to the usual books, chairs and tables, the room contained a placard put there by Professor Wheeler, with a quatrain directed against careless borrowers of books.

I was made secretary of the society, the only political position I have ever filled, and also assistant editor of *Psyche,* the entomological magazine published by the club. Brues was editor. The duties of the assistant editor included putting each issue into envelopes and addressing them to the subscribers, and also helping raise funds to make up the deficit of publishing. This was the beginning of my real acquaintance with Tom Barbour, whom I had met before only on occasional visits to the Museum to see its director, Samuel Henshaw. Brues suggested I drop into Barbour's office and ask for a fifty-dollar donation. This is not easy to do even when one knows a person well, but I had heard so much about Barbour, his vast wealth and his great generosity, that I walked into his office without hesitation. He was sitting at his desk, a tremendous man, swarthy, with thick curly black hair. When I told him what I was there for, he turned and looked me straight in the face with a most pathetic expression; this changed into a ferocious glare as he said:

"Here I sit in my office, wondering where my next meal is coming from, and you, *you* come in and ask me for fifty dollars!"

He used curse words in those days, and he punctuated this remark with several of them. Abashed, I was bowing my way out, when he roared, "Stop! Did you say fifty dollars? Wait a moment till I get my checkbook"[9]

Dr. Mann's book and much of his life concerned his varied explorations throughout the tropics of the world. In 1917 he received an appointment with the Department of Agriculture and the National Museum to study ants and to help in the survey of insect pests in the American tropics—which prompted him to wire Professor Wheeler: "The University's loss is the government's gain." In 1923 he participated in a nearly disastrous expedition to remote parts of the Amazon Basin, an expedition described vividly by Gordon MacCreagh in *White Waters and Black,* where Mann is identified simply as "The Entomologist."[10] In 1925 Mann became director of the National Zoological Park, and at this he was enormously successful—so much so that even Wheeler forgave him for leaving a full-time career in entomology. He did not abandon his ants, however. And his wit and charm added much to the lives of those around him, including Dr. Wheeler, who in 1937 was looking forward to accepting an invitation extended by Dr. and Mrs. Mann to accompany them on an expedition to Sumatra—a trip Wheeler did not live to experience.

Wheeler's students did not confine their subjects of research to flies or ants or the gypsy moth, but were apt to work with representatives of any insect order and from almost any point of view. In a school which was supposed to be concerned mainly with economic problems, Bussey students were unusually versatile and original in their research and, in fact, in their outlook toward biology in general. Howard M. Parshley (S.D., 1917) wrote a thesis on "The Hemiptera-Heteroptera of New England," and continued to publish on that group throughout his life while a professor of zoology at Smith College in Northampton, Massachusetts. But gradually over the years Parshley became more and more interested in sociology (a subject which also interested Professor Wheeler intensely), and he contributed a number of papers and two books in this field.

Much of the entomological research at Bussey was taxonomic, but life history and developmental studies were often included, too. The illustrations of these papers were often reminiscent of Wheeler's earlier cytological drawings: exquisite in minutest details. The Philippine L. B. Uichanco (S.D., 1922), working on aphids, and R. F. Hussey (S.D., 1923), working on aquatic bugs, were involved with embryological studies. Bussey had an occasional woman student, too, although their degrees officially came from Radcliffe. One was Esther W. Hall (Ph.D., 1921), who worked on braconid wasps parasitizing aphids, another Priscilla Butler Hussey (Ph.D., 1925), who studied certain aspects of embryonic development in the same aquatic bugs that interested her husband.

Of all of the hundreds of students and research associates who were a part of the Bussey, probably none received more recognition or lasting fame than did Alfred C. Kinsey (S.D., 1920). For a man who was to find himself on the forefront of biological advance, Dr. Kinsey (1894–1956) was almost an anachronism, for he was basically an old-fashioned naturalist who was interested in learning all he could about the biological world around him. As a teenager he had enjoyed watching birds and studying flowers, and one of his later, lesser known publications was a book called *Edible Wild Plants of Eastern North America,* coauthored with Professor M. L. Fernald of Harvard. As revised by Reed C. Rollins of Harvard in 1958, this book is still the standard work on the subject and is now in its fifth printing. As a teacher Kinsey felt that his students, and all students, should be concerned with understanding the basic principles fundamental to all organisms. Consequently, he wrote a textbook for high school biology, *An Introduction to Biology,* which was widely used throughout the United States for many years after its publication in 1926. He sent Dr. Wheeler a copy of this, and in a covering letter (November 4) wrote: "May I further express my appreciation because I find, upon searching for the origin of my present conception of Biology as distinct from a straight combination of Zoology and Botany, that I owe a great deal to you and your continued treatment of questions in a broad, biological way."[11] In the second edition Kinsey had several portraits of living scientists, including Wheeler, to show that the best biological work had not necessarily been done in the past.

In the *American Men of Science,* Dr. Kinsey listed his specialties as, first, elementary biological curriculum, and next, life histories, gall formation, evolution, geographical distribution, taxonomy, and the nature of species of gall wasps—all this before his last great study on human sexual behavior. His work on the gall wasps, or Cynipidae, was begun at Harvard under Professor Wheeler, and continued at Indiana University, where Kinsey went to join the faculty in 1920 and where he remained for the rest of his life. These wasp studies were primarily taxonomic, but were done in an unusually thorough manner, including a detailed analysis of many biological aspects of populations sampled over a wide geographical range. He eventually came up with some interesting though somewhat controversial ideas on the origin of species.

Dr. Kinsey's studies of the sexual behavior of human males and females was begun about 1938, apparently as the result of his work in counseling students, who asked many questions for which there seemed to be no adequate information on which to give helpful advice. And so Kinsey, first alone, and then with help from other men interested in statistical analysis and with money from various sources to aid in the mechanics of this tremendous undertaking, began his studies of humans. In 1948 the first in a series of intended nine volumes was published. In his long and excellent review of this book Bentley Glass began:

> Great scientific books are of two chief sorts. The more frequent kind is that which surveys an active field of science, summarizes the work of perhaps hundreds of scientific lives, and with masterly grasp achieves a synthesis of scattered and ill assimilated materials. Such were Wilson's *The Cell* and Bayliss' *General Physiology,* and in more recent years, Dobzhansky's *Genetics and the Origin of Species.*
>
> But once in a great while a book comes along which marks the commencement, not the climax or end, of a scientific era. These are the books that virtually begin a new science, that blaze a trial into the unknown. Such a book is *Sexual Behavior in the Human Male.*[12]

Dr. Kinsey died in 1956, at the age of sixty-two, but he did live long enough to see his work well launched with the publication in 1953 of the second volume, *Sexual Behavior in the Human Female* and to realize that these two best sellers were providing from their sales further financial assistance (to say nothing of interest) for expanding this much needed research. As Professor Wheeler had once said in a review (in 1927) of several books on human reproduction and the family: "What we need, before we begin to apply our still very incomplete biological knowledge of sex and the family to human society, is a much more comprehensive and detailed knowledge of the actual sexual situation in civilized societies at the present time. What we actually know about this matter is . . . extremely meager and unsatisfactory." Wheeler then added his own facetious touch to the matter: "Evidently a very considerable proportion of the population is over-sexed, under-sexed, intersexed or no-sexed, and, therefore, not well-suited to family life of the old-fashioned, rural, or garden variety. It is, of course, very easy to tell all these people to go to hell, but many of them are so devilishly attractive and, apart from their sexual behavior, so very efficient socially, that they are constantly being married by those who are blessed with normal sexual proclivities and ideals." Wheeler presented a number of challenges to future students of human sexual behavior, but how much he actually influenced Kinsey in his work is difficult to say.

Among Wheeler's many students at the end of the teens and in the early nineteen-twenties were the brothers, C. L. Metcalf (S.D., 1919), and Z. P. Metcalf (S. D., 1925). Both men were born and raised in Ohio and did their undergraduate work at Ohio State University. C. L. Metcalf (1888–1948) came to Harvard first to work under Wheeler and Brues and did a taxonomic study on male syrphid flies, using their genitalia to distinguish the species. He later taught at the University of Illinois. Today he is remembered chiefly for his work in economic entomology. His best known book, written with Dr. W. P. Flint, is *Destructive and Useful Insects,* first published in 1928. Z. P. Metcalf (1885–1956) taught at North Carolina State College in Raleigh and devoted his life to the study of Homoptera, beginning with his doctor's thesis on fulgorids or plant hoppers. At the time of his death he was engaged in preparing an extensive *Catalogue of the Homoptera of the World* and had fifteen volumes either published or in production.

J. G. Myers (S.D., 1926) was a student at the Bussey at about the same time as Z. P. Metcalf and was working in the same insect order, but there the resemblance ends. Myers (1897–1942), though born near Rugby, England, had lived in New Zealand since 1911. After returning from World War I, he completed his B.S. and M.S. degrees at University College, Wellington, and for about five years worked as entomologist for the Biological Division of the New Zealand Department of Agriculture. Then in 1924 he won the coveted honor of an 1851 Exhibition Scholarship, which he decided to use to come to Harvard to study. Although Myers worked on cicadas and not ants, he and his methods of study must have been very pleasing to Wheeler. Myers' thesis, published in 1929 as a book, *Insect Singers,* is filled with references and quotations concerning the role these insects have played throughout the course of human history and at the same time presents his scientific observations, which obviously were made with great care. His thesis was unusual and his style of writing delightful—something that can seldom be said of graduate students in biology. And even the acknowledgments in his thesis provide a relief from the usual clichés: "To thank specifically all who have been a source of inspiration or assistance in the work would be almost impossible in this place. For the first I am indebted above all to Dr. William M. Wheeler. To say that I have tried to accord to his work, 'the sincerest form of flattery,' would be to admit that I have attempted the impossible." But he did attempt the impossible—even to the point of challenging his mentor's high estimation of the ants. Here is the concluding paragraph of the Introduction to his book:

> There is most ancient precedent and high modern philosophical authority for looking to the ants as in many respects the paradigm of social beings. "Folk-lore and primitive poetry and philosophy show the ants as an abiding source of similes expressing the fervid activity and cooperation of men." They command a cool intellectual admiration for the organized thoroughness with which the individual is subordinated to the needs of that super-organism—the community as a whole. But when men shall have been cured of the disease which some call metaphrenia and others civilization, when they shall have ceased their "special and devout worship of a strange god whom they call Progress" and shall have turned once more to Pan and Orpheus and the art of living, then will they see in the cicadas living exemplars of the otherwise fictitious "noble savages" of Rousseau and will understand why the Greeks hailed these insects as the favorites of the Muses.[13]

After leaving Harvard, Myers worked mainly on problems of biological control, and he traveled around Europe, Africa, Australia, New Zealand, the West Indies, and South America searching for possible parasites. His publications were numerous, and his untimely death in an automobile accident before he was forty-five was a great loss to biology.

William Morton Wheeler's major interest in the Hymenoptera was reflected by the large number of students he had working on insects of this order and the undoubted success of these men in their fields. Theodore B. Mitchell (S.D., 1928) and Otto E. Plath (S.D., 1928) worked on bees, George Salt (S.D., 1927) on parasitic wasps, and Richard P. Dow (Ph.D., 1935) on behavior in sphecoid wasps. George Salt (b. 1903), although a Canadian by birth, has long been associated with the Cambridge University in England, where he has done outstanding work on the ecology and behavior of this most complex and difficult group of insects. Plath (1885–1940) spent his life in the Boston area (he was at Boston University) and wrote a book called *Bumblebees and Their Ways,* published in 1934, for which Wheeler wrote a brief introduction. This book was based primarily on research done at Bussey on the genus *Bombus,* and one aspect of his study had to do with life in the nest. To aid in this, Plath had nailed together a couple of old packing cases behind Bussey and made an adequate but rather unsightly home for his bee colony. "One day," so the story goes, "Wheeler was escorting some distinguished entomological visitors around the place, and, when they came to this somewhat ramshackle sight, he calmly informed them that this was the 'Bussey Bombarium.' He did this in a purely matter-of-fact way, and was probably unaware that his erudite language had conferred a dubious distinction on a utilitarian but quite unprepossessing eyesore." Thus did Herbert Friedmann, also then at Bussey, summarize the incident in retrospect.

The other student of bees, Theodore B. Mitchell (b. 1890), like Z. P. Metcalf, was located at North Carolina State College in Raleigh, and his major contributions to entomology are his taxonomic studies of leaf-cutting bees of the genus *Megachile* (done in part at Bussey), and his two volume classic, *Bees of the Eastern United States,* published in 1960 and 1962. In the foreword to the first volume of this latter work he includes the following paragraph:

I joined the teaching staff of the Department of Zoology and Entomology at North Carolina State College in 1925. The late Z. P. Metcalf, Head of the Department, was an authority on the Order Homoptera and an enthusiastic taxonomist with a world-wide reputation. To him I am deeply indebted for the assistance and words of advice, for encouragement and support, and for new ideas and techniques. The association lasted for more than thirty years, until his death, and throughout this period complete freedom was enjoyed in the performance of teaching duties and research activities. In this assignment the summer months were available for research and for advanced study. This included studies at Bussey Institute of Harvard University where the research was carried on under the direction of W. M. Wheeler and C. T. Brues. This association with these two outstanding teachers served to augment my interest and enthusiasm for taxonomic research.[14]

Without doubt the students who must have been most appreciated by Wheeler were those who studied ants. First was William Mann, but then came George C. Wheeler (S.D., 1921), who worked on ant larvae, Frank M. Carpenter (S.D., 1929) with a thesis on fossil ants, George S. Tulloch (Ph.D., 1931), a student of ant morphology, William S. Creighton (S.D., 1928) and Neal A. Weber (Ph.D., 1935), both of whom worked on taxonomic revisions of certain ant genera. (The difference in the doctoral degrees was due to administrative changes at Harvard, which will be explained later.) George Wheeler (who was not related to William Morton) had a long career as professor and chairman of the department of biology at the University of North Dakota and continued to publish extensively on ant larvae. Tulloch was for many years professor of biology at Brooklyn College, while Creighton held a similar position at City College of New York (joining A. L. Melander, who had moved to New York from Washington state in 1926). Frank Carpenter, as we have mentioned, remained at Harvard, where, as Alexander Agassiz Professor of Zoology, he became a world authority on fossil insects. Neal Weber, Wheeler's last student of ants, specialized in leaf-cutting ants and the fungi they cultivate. Weber has published extensively on ants and has traveled widely throughout the globe; since 1947 he has been professor of zoology at Swarthmore College in Pennsylvania.

Two others of Wheeler's later students were to make major contributions, though not to the study of ants. These were Philip J. Darlington, Jr. and Marston Bates, both of whom spent extended periods of time in the tropics shortly after receiving their degrees. Darlington (Ph.D., 1931) wrote a thesis entitled "The Cicindelid and Carabid Beetles of Central New Hampshire with Special Reference to the Geographical Relationships of the Mountain Fauna." Darlington became curator of insects and later Alexander Agassiz Professor of Zoology at Harvard. His book *Zoogeography* was awarded the Elliot Medal of the National Academy of Sciences for 1957 (an award Wheeler had received in 1924). Marston Bates (Ph.D., 1934) wrote his thesis on "The Butterflies of Cuba," but since then has become an authority on mosquitoes as well as a highly successful writer of popular books on natural history, two of which are *The Nature of Natural History* and *The Forest and the Sea.* He is currently a professor at the University of Michigan in Ann Arbor. One cannot help but note that almost every one of Wheeler's students included among his many publications at least one and often several books.

Wheeler had a number of candidates for master's degrees, and even though several later received doctorates elsewhere, they often referred to the help and inspiration they had received from their brief association with Wheeler at Harvard. P. H. Timberlake (M.A., 1910) had been at Harvard when Wheeler first arrived, and considered himself lucky to have been in his first course. Timberlake has long been at Riverside, California, doing basic work in biological control and on the taxonomy of bees. E. H. Strickland (S.M., 1912), an Englishman, became head of the entomology department at the University of Alberta, and an outstanding teacher. Harold R. Hagan (S.M., 1917), who was for some years in Hawaii and later at the City College of New York, published *Embryology of the Viviparous Insects* in 1951. This book, which received the A. Cressy Morrison Prize in Natural Science from the New York Academy of Sciences, is dedicated to two men, one of them "The Late Professor William Morton Wheeler." In the preface, Dr. Hagan said that Professor Wheeler had suggested the title and scope of the work. Two other men, Guy C. Crampton (S.M., 1921) and William J. Clench (S.M., 1923) were also master's candidates at Bussey. Dr. Crampton (1881–1951) worked in the field of insect morphology and was at the University of Massachusetts for many years. Dr. Clench (b. 1897) has just retired as curator of mollusks at the Museum of Comparative Zoology at Harvard University.

P. W. Whiting (S.M., 1912), long at the University of Pennsylvania and now at Oak Ridge, Tennessee, was an early student at Bussey, and still recalls vividly Professor Wheeler's stimulating lectures so effectively illustrated with blackboard drawings made as he talked. Dr. Whiting said that at the end of the hour he would frequently find that he had been so entranced that he had failed to take notes. He recalls, too, that at the Bussey seminars, Professor Wheeler always joked with a straight face and never laughed or smiled at his own jokes. In the late twenties, Dr. Whiting, along with his wife, Dr. Anna R. Whiting, did some research at Bussey one semester, and Mrs. Whiting says that she does not recall that Dr. Wheeler ever spoke to her: "He was not enthusiastic over women in the laboratory apparently."[15] There were other students in the late twenties who also felt that although Dr. Wheeler's writings continued to be models of excellence and of great interest, student seminars and perhaps even his course lectures did not reflect Wheeler at his best but tended to be dull. As previously noted, he seemed to feel very deeply the weight of his academic responsibilities.

Whether or not his students agreed, Wheeler did have definite opinions on how they should be educated, as is shown in the following brief quotation from his essay, "The Dry-rot of Academic Biology," first given in 1922:

> It seems to me that there are two periods when the young biologist is most susceptible to lethal infection by the [dry rot fungus] spores that are continually being thrown off by his professors. One is his freshman year, when he should be stimulated to develop an enthusiastic, receptive attitude, the other his graduate year or years, when he may be expected to adopt an independent, adventurous, and creative attitude toward his science. Of course, the treatment of advanced students is easy for any professor who will follow the excellent example of the late Professor Roland of Johns Hopkins. The story is told that he was once presented with a list of rules for teaching graduate students and that he crossed out all the items and wrote beneath: "Neglect them!" Despite this very convenient precept, many of us coddle our graduate students till the more impressionable of them develop the most sodden types of the father-complex. Some of us even wear out a layer of cortical neurones annually, correcting their spelling and syntax. One fussy old duru of my acquaintance has destroyed both of his hemispheres, his corpus callosum, and a large part of his basal ganglia hunting stray commas, semi-colons, dashes, parentheses, and other vermin in doctor's dissertations.
>
> Not only do many of us wear out our most valuable tissues converting the graduate students into mere vehicles of our own interests, prepossessions, and specialties, but nearly all of us fail to excite in them that spirit of adventure which has in the past yielded such remarkable results in the development of our science. The finest example of this lack of vision is seen in the stolid indifference to exploration and research in the remoter portions of our own country, in foreign lands, and especially in the tropics.

There is no question that Wheeler left his graduate students pretty much to their own devices, although he often welcomed them in his office and took time to discuss his research and theirs. His ant collection was kept in wooden boxes on the shelves of his office at Bussey, and trusted students were given a key and allowed use of his collection and of his library. Visitors were also often treated generously of his time and collections. But now and then, when he was especially absorbed in his work or bored with a visitor, he would resort to a ruse. On one occasion he received a visit from the California hemipterist E. P. VanDuzee, who was anxious to take some representative ants from Wheeler's collection. After greeting his visitor, Wheeler saw to it that he was treated to lunch at the dormitory. Then, knowing that VanDuzee was to catch a bus at four o'clock, he asked his student William Creighton to show VanDuzee some of the good collecting areas around Boston. VanDuzee was not overly keen on this, and when Creighton's ancient vehicle began sputtering on the slopes of the Blue Hills he became increasingly impatient. They finally made it back to Boston barely in time to catch the bus—and much too late for Dr. VanDuzee to take any of Wheeler's ants back to California!

Another visitor who presented a slightly different problem was R. J. Tillyard, a distinguished Australian entomologist who was especially partial to dragonflies, living and fossil. As a young man, Tillyard had made brilliant contributions to entomology, but later in life he had tendencies to lose contact with reality from time to time. In 1928 he wrote to F. M. Carpenter regarding his plans to visit Harvard. He asked if it might be possible to arrange an appointment with Margery Crandon, a well-known medium of Lime Street, Boston, and to take with him a specimen of the giant fossil dragonfly *Megatypus* (known only from a wing). He hoped that she might be able to cause the dragonfly to materialize in all its original splendor. While it was still flopping about, perhaps he could make a wax impression of it! On hearing of Tillyard's proposal, Wheeler is reported to have remarked: "What if the damned thing flew out the window?" But it turned out that Margery was unwilling to take on a dragonfly.

From time to time Bussey Institution had working in it, or from it, various research associates or visiting investigators who were not candidates for degrees. For example, Theodore H. Hubbell, who went on to a distinguished career as professor and director of the Museum of Zoology at the University of Michigan, held a research associateship in 1922–23. Another was Herbert Friedmann, who is now director of the Los Angeles County Museum of Natural History and one of the country's leading ornithologists. In a letter written in 1967, Dr. Friedmann explained his association with Dr. Wheeler:

I took my doctorate in June, 1923, at Cornell University, on a thesis devoted to the brood parasitism of the brown-headed cowbird. That summer the National Research Council Awarded me one of their first batch of post-doctoral fellowships, with the problem of studying the South American cowbirds in Argentina to round out my work in North America. They assigned me to Professor Wheeler, as he had long been interested in parasitism, inquilinism, and related problems in social insects, especially ants. I was delighted with the prospect of this assignment as I had read many of Wheeler's papers and had enjoyed many of his judgments and evaluations of what then passed for modern ethology.

I first met Wheeler at the Bussey Institute in Jamaica Plain . . . and I had only a short talk with him prior to my departure for Buenos Aires. His interest in the cowbird problem was, however, heartening and I looked forward to telling him on my return the following year of the new data I hoped to gather. At that time Boston, like the rest of the country, was indulging in the cozy dissipation of the Prohibition era, and my half hour with Wheeler was interrupted by a telephone call that could only have come from a bootlegger. At least, hearing only Wheeler's replies, I had to imagine the questions. When Wheeler said with professional authority "I am supposed to know something about insects, so don't try to fool me with false cobwebs on the bottles; save them for someone else," I felt pretty satisfied with my interpretation. Ten days later, the steamer I took to Buenos Aires put in at Bermuda to take on liquor for the thirsty passengers. Looking over the rail with my fellow travellers, I heard someone say "This stuff is for real, no cobwebs on the bottles."

When I returned next to Boston it was only for a week as I was then going to the Texan-Mexican border to study the remaining cowbird species, and shortly thereafter, to South Africa to begin a long field study of the other brood-parasitic birds, the cuckoos, honey-guides, and weavers. Wheeler had been away at a sanatorium . . . for a nervous breakdown, but had just returned and appeared interested in what I had to tell him about my results, and he gave me a piece of advice that I have since passed on to others. "Don't get so wrapped up in the new field experiences ahead that you will leave the older ones unwritten. They will still be new to their readers."[16]

In 1929 Wheeler wrote a three-page review for *Science* of Friedmann's first book, which was a summary of his studies of the cowbirds, part of which Wheeler had "supervised." It began as follows:

> Although this book contains the results of only a portion of Dr. Friedmann's researches on parasitic birds, it probably represents the most important contribution to ornithology within recent years. It is, moreover, of no little interest to the parasitologist and entomologist. The parasitic behavior which consists in ovipositing in the nests of alien species and leaving them to rear the resulting young has been observed in members of no less than five natural families of birds: the cuckoos (Cuculidae), cowbirds (Icteridae), weaver-birds (Ploceidae), honey-guides (Indicatoridae) and ducks (Anatidae). The present work, which is confined to a detailed field-study of the cowbirds, with an account of their geographical distribution, taxonomy and ontogeny and an extensive citation of the pertinent literature, is an admirable demonstration of the kind of ethological investigation that has to be accomplished before the physiologist or experimentalist can even approach the fundamental problems of parasitism.

Most of us who are teachers hope that among our students will be a few who will be sufficiently inspired by the subjects which interest us, either in or out of class, to wish to devote their own lives and energies to work in these same areas. We may even wish them the success and fame we never had, for it is nice to bask even in reflected glory. We have not mentioned all of the students of Dr. Wheeler, nor discussed with any consistency his better known ones. Perhaps Wheeler was not always at his best in the classroom, but there can be no doubt that his brilliant mind and his great industry infected all who came in contact with him. There are few if any professors of entomology who have produced such a successful and distinguished group of students.

10 / Student of Animal Behavior

Wheeler's years at Harvard were busy ones. He was not only a teacher and administrator, but he lectured widely and his research carried him to many parts of the world. Rather than trying to summarize this period on any sort of year-to-year basis, we prefer to discuss them by topics and areas of interest. In the preceding chapter we discussed some of his students, and in the chapter before that we tried to summarize some of his research on ants. In the present chapter we wish to examine Wheeler's contributions to the study of animal behavior. His interests in this field reached their culmination in a lecture, "On Instincts," presented before the Royce Club of Boston in 1917 and later published in the *Journal of Abnormal Psychology*. Professor W. H. Thorpe of Cambridge University has recently remarked that Wheeler, "in his short but brilliant essay on the subject [of instincts] has given an indication of what might be done by a scholar with the right and rare combination of abilities."[1] Before examining this essay in detail, we would like to return briefly to Wheeler's formative years and attempt to trace the development of some of his thinking in this field. We believe that Wheeler deserves more credit for the priority of certain terms and concepts in behavior than he has generally received.

Wheeler's approach to behavior was that of a field naturalist, an inclination that is traceable to his early associations with Dorner, Rauterberg, and other members of The Wisconsin Natural History Society. Above all he was influenced by George and Elizabeth Peckham, for whom he developed, he said, "an almost filial piety." In the 1880s the Peckhams were conducting their pioneering studies on the mating behavior of spiders, and Wheeler shared in their observations and prepared some of the sketches. The Peckhams felt that their studies supported Darwin's theory that the females actually select those males of their species that possess the most conspicuous colors and courtship displays. This theory had been attacked by A. R. Wallace and others, but it is now commonly accepted that sexual selection has been a major factor in the evolution of brilliant colors and elaborate structural features of the males of many groups of animals. The Peckhams' careful marshalling of descriptive data, their frequent references to published accounts, and their thoughtful argumentation: all of these came to be characteristic of Wheeler's writings. A descriptive, comparative approach similar to that of the Peckhams is already apparent in short papers of Wheeler's on courtship in midges and oviposition in grasshoppers published in 1889 and 1890.

Wheeler's early interest in mimicry can also be traced to the Peckhams. Mrs. Peckham's paper on protective resemblance in spiders appeared in 1889, and during the same year Wheeler, then twenty-four, published a note on two cases of mimicry he had observed. He concluded his paper with a remark that many a later worker might well have heeded: "Only field study can be of any service in deciding whether a particular insect is mimetic." To this he added a statement from Drummond to the effect that "it is not only the form but the behavior of the mimetic insect, its whole habit and habitat, that have to be considered; so that mere museum contributions to mimicry are almost useless without the amplest supplement from the field naturalist." This was shortly before the appearance of E. B. Poulton's important book *The Colours of Animals,* which Wheeler reviewed for *Science*. Wheeler's first venture into the field of popular writing was a survey of some examples of mimicry, protective coloration, and "scare organs," published under the title "Impostors among Animals" in *Century Magazine,* July 1901.

At Clark University, from 1890 to 1892 Wheeler was exposed to lectures and seminars by Clark's philosopher-psychologist president, G. Stanley Hall, a man of "assimilative erudition" whose "thoughts came in such a volume of suggestion that it seemed as if he had three sets of vocal organs instead of one, they could not have given expression to his thought."[2] Hall was primarily a humanist, with little taste for experiment, quantitative treatment of results, or rigorous analysis. During the year 1890 William James' *Principles of Psychology* was published and in 1894, C. Lloyd Morgan's *Intro-*

duction to Comparative Psychology. During this same period, in Germany Wilhelm Wundt's *Grundzuge der Physiologischen Psychologie* was going through successive editions, and provocative papers on the behavior of lower organisms were appearing from the pens of Jacques Loeb and J. von Uexküll. Von Uexküll stressed the "combat to the end," as he called it, between comparative psychology and comparative physiology, placing himself firmly in the second camp with his credo that the scientist has to do "only with processes that can be objectively demonstrated," yet calling for "the continued and accurate observation of the living animal in its environment."[3] The publications of these men, and many others, were familiar to Wheeler, throughout his life an avid reader in the fields of psychology and animal behavior.

But of these men, it was Jacques Loeb who was to have the most immediate impact upon Wheeler, as indeed he did upon many American biologists. Irritated by military and political conditions in Germany and encouraged by a letter of praise from William James and by personal contacts with Henry B. Ward and others at Naples, Loeb had moved to the United States in 1891. At Würzburg, Loeb had become convinced, through his studies under Adolf Fick, professor of animal physiology, and Fick's close friend and associate, Professor Julius Sachs, that physical chemistry was the key to biology. He had been introduced at Naples to sea urchins as experimental animals, and to the mechanical techniques of embryo manipulation, and he was convinced that the developmental mechanics of Wilhelm Roux would make it possible to explain all of the activities of organisms according to the exact laws of physics and chemistry—a "mechanistic conception of life."

In 1892 C. O. Whitman invited Loeb to join his newly formed department of biology at the University of Chicago and to give a summer course in physiology at Woods Hole. Loeb was not the sort of person one accepted casually. His was "a spirit zealous, quick, and full of youthful fire," and he had "a mind always alert, poised to turn easily in any direction, and operating with bewildering speed and certainty.

"His intimate talks in the laboratory were at once the joy and despair of fellow workers. His mobile features, his expressive, eager eyes, alight with enthusiasm, were a fascinating study as he flashed from mood to mood, smiles and frowns following in rapid succession. Quick to wrath, he was also quick to feel the folly of anger and in the midst of a tempest he would suddenly stop, then smile, and at length burst into laughter as the incongruity of the situation dawned upon him . . . In conversation the emotional character of his thought, with its sudden flashes, might sometimes prove exhausting or even bewildering to more phlegmatic natures. A visitor to his laboratory was quite apt to leave in a somewhat breathless state."[4]

As a young man, Loeb had addressed himself to no less a problem than an experimental approach to free will. Sachs had demonstrated that plants respond to light and certain other stimuli in the manner of simple machines, and Loeb, beginning in 1888, had extended Sachs' tropism theory to animals. As Loeb explained it:

> Normally the processes inducing locomotion are equal in both halves of the central nervous system, and the tension of the symmetrical muscles being equal, the animal moves in as straight a line as the imperfections of its locomotor apparatus permit. If, however, the velocity of chemical reactions in one side of the body, e.g., in one eye of an insect, is increased, the physiological symmetry of both sides of the brain and as a consequence the equality of tension of the symmetrical muscles no longer exist. The muscles connected with the more strongly illuminated eye are thrown into a stronger tension, and if new impulses for locomotion originate in the central nervous system, they will no longer produce an equal response in the symmetrical muscles, but a stronger one in muscles turning the head and body of the animal to the source of light. As soon as the plane of symmetry goes through the source of light . . . the animal will move in a straight line to the source of light until some other asymmetrical disturbance once more changes the direction of motion.
>
> What has been stated for light holds true also if light is replaced by any other form of energy. Motions caused by light or other agencies appear to the layman as expressions of will and purpose on the part of the animal, whereas in reality the animal is forced to go where carried by its legs. For the conduct of animals consists of forced movements.[5]

The wide acceptance of Loeb's ideas led to the description of many apparent tropisms and to the development of an elaborate terminology for "forced movements." Wheeler contributed the word anemotropism (orientation with respect to air currents) in 1899. He cited several examples from among flies he had observed near Chicago. "If a poising swarm . . . be watched when there is a gentle, constant wind, all the flies will be seen to head directly towards the wind and not to deviate from this position while they are on the wing and the wind continues. And in walking some distance one still finds all the swarms of a given locality oriented in the same direction like the weathercocks on the barns and churches. If the wind shifts the insect at once changes its position so that it again faces to windward." Like a true disciple of Loeb, to whom he pays tribute, Wheeler believed that "the organism naturally assumes the position in which the pressure exerted on its surface is symmetrically distributed and can be overcome by a perfectly symmetrical action of the musculature of the right and left halves of the body."

Fraenkel and Gunn, in their book of 1940, *The Orientation of Animals,* recognize Wheeler as the first person to describe orientation with respect to air currents. However, like other contemporary workers, they prefer *taxis* to *tropism* for directed orientation movements, and they further question whether the prefix *anemo-* is desirable, since there is now evidence that at least in some flies it is not a question of the detection of air pressure but of optical orientation, that is, the insects fly into the wind so that their forward motion balances the force of the wind to the extent that they are able to maintain a constant position with reference to the visible background.[6]

In his 1899 paper, Wheeler reviewed examples of other kinds of tropisms reported in insects and suggested that various tropisms "either singly or together will explain many of the instincts of insects." However, he concluded that despite the value of the study of forced movements, this study may not satisfactorily explain all aspects of the behavior of animals.

We know that the insect responds not only to known external stimuli but also to certain unknown internal stimuli originating within the cells of the alimentary tract, reproductive organs, etc., and that the responses to these stimuli are often remarkably complex, as e.g. in the elaborate feeding and nesting instincts of ants, bees, and wasps. Nor does the complication of the problem of instinct end here. It is greatly increased by two further considerations, first by our complete ignorance of the protoplasmic changes, chemical and physical, which precede or accompany these tropisms or the responses to stimuli in general; and second, by the difficulty of explaining why all these responses are so marvellously adaptive. I venture to assert, nevertheless, that it is better to face these difficulties, insuperable as they appear, than to continue investigation in that spirit of anthropomorphism which has been such a fruitful source of misinterpretation in the comparative study of habits and instincts.

Others besides Wheeler soon came to realize that Loeb's theories were not as all-embracing as he believed. H. S. Jennings and S. O. Mast, both at Johns Hopkins, showed that there were many shortcomings in the tonus theory as applied to various Protozoa and insects. In 1909 Jennings told the Sixth International Congress of Psychology that tropisms "form a part (a very minor part, I should say) of the behavior of many organisms. The tropism is not a uniform phenomenon; the various reactions classified under the concept show diversities in practically every possible category . . . For the real elements of behavior we must look in an entirely different direction, toward such analyses of the complex activities of organisms as are attempted by von Uexküll and Sherrington."[7]

Nevertheless, Loeb's concepts continued to prove useful. In his book *Ants* (1910), Wheeler cited Loeb's explanation of the mating flight of ants as a sudden reversal of phototropism; that is, the males and females before maturity are photonegative, but as the time for the nuptial flight arrives, they become photopositive; following the flight the females once again become photonegative as they seek out a place to found a colony. In 1919 A. Kühn published a new classification of orientation movements which incorporated many of Loeb's ideas as well as some of those of his antagonists.[8] It was Kühn who first clearly restricted the word *tropism* to the curvatures of plants and fixed animals, the word *taxis* to orientation movements of motile animals. With modifications by Fraenkel and Gunn (1940) and others, Kühn's terminology still finds widespread use in animal behavior today.

In the final analysis, Loeb was to have less influence upon Wheeler than the Peckhams, C. O. Whitman, and that remarkable Swiss myrmecologist and psychiatrist, Auguste Forel. Forel's long series of publications began in 1869 and did not conclude until his death in 1931. Forel summed up many of his ideas on insect behavior in a perceptive lecture presented at the Fifth International Congress of Zoology in Berlin in 1901—a lecture translated by Wheeler for *The Monist* in 1903.

In the meantime Wheeler's most direct contacts were with Whitman, and although Whitman's early work was in the field of embryology, and Wheeler's doctor's thesis under Whitman at Clark was also in embryology, both men had a strong predilection for comparative, phylogenetic studies of behavior. Whitman is said to have kept pigeons as a boy and "was fascinated by them and sat and watched them by the hour, intensely interested in their feeding, their young, and everything they did."[9] Whitman's classic studies of pigeon behavior were not published until several years after his death. But his views were made known much earlier, especially in his lecture "Animal Behavior," given at Woods Hole in the summer of 1898.[10] This essay is now recognized as something of a milestone in the zoological study of behavior. For example, the psychologist E. H. Hess began his recent (1962) survey of this field with the comment that in 1898 Whitman "wrote the sentence that initiated the birth of modern ethology: 'Instincts and organs are to be studied from the common viewpoint of phyletic descent.' "[11] And K. Z. Lorenz, who more than anyone else is responsible for the current surge of interest in this field, has acknowledged that "endogenous movements [that is, instincts in Whitman's sense] were not only discovered and recognized as a very distinct phenomenon by C. O. Whitman as early as 1898, but also systematically studied and evaluated as taxonomic characters."[12] Devotees of the French naturalist Jean Henri Fabre may feel that too much credit is here granted to Whitman; but Whitman's view of instinct was far more balanced and more profound than that of Fabre, owing more to the influence of Darwin and the Peckhams, as he acknowledged. In contrast to Fabre, Whitman maintained that: "The clock-like regularity and inflexibility of instinct . . . have been greatly exaggerated. They imply nothing more than a low degree of variability under normal conditions . . . Close study and experiment with the most machine-like instincts always reveal some degree of adaptability to new conditions. This was made clear by Darwin's studies on instincts, and it has been demonstrated over and over again by later investigators, and by none more thoroughly than by the Peckhams in the case of spiders and wasps."

Whitman's essay is developed around three groups of organisms he knew well: the leech *Clepsine,* the mudpuppy *Necturus,* and, of course, pigeons. The behavior of the leech in rolling into a passive ball following engorgement he considered to be instinctive, "since it is performed by the young after the first meal as perfectly as by the adult"; it is highly adaptive, since to drop off the host fully extended "would retard descent and increase the chances of capture by fish."

Looking more closely at the nature and origin of this instinct, it will be seen to be quite a natural performance, in keeping with the most fundamental features of the animal's organization, and only a special application of a more general act that is primary and organic as much as tasting, seeing, or sleeping. The more general act consists simply in tucking or rolling the head under, as often happens when the animal is resting . . . The same act, carried a little further, gives the half-rolled condition . . . often assumed if the leech is sick or has been injured . . . It is only up a step futher to realease the sucker and fold it over the part already rolled up, thus completing the part ball to a whole ball [as occurs in several situations outlined] . . . From beginning to end we have only one act, in different stages of completion, simply different degrees of one and the same process. Having the general act to start with, it is easy to see how it might be made use of for particular purposes; in other words, how special adaptations of a useful kind might arise . . . Natural selection would steadily improve upon the results, and the special adaptation, in different stages of development in different species, as we find it today in different *Clepsines,* would lie in the direct line of progress. This view does not of course presuppose intelligence as a guiding factor, and therefore lends no support to the theory of instinct as "lapsed intelligence," or "inherited habit."

Reasoning in this manner from his knowledge of the behavior of leeches, mudpuppies, and pigeons, Whitman developed a series of maxims which (with some changes in wording) still serve well those persons interested in the comparative behavior of animals in their natural environment: (1) "The point of special emphasis here is that instincts are evolved, not improvised, and that their genealogy may be as complex and far-reaching as the history of their organic bases." (2) "The first criterion of instinct is, that it can be performed by the animal without learning by experience, instruction, or imitation. The first performance is therefore the crucial one." (3) "The main reliance in getting at the phyletic history must be comparative study." (4) "Plasticity of instinct is not intelligence, but it is the open door through which the great educator, experience, comes in and works every wonder of intelligence."

We do not know Wheeler's initial reaction to Whitman's lecture. Undoubtedly much of it was "old hat"; for Wheeler had been associated with Whitman for more than ten years, and had surely discussed many of these points with him. In his paper on ponerine ants, published in 1900, Wheeler commented that "it can hardly be doubted that there is a phylogeny of instincts"—essentially a paraphrasing of his mentor's remarks. Comparative, phylogenetic studies were to occupy Wheeler's attention for most of his life.

In several short but important papers published between 1902 and 1905, Wheeler sounded a clarion call for the field of ethology—a word Whitman had never used, even though he is now judged to have written "the sentence that initiated the birth" of that field.

"A study of recent literature," wrote Wheeler in *Science* in June 1902, "reveals the fact that zoologists are much in need of a satisfactory technical term for animal behavior." The Germans tended to use the word *biologie* in a special, restricted sense for this field, but it made little sense to use the same word for both a major field and a subdivision of that field. Ernst Haeckel had coined the word *oekologie* in 1866, basing it on the Greek word *oikos,* meaning home. Ecology, as the field come to be known, stressed the habitats of plants and animals and the interactions between organisms and their environment. "The only term hitherto suggested which will adequately express the study of animals, with a view to elucidating the true character as expressed in their physical and psychical behavior toward their living and inorganic environment," wrote Wheeler, "is *ethology*. This term has been employed to some extent by French zoologists and . . . attempts have already been made to establish its English usage. Dahl has advocated its introduction into Germany [but] the retention of 'Biologie' has been ably defended by Wasmann." The Greek word *ethos,* as Wheeler went on to explain, refers to character or manners; "certainly no term could be more applicable to a study which must deal very largely with instincts and intelligence as well as with the 'habits' and 'habitus' of animals." The central position of behavioral study among the biological disciplines was clearly appreciated: "Whenever we undertake the detailed or exhaustive study of an ethological problem we are led imperceptibly into the details of physiology, psychology, morphology, embryology, taxonomy or chorology, according to the particular aspect of the subject under consideration. On the other hand, the interests of all these various sciences are slowly but surely converging to a point which is not far from the center of gravity of 'ethology.' "

As Wheeler pointed out in a later paper, John Stuart Mill, in 1843, was apparently the first to use the word *ethology,* referring to the science of human character. In 1859 Isidore Geoffroy St. Hilaire used the word in the sense of scientific natural history, and through

St. Hilaire's influence the word became well established in France, although it was not until close to the turn of the century that it was accepted in other major languages. The *Zoological Record* of London began using ethology as a heading for studies of "habits, instinct" in volume 38 for the year 1901. There is nothing in the background of the term to justify its use for anything other than the study of natural behavior in its broadest sense. As for the subject matter of the field, one must of course go back at least to Réaumur (1683–1757), whose contributions were reviewed by Wheeler in 1926 in his publication and translation of *The Natural History of Ants,* based on a previously lost manuscript of Réaumur. And one must certainly include the work of Charles Darwin, especially his book *The Expression of the Emotions in Man and Animals* (1872), as well as the work of Spalding (1873) on the instincts of birds,[13] to say nothing of Fabre and others.* Thus the statement by Lorenz in 1955 that ethology is "that branch of research started by Oskar Heinroth [1911]"[14] seems hasty,† and Thorpe's 1956 essay "Ethology as a New Branch of Biology"[16] seems absurdly titled, as does Baerends' 1959 review "Ethological Studies of Insect Behavior"[17] (why not physiological studies of insect function?). These men are, of course, reflecting the current usurpation of the word ethology by a particular school of European workers—a school which is "modern" at least in the sense that it disavows most of the history of its field. Lorenz at other times has in fact acknowledged the influence of Whitman (as in the quotation presented earlier), and the views of Whitman's student Wallace Craig have become doctrine in contemporary ethology. Craig's paper "Appetites and Aversions as Constituents of Instinct"[18] was actually published a year before

* Shortly after Fabre's death in 1915, at the age of ninety-two, Wheeler reviewed his life's work for the *Journal of Animal Behavior.* He concluded that although Fabre "had a strong tendency to schematize his observations and to ignore the variability of instinct," he was "the first to realize the full importance of a scientific study of animal behavior" and "the first consistently to apply the experimental method to the investigation of the animal mind . . . [His experiments were] not the less illuminating and conclusive because they were carried out with crude, homemade apparatus. It is as instructive as it is humiliating to read his results and to reflect on the mountains of complicated apparatus in our modern laboratories and the ridiculous mice in the form of results which only too frequently issue from the travail of 'research.' "

† But in a recent interview in *Harper's Magazine* (1968) Lorenz remarks that (about 1935) "I collaborated for a time with Professor Niko Tinbergen, and between the two of us we founded the science of ethology."[15] Even poor Heinroth is now left holding the bag!

the posthumous appearance of Whitman's major work on the behavior of pigeons in 1919. Wheeler made frequent reference to Craig's ideas well before they became incorporated into the doctrines of Lorenzian "ethology." Note, for example, these remarks in *Social Life Among the Insects* (1923): "The appetites of hunger and sex arise from internal stimuli which compel the organism to make random or trial and error movements till appropriate, specific external stimuli are encountered. Then a sudden, consummatory reaction occurs and the relieved organism lapses into quiescence till the internal stimuli again make themselves felt."

Wheeler did much to establish the word *ethology* in its broader sense, not only through his 1902 paper in *Science*, but also through various other lectures and articles in the years that followed (although the word *biology* in its narrow, German sense has by no means completely disappeared even today, and its use has been encouraged by the attempt to narrow *ethology* to one particular approach). In an address before the American Society of Naturalists in 1904 (published in *Science* in April 1905), Wheeler eloquently described the importance of ethology in evolutionary biology:

> Inasmuch as ethology deals with processes, or phenomenal diversity in time, whereas morphology deals with the spatial diversity of phenomena, it is evident that ethological must be very different from morphological characters. It might even be said that the ethologist has no right to speak of a process as a character or characteristic, and the original Greek meaning of these words would seem to limit their use to the structural configurations resulting from specific acts or processes. This need not prevent us, however, from extending the meaning of the terms to include also the typical and specific reactions of the organisms to their environment. Certainly in the case of the human species, which is best known ethologically, the terms character and characteristic are hardly used of physical structures, but almost exclusively of typical modes of activity.
>
> In its application to ethology the mutation theory* can only mean that organic species must differ from one another by discrete idiosyncrasies of behavior. Most biologists would probably regard any discussion of mutation from the ethological standpoint either as superfluous or as necessarily and merely confirmatory of the results of morphological study. In their opinion it would follow as a matter of course that the

* Hugo DeVries' mutation theory was at this time only a few years old, and Wheeler's paper was part of a symposium on this subject.

functional and ethological characters of organisms must fluctuate or mutate accordingly as the structural characters vary continuously or discontinuously. In my opinion this is not so self-evident as it would appear to be at first sight.

It is true, of course, that the various structural categories from the phylum down to the species, subspecies, variety, sex and individual—all show what may be regarded as correlated or corresponding ethological characters, although this correspondence is often very loose, vague and irregular, for it is evident that slight morphological may be correlated with complex ethological characters, and conversely. Some such correspondence may also be observed in hybrid forms. All this is usually taken for granted . . . If we follow up the matter, however, we soon find that in the field of possible observation the ethological tend to outstrip the morphological characters. We observe great differences in habits and behavior between genera of the same family, between species of the same genus, and what is most significant, between individuals and even twins of the same species. At the same time we may be utterly unable to point out the corresponding structural differences, which, according to any theory of parallelism, should accompany such pronounced ethological distinctions . . . It is undoubtedly a matter of considerable theoretical and practical importance that we are able to detect ethological where we can not detect morphological differences or characters.

We may, in fact, be permitted to reverse the matter and take the point of view of the psychologist and metaphysician rather than that of the morphologist. In other words, we may start with behavior or the dynamic, *i.e.,* physiological and psychological processes of the organism, and regard the structure as their result or objectivation. The organism makes itself—the *ethos* is the organism. In this sense the honeycomb is as much a part of the bee as is her chitinous investment, and the nest is as much a part of the bird as her feathers, and every organism, as a living and acting being, fills a much greater sphere than that which is bounded by its integument.

Although the time is so very limited, permit me to digress somewhat further on a more practical consequence of the view here advocated. We are certainly justified in regarding ethological characters as very important, as belonging to the organism and as being at least complementary to the morphological characters. If this is true, our existing taxonomy and phylogeny are deplorably defective and one-sided. To classify organisms or to seek to determine their phylogenetic affinities on purely structural grounds can only lead, as it has led in the past, to the trivialities of the species monger and synonym peddler. This has been instinctively felt by all biologists whose development has not been arrested in the puerile specimen-collecting stage . . . Every field naturalist knows that he is frequently guided to the more delicate specific and varietal distinctions, not so much by the structural differences between the organisms he is observing, as by differences in their habitat or behavior. Then closer scrutiny may often, although not always, reveal correlated structural differences. When such structural differences are not to be detected we speak of ethological species, and the number of these is undoubtedly much greater than was formerly supposed . . .*

The fact that the morphologist has so consistently either neglected or opposed the use of ethological characters in classification shows very clearly that in his heart of hearts he has never very earnestly concerned himself with the parallelism of structure and function. He is inclined to regard function, especially psychical function, as something utterly intangible and capricious. For does it not seem to make its appearance in the embryo or young *after* structure has developed, and to depart at death before the dissolution of visible structure? And are not our museums largely mausoleums of animal and plant structures which we can forever describe and redescribe, tabulate and retabulate, arrange and rearrange, without troubling ourselves in the least about anything so volatile as function?

It is, indeed, not only conceivable, but very desirable, that a taxonomy should be developed in which the ethological will receive ample consideration, if they do not actually take precedence of the morphological characters. It is certainly quite as rational to classify organisms as much by what they do as by the number of their spines and joints, the color of their hairs and feathers, the course of their wing nervures, etc. To regard our existing purely structural classifications as anything more than the most provisional of makeshifts, is to ignore the fact that the vast majority of organisms which they are designed to cover are known only from a few dead exuviae. There are, of course, enormous difficulties in the way of constructing ethological classifications, quite apart from the fact that our knowledge of behavior is even more fragmentary than that of structure, as any one will realize who tries to write an ethological description of some common animal or group of animals. In morphology the elements of description can be treated as parts of an orderly and traditionally respected routine, but in ethology we still lack the necessary preliminary analysis of the more complex instincts, and are therefore unable to construct uniform and mutually comparable descriptions. One great desideratum in ethology at the present time is a satisfactory

* A statement amply borne out by recent studies of complexes of sibling species in, for example, leaf beetles, crickets, and fireflies. For discussion, see E. Mayr, 1963, "Morphological Species Characters and Sibling Species."[19]

and sufficiently elastic working classification of the instincts and reactions, like that of the organs and organ systems of the morphologist. Such a classification can be developed only by comprehensive, comparative study of behavior in a number of genera and families and not by any amount of intensive study of a few reactions in a few species.

In the meantime Wheeler himself put his own dicta into practice in his studies of ants, and later in his compendia on social insects generally. A further contribution to the theoretical aspects of this subject was provided by his address before the New York Academy of Sciences in November 1907, "Vestigial Instincts in Insects and Other Animals," which was published in 1908 in the *American Journal of Psychology*. Here he defined instinct as "an action performed by all the individuals of a species in a similar manner under like conditions" and proceeded to review various examples of instincts which appear to have lost their former function. He cited Darwin's example of dogs which, before lying down, turn in a circle, even inside a house, the vestige of a behavior pattern once useful in treading down the grass. He also cited a "negative case" from C. O. Whitman: "When we ask . . . why our domestic pigeons no longer alight or nest in trees, or do so only in very rare instances, no amount of experimentation on the nesting habits of these birds can assist us in answering this question. A comparative study, however, shows that the domestic bird is in all probability derived from the rock-pigeon, a form that had developed an instinct to alight and nest only on cliffs or open ground, and this peculiarity, granting the wonderful conservatism and other peculiarities of hereditary transmission, accounts for the negative behavior above mentioned."

After presenting a number of similar examples, several from among ants and other social insects, Wheeler drew several conclusions which may be worth quoting in part:

1. The vestigial instinct action presents itself as an act of racial or phyletic recollection . . . vestiges, both instinctive and structural, often remain latent for generations and then suddenly manifest themselves under the stress of extraordinary stimuli . . .

2. The vestigial instincts obviously represent a part of the animals' endowment, and their manifestation shows that the capacities of even the lower organisms are greater than their ordinary routine behavior might lead us to suppose . . . Vestigial organs and instincts may be useful as starting points for entirely new adaptations, for there are not wanting cases of vestiges that have acquired new functions.

3. The foregoing and similar considerations lead me in conclusion to a few remarks on the method of studying vestigial instincts and instincts in general . . . I am convinced that our knowledge . . . would gain immensely by comparative study of whole genera or families of closely related organisms, for we know of no case in which an instinct is peculiar to a single species . . . and no cases in which two species manifest an instinct in precisely the same manner.

These were matters that Wheeler was to return to many times, not only in various short articles and in introductions to books such as Phil and Nellie Rau's *Wasp Studies Afield* (1918) and O. E. Plath's *Bumblebees* (1934), but also in his own books, especially *Ants* (1910) and *The Social Insects* (1928). However, his major contribution to ethological theory is to be found in his 1917 lecture and article "On Instincts." This essay is a somewhat personal view rather than a full review of the field or a proposal of new conceptual schemes. His defense of anthropomorphism ("the eighth deadly sin," as he put it) and of Lamarckism ("the ninth mortal sin") seem repellent to some modern students of behavior, and the one new term he proposed (*hapaxoraeic instincts* for those behavior patterns which are "manifested only once and with great adaptive perfection during the life of the individual organism") has not proved useful. Nevertheless the essay is very readable and filled with typical Wheeler witticisms; it must have elicited many guffaws from his audience, as intended. He began as follows:

The economic entomologist is primarily engaged in a study of the relations of insects to one another, to other organisms, and especially to man. All such relations, in so far as they involve the human race, may be distinguished as either actually or potentially beneficial or injurious, and are, of course, due to peculiarities of behavior. And since behavior, both in insects and in man, is fundamentally of the type called instinctive, the economic entomologist, far from having to apologize for an interest in the perennial problem of instinct, would be deserving of censure if he failed to keep it constantly in mind. Moreover, no class of organisms offers such a marvelous field for the study of instincts as the Insecta. No other class, with the possible exception of birds, shows anything like the diversity and complexity of these phenomena and, owing to the great number of species, genera, and families that have survived the vicissitudes of geologic time—a number far in excess of that of all other organisms on the planet—no other class exhibits such a complete representation of the historical or phylogenetic stages of many instincts.

As mere phenomena the instincts are well known, and there is practical unanimity among authors in regard to their peculiarities. Any behavior is designated as instinctive which originates in an impulse, but the nature of impulse cannot be defined further than to say that it has both a conative and a cognitive aspect. Those who lay greater emphasis on the conative aspect prefer to use such terms as impulse, *Trieb, horme,* life urge, *élan vital,* etc., whereas those who wish to suggest the cognitive aspect use such terms as craving, appetite, desire, interest, libido, etc. The impulse is evidently the center or core of the instinctive activity, which is peculiarly fixed and mechanized, very rigidly dependent on inherited structure or organization and therefore very uniform, or variable only within rather narrow limits, in all the individuals of one or both sexes of a species. Behavior of this kind has the attributes of compulsion or necessity and is at the same time highly adaptive or purposive, though the organism manifesting it is unaware of any purpose, or at any rate is usually aware only of an immediate purpose, even when the behavior is accompanied by consciousness.

Wheeler then proceeded to provide a valuable summary of three major currents of opinion regarding behavior, which he called the theological or teleological, the physiological or mechanistic, and the psychological or anthropomorphic. Each of these he was able to trace back to certain Greek philosophers, including Aristotle, Democritus, and Heraclitus. He placed himself in the third school, with the following comments:

The psychological view of instinct has certain great advantages. It is naturally genetic and favorable to the interpretation of organisms as historical beings. Instinct and intelligence are not regarded as separate faculties but as extreme phases of one psychical process which in our individual experience is continually lapsing from conscious and intelligent performance to the mechanized status of habit. The same process is supposed to have gone on throughout the phylogenetic history of the organic world and to have resulted in all the characteristic structures and behavior of existing organisms. In other words, instinct is essentially inherited habit. Hence individual experience, which is rejected as of no value by the Neo-Darwinians in comparison with the fortuitous concourse of accidental germ plasma variations, must affect, at least in some measure, the constitution of succeeding generations. Expressing such a view means of course committing the ninth mortal sin, known as Lamarckism, which is faith in the inheritance of acquired characters and in the opinion that the function creates the organ . . . The completed instinct is merely the congealed result, so to speak, of more fluid or unstable activities of the random, trial and error, or perseverance type initiated by a feeble intelligence. The argument used by my old teacher, Prof. C. O. Whitman, and repeated by Holmes, that lapsed intelligence cannot account for instinct, because the phylogenetic sequence in the animal kingdom is instinct in the lower, followed by intelligence in the higher forms, seems to me easily answered, if we admit what the researches of Jennings and others seem to compel us to admit, that even the lowest organisms have a glimmer of intelligence."

Clearly Wheeler used the word *intelligence* (as was common in his time) for any departure from strict stereotypy, whether it be innate plasticity, physiological adaptation, or simple forms of learning and memory. What Wheeler is saying is that individual experience in the form of repeated performance of an act in one of several possible ways can, in a constant environment, result in the formation of more stereotyped, more strictly genetically determined behavior. Or, as Forel had expressed it, "instinct . . . is . . . phylogenetically inherited intelligence which has gradually become adapted and crystallized by natural selection." As Wheeler pointed out in reviewing a translation of Forel's major work on ants in 1930, Forel used the word *intelligence* "in the modern sense of 'behavioristic adaptability' and not in the scholastic sense as a synonym of 'ratiocination,' or drawing conclusions from premises."

These ideas are very much what James Mark Baldwin had in mind when he developed his concept of "organic selection,"* although it would have strengthened Wheeler's essay considerably had he developed this theme more fully and been less eager to commit that ninth mortal sin, for in fact Baldwin thought of organic selection as something of a compromise between Darwinism and Lamarckism. In Baldwin's words: "If it be true that those variations which can accommodate, either very much or very little, to critical conditions of life are the ones to survive, and that such variations will be accumulated and will in turn progressively support better accommodations, then it is the accommodations which set the pace, lay out the direction, and prophesy the actual course of evolution. This meets the view of the Lamarckians that evolution does

* Baldwin first proposed his theory in 1896 but developed it more fully in his book *Development and Evolution* in 1902. Wheeler reviewed this book very favorably for *The Psychological Review* in January, 1903.

somehow reflect individual progress; but it meets it without adopting the principle of Lamarckian inheritance."[20]

The term *organic selection* has largely been abandoned, but as the so-called "Baldwin effect," the concept still survives in the modern synthetic theory of evolution. G. G. Simpson has presented an excellent review, defining the Baldwin effect as the view that "characters individually acquired by members of a group of organisms may eventually, under the influence of selection, be reinforced or replaced by similar hereditary characters." Simpson believes that the Baldwin effect is fully plausible in current evolutionary contexts, although we know of few if any examples in which it can be said that it has indubitably occurred. As Simpson says:

> Genetical systems do not directly and rigidly determine the characteristics of organisms but set up reaction ranges within which those characteristics develop. An "acquired character" or specifically an adaptive modification (that is, an accommodation) necessarily occurs within a genetically determined reaction range. The range may be relatively broad or extremely narrow. In any case an accommodation has genetical limits and develops only in the framework of the genetical system, but in a labile reaction range the particular form taken by a developing organism depends also on interaction with the environment. The genetical system evolves and the reaction range correspondingly changes. The range may come to cover different possibilities or it may become broader or narrower. If it becomes narrower, the possibilities for individual modification of characteristics become fewer. An accommodation that in a broader range occurs only as a specific response to a particular interaction with the environment may as the range narrows become the only developmental possibility. Then the Baldwin effect may occur: a response formerly dependent on a combination of genetical and environmental variables may become relatively or even absolutely invariable.[21]

Simpson has here phrased the old statement that "instinct is inherited habit" in modern and thoroughly acceptable terms—but in fact it is not Lamarckian at all. Even Mayr, who (like many others) tends to reject the Baldwin effect in the origin of structural characters, believes it possible that a new behavior which is "at first a nongenetic modification of an existing behavior [may], as a result of learning, conditioning or habituation, [be] replaced (by an unknown process) by genetically controlled behavior."[22]

It would, of course, be ridiculous to credit Wheeler with saying things he did not say. Working with ants and other social insects as he was, he simply found it difficult to believe that random mutations were capable of creating all of the exceedingly complex and highly adaptive behavior patterns of these insects and their social parasites. It is very probably true that "habits," in the sense that in a given environment one of several modes of behavior permitted by the genetic system is found to work most effectively, may be "inherited," in the sense that there may be selection for genes rendering this mode of habit more automatic. However, such limitation may be followed by a broadening of behavioral response around this particular mode, and this in turn may be followed by ritualization of some mode within the new spectrum of variation. Some such pseudo-Lamarckian mechanism may in fact be very important in molding complex behavior patterns, particularly in social insects, although we still know very little about it. Wheeler seemed to sense this in 1917, and it is natural enough that he couched his impressions in terms in vogue in his day.

Wheeler often expressed the belief that behavioral evolution may precede structural evolution—that once an instinct has been formed it sets the stage for structural change assisting in the performance of this behavior. As he put it in 1905: "we may . . . regard the structure as [the] result or objectivation [of behavior]"; and in 1917: "the function guides and modifies and builds up the organ according to the principle of *Funktionswechsel,* first elucidated by Anton Dohrn. The function is continually changing, shifting, and dichotomizing in obedience to the needs and experience of the organism, and the organ merely reflects these changes in the development of its various parts." Similar ideas have been expressed quite recently by such biologists as Sir Alister Hardy,[23] although they have still to be fully appreciated by most evolutionists.

The applications of these ideas to mimicry and protective coloration were well brought out in a 1912 lecture sponsored by the Boston Society of Natural History and alluded to in the preceding chapter. "In my opinion," Wheeler remarked, "insects which resemble leaves must have developed this resemblance by a process of active, sympathetic adaptation of their color and form to that of the leaves, the warning types must have developed their garish colors and threatening attitudes in intimate correlation with their stinging powers or nauseous body-fluids and the mimics, though lacking these latter characters, must have actively, though of course unconsciously, copied their models. This sounds like the mere substitution of a more mysterious for a less mysterious explanation, and will undoubtedly be so regarded by those who reject all explanations that are not purely mechanical. But it should be noted that natural selection offers no explanation of the origin but merely of the survival of adaptations."

It can readily be appreciated that Wheeler came to be labeled "Lamarckian" and "vitalist" by those addicted to such name-calling. In point of fact, his researches on social insects had taught him to avoid simplistic explanations of any kind, from "forced movements" and the theory of the gene to Bergson's "élan vital" and Driesch's "entelechy." His occasional defense of Lamarckism (or what he called Lamarckism) was a reaction to Darwinism in its popular, oversimplified form. It is true that Wheeler found Bergson's writings attractive. When he received an honorary degree at the University of Chicago he lectured on "Bergson's Philosophy of Instinct as Viewed by an Entomologist." But when Alfred Kinsey asked him to lecture on the same subject at Indiana University in 1923, Wheeler declined and remarked that he now departed considerably from Bergson in his views. And in an address to the Sixth International Congress of Philosophy in 1926, he remarked that "the resort to such metaphysical agencies [as élan vital and others] has been shown to be worse than useless in our dealings with the inorganic world and it is difficult to see how they can be of any greater service in understanding the organic." By 1928 he was prepared to label élan vital and entelechy as "little more than fetishes," while at the same time remarking that many biologists "were prostrating themselves before mechanism [while] some of the more bolshevistic physicists very stealthily carried it off and dropped it into the sea."

The final section of Wheeler's 1917 essay "On Instincts" is a treatment of some of the major methods of studying behavior. He remarks that "the theologians, metaphysicians, and sociologists have, of course, developed no peculiar methods of investigating biological phenomena. They take over, manipulate, and interpret the output of investigation and, to judge from the result, some of them ought to be forbidden by law to indulge in such practices." Basically, according to Wheeler, there are three methods of studying behavior worthy of consideration: the experimental, the historical, and the psychopathic. Regarding the first, he remarks that:

The experimental method, so universally applicable and successful in physics and chemistry, is certainly of much more limited service in the departments of biology that deal with the living organism . . . In the study of animal behavior important results have been achieved by experimentation in detecting the limits of the variations of instincts, in disposing forever of the notion of their infallibility, and in elucidating the relations between stimuli and responses. The serious limitations of the method lie in the fact that the living plant or animal is not a mere mechanical system but a creative organism, a being that cannot be isolated from its environment like a material system and one which has the ability to epitomize its whole past in its structure and behavior. It will always be necessary, therefore, to supplement experimentation with the historical method.

Wheeler's examples of the use of the historical method were drawn from the defensive behavior of *Formica* ants, the nuptial dances of empidid flies, and the spinning behavior of moths and butterflies. After describing the mechanism by which *Formica* ants spray their enemies with formic acid, he remarks: "We now know the machine and how its works and my friend Dr. Loeb would undoubtedly ask me: 'What more do you want?' I would reply that I am one of those absurdly inquisitive people who might like to know how the pismire came into possession of its atomizer."

He then proceeds to discuss the evolution of the hymenopterous ovipositor and its associated behavior, its conversion into a sting, and finally the reduction of the ovipositor in the higher ants and its replacement by a large, contractile reservoir of poison that can be sprayed some distance. The behavior of dance flies (Empididae) shows even more clearly the value of comparative studies. The males of certain species are known to carry about a balloon of frothy material and to present it to the female prior to courtship. This behavior makes sense only in a comparative context, for it is known that some species include a minute dead insect in the frothy mass, that others capture larger prey and envelop it in saliva before presenting it to the female, and that still others (the apparently most primitive members of the group) merely capture an insect and present it to the female, which feeds upon this prey during copulation. Still other empidid males pick up petals or bits of wood and present them to the females in lieu of prey. In the moths and butterflies the trend is actually from simple to complex then back again to simple.

The larvae of some primitive moths remain enveloped in simple cases, often containing foreign matter, throughout their lives. In more specialized moths the silk is retained in the silk glands until the larva is mature, when a complex cocoon is formed. But some skipper caterpillars "spin a flimsy and degenerate cocoon," while butterflies make no cocoon at all, but simply suspend themselves from a disc or saddle of silk. Wheeler sums up this section of his essay as follows:

Of course, I do not pretend that the historical method of studying instincts, as I have endeavored to illustrate it, is capable of yielding results of great precision or certainty. It has serious limitations, some of which are inherent in the limitations of the living and fossil faunas accessible to us. Many of the most extraordinary instincts are exhibited only by isolated and specialized groups of species, and though we may be able to detect certain developmental tendencies within a group, it is sometimes impossible, and may always be impossible, owing to the extinction of the more primitive allied forms, to form any satisfactory conception of the origins or early stages of a particular instinct . . . Other limitations are inherent in the method itself, which is of such a character as to require constant revision and considerable restraint and taxonomic information on the part of the one who employs it. It is, nevertheless, sufficiently valuable to merit more attention on the part of modern biologists, and especially of some of our students, whose intellects are obnubilated by the notion that biology begins and ends in physics and chemistry and that it is bad form to be able to recognize at sight more than fifteen animals and ten plants.

The final section of "On Instincts" is a consideration of the psychopathic approach to behavior. Here we find Wheeler in his best after-dinner-speaking form, and we quote this section at length in the hope of putting the reader of this chapter in as jocund a mood as Wheeler must have left his audience at the Royce Club on the twentieth of May in 1917. Wheeler states that as a result of his attraction to this approach to behavior, he was led "to read some twenty volumes of psychoanalytic literature comprising the works of Freud, Jung, Brill, Adler, Ernest Jones, Ferenzci, Bjerre, and W. A. White, with the result that I feel as if I had been taking a course of swimming lessons in a veritable cesspool of learning." Then he goes on:

Having had a fling at nearly all the types of biologists and at the nonbiologists who have handled instinct, I now see my opportunity to get under the skin of the psychologists. After perusing during the past twenty years a small library of rose-water psychologies of the academic type and noticing how their authors ignore or merely hint at the existence of such stupendous and fundamental biological phenomena as those of hunger, sex, and fear, I should not disagree with, let us say, an imaginary critic recently arrived from Mars, who should express the opinion that many of these works read as if they had been composed by beings that had been born and bred in a belfry, castrated in early infancy, and fed continually for fifty years through a tube with a stream of liquid nutriment of constant chemical composition. To put it drastically, most of our traditional psychologies are about as useful for purposes of understanding the human mind as an equal number of dissertations on Greek statuary would be to a student eager for a knowledge of anatomy. Such a student at once learns that the object of his investigation, the human animal body, is very largely composed of parts offensive to the aesthetic sense, but this does not deter him from studying them as thoroughly as other parts. The typical psychologist, who might be expected to study his material in the same scientific spirit, does nothing of the kind, but confines his attention to the head and upper extremities and drapes or ignores the other parts.

Now I believe that the psychoanalysts are getting down to brass tacks. They have discovered that the psychologist's game, which seems to consist in sitting down together or with the philosophers and seeing who can hallucinate fastest or most subtly and clothe the results in best English, is not helping us very much in solving the terribly insistent problems of life. They have had the courage to dig up the subconscious, that hotbed of all the egotism, greed, lust, pugnacity, cowardice, sloth, hate, and envy which every single one of us carries about as his inheritance from the animal world. These are all ethically and aesthetically very unpleasant phenomena but they are just as real and fundamental as our entrails, blood, and reproductive organs. In this matter, I am glad to admit, the theologians, with their doctrine of total depravity, seem to me to be nearer the truth than the psychologists. I should say, however, that our depravity is only about 85 to 90% . . .

In nothing is the courage of the psychoanalysts better seen than in their use of the biogenetic law. They certainly employ that great biological slogan of the nineteenth century with a fearlessness that makes the timid twentieth-century biologist gasp. But making all due allowance for the extravagant statements of Freud and Jung and their disciples, any fair-minded student of human nature is compelled to admit that there is a very considerable residuum of accurate observation and inference in their accounts of the dream, of the perversions of the nutritive and sexual instincts, of the erotic conflicts and repressions, and of the surviving infantilisms. If Freud told us, as he probably would if he were here, that all of us who have been smoking this evening have merely been exhibiting a surviving nutritional infantilism with the substitution of cigars for our mothers' breasts, we should, of course, exclaim, like some New England farmer confronted with a wildly improbable statement, *Gosh!* But after all, is the substitution by a man of a roll of dried *Nicotiana* leaves for a woman's breast any more preposterous than the Empidid's substitution of a balloon of salivary bubbles for a juicy fly . . .?

To me one of the most striking indications that the psychoanalysts are on the right road is the fact that many of their theories have such a broad biological basis that they can be applied, *exceptis excipiendis,* to a group of animals so remote from man as the insects. This has not escaped Jung, who calls attention to the striking analogies between the nutritive caterpillar stage and human infancy, the chrysalis and the period of latency, and the imaginal butterfly and puberty in man. There are even cases of repression and sublimation as in the workers of social insects, and did time permit I could cite examples of multiple personality or of infantilisms, that is, larval traits which survive or reappear in the adults of many species. Insects undoubtedly sleep. Do they dream? If they do, what a pity that we shall never be able to apply the Freudian analysis to the dreams of that symbol of sexual repression and sublimation, the worker ant!

But these are trivial considerations. The great fact remains that the work of the psychiatrists is beginning to have its effect even on such hidebound institutions as ethics, religion, education, and jurisprudence, and that the knowledge that is being gained of the workings of our subconscious must eventually profoundly affect animal no less than human psychology, since the subconscious *is* the animal mind.

Nine years later, Wheeler had another fling at the psychologists in his review of John B. Watson's *Behaviorism* for Raymond Pearl's *Quarterly Review of Biology*. "Psychology," wrote Wheeler, "is universally admitted to be a queer science, and one can hardly fail to observe that its queerness . . . is due in no small measure to the mental peculiarities of its devotees. When we consider that, with very few exceptions, the men who have developed psychological science have been either philosophers, priests or paedagogues, or hybrids of all three, its strangely inhibited, not to say eunuchoidal, aspect is easily understood . . . Their professions require that they shall give the impression that they subtly vegetate on cloud-banks of ratiocination and pure asexual love and that they possess little or no first hand acquaintance with the instincts, emotions or appetites, not to mention uglier propensities."

However, Wheeler pointed out, a number of younger men were attempting to break away from "anaemic psychic processes" and to approach an understanding of the mind by describing and analysing behavior. "Watson, as one of the most obstreperous of the youngsters, has been so frequently spanked that he has by this time undoubtedly developed ischial callosities of some thickness. Since henceforth the application of the ferule in the same *locus* would be ineffectual, a more sensitive spot might be sought. To the biologist the whole situation is replete with fun and confusion."

According to Wheeler, Watson's major shortcoming was that "his lack of interest in the history of science has led him to regard himself as the Messiah of a new scientific dispensation dating from the year of grace 1912." In fact, the biological study of behavior can be traced to Réaumur, if not to Aristotle, and long before Watson's time "the methods of ethological, or behavioristic science" had

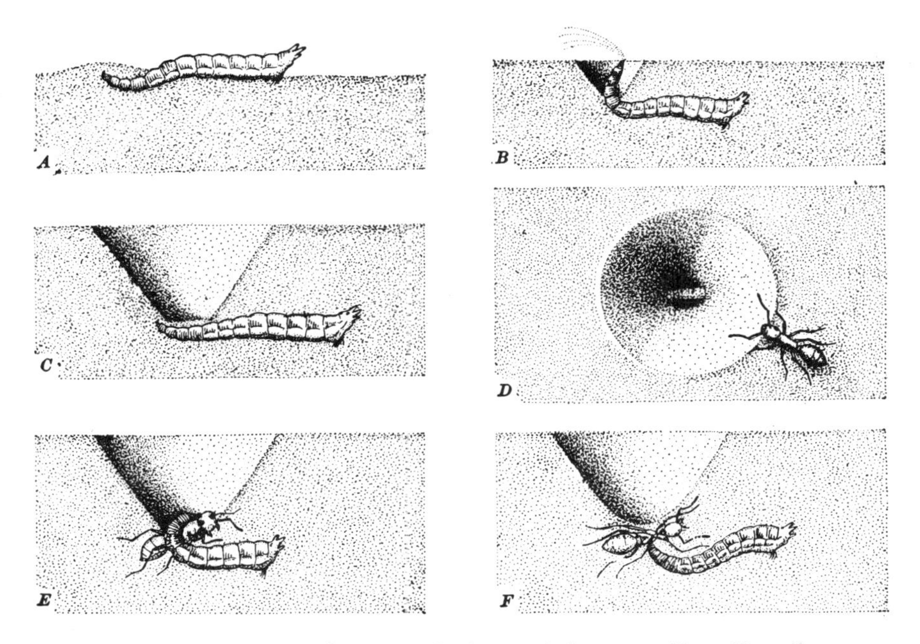

A figure from *Demons of the Dust:* the larva of the worm lion, *Vermileo comstocki,* building its trap and capturing an ant. Wheeler worked out the life history of this insect on the basis of material collected in California in 1917. (W. W. Norton & Company, Inc., 1930)

proved illuminating. It was obvious to Wheeler that "there is nothing startlingly novel in the conceptions and methods employed by Watson" and that "both he and the psychologists seem to have been duped by a new word for an old thing." However, Wheeler felt that Watson was too ready to reject all data arrived at by introspection and much inclined to overstress the effects of environment and to underestimate the role of heredity in determining behavior. The stir caused by Watson's behaviorism has long since passed, and in retrospect Wheeler's appraisal of his work seems basically sound.

In the last two decades of his life, Wheeler continued to apply his biological approach to behavior in his studies of social insects as well as in his studies of ant lions and worm lions (in *Demons of the Dust,* 1930). His description of pit construction by ant lion larvae reads like something out of the pages of modern ethology. After rejecting Doflein's characterization of this insect as a "reflex automaton," Wheeler remarks that pit-digging "is clearly appetitive, or a 'goal-activity,' initiated and guided by a constellation of internal and external stimuli . . . It is at first often executed by random movements but soon becomes a more definitely organized preparation for securing the prey, culminates in the consummatory acts of its seizure and digestion and concludes with a state of satiety and quiescence."

Demons of the Dust, The Social Insects, and Wheeler's other books and papers were widely read in their day, and in our opinion had a good deal more to do with the development of current behavioral theory than is commonly appreciated. That most modern treatments of animal behavior concern themselves almost wholly with birds and mammals, minor groups of animals in terms of behavioral diversity and numbers of species, is a situation that hopefully will be corrected in the course of time.

11 / Colleagues, Credos, and Crises

William Morton Wheeler, in a characteristically thought-stimulating and smile-provoking lecture delivered in 1911 at the Marine Biological Laboratory, Woods Hole, Massachusetts, discussed insect parasitism (Wheeler, 1911). One sentiment expressed by this outstanding entomologist, and shared by many biologists today, was that of apology for being, not an experimentalist, but an "old-fashioned" student of insect life, constrained by the extraordinary intricacy of his science to maintain the closest and most sympathetic cooperation with taxonomists, morphologists, and students of geographical distribution.

This monograph will deal with biology, sensu stricto, and with experimental work that is largely in the fields of behavior, ecology, and genetics. Wheeler, to judge from his opinions in 1911, would express criticism of the genetical section, orthodox as it is, for he had little appreciation of this field of research. Perhaps, had genetics dealt in its early days with a hymenopterous insect instead of "that stupid little saprophyte" (his description of *Drosophila* as expressed to his students), he might have been convinced of the validity of Mendelism.[1]

Thus does Anna R. Whiting begin her paper "The Biology of the Parasitic Wasp *Mormoniella vitripennis* [=*Nasonia vitripennis*] (Walker)," which was published in the *Quarterly Review of Biology*, September 1967. This feeling about Professor Wheeler, which is so well expressed by Dr. Whiting, has been widespread among scientists and friends who knew him and his work and has caused considerable speculation as to why there should be such an apparent blind spot in a man who was obviously such an intellectual man of the world. Did he really fail to appreciate the field of genetics? Was he serious, or was he merely twitting his colleagues, when he spoke of genetics to the American Society of Naturalists in 1922 as a "dear little bud" from the stolon of natural history, "so promising, so self-conscious, but, alas, so constricted at the base?"

The history of genetics has been written by several of the people who helped make it, and their observations placed against the facts of Dr. Wheeler's life make an interesting study. In the 1890s in Europe, cytologists and embryologists such as Boveri and van Beneden were finding more and more morphological details in the cells they were studying. These men and others had come to the conclusion that the nucleus contained the physical basis of inheritance, and that a specific component, chromatin (a complex organic acid), probably was involved in the actual transmission mechanism of heredity. This was the conclusion of Edmund Beecher Wilson in 1900 in *The Cell in Development and Inheritance,* but at best it was only speculative. And, as Wilson stated, the fundamental problems remained: "How do the adult characters lie latent in the germ cell, and how do they become potent as development proceeds? This is the final question that looms in the background of every investigation of the cell."[2]

In Freiburg, Germany, August Weismann (1834–1914), in *Das Keimplasma,* presented his concept of the continuity of germ plasm, saying that the offspring inherits its characters not from the body of the parent, but from the germ cells, which in turn have derived their properties from the ancestral germ cells which preceded them. This idea of germ plasm led to even greater effort of the cytologists, and first chromosomes were discovered, then mitosis, and soon meiosis (predicted first by Weismann). As these ideas were presented, Lamarck's theory of the inheritance of acquired characteristics came under increasing disfavor, and unfortunately along with it went almost all immediate concern with the possible effect environment might have on development.

Meanwhile, a Dutch physiologist, Hugo De Vries (1848–1935) (who had been a student of Julius Sachs at Würzburg), came up with some interesting ideas on the mechanism of inheritance. He postulated that the hereditary substances were in the form of self-replicating units, which he called *pangenes.* (Darwin had used this word in a different sense, so it was fortunate that the Dane W. L.

Johannsen soon proposed the shorter word *gene.*) Thus De Vries thought of species characters as being "composed of separate more or less independent factors," the changing combinations of which "determine the extraordinary variety of the world of organisms"—this a decade before the rediscovery of the work of Gregor Mendel. Furthermore, De Vries' work on the breeding of evening primroses led him to propose his *Mutationstheorie* in 1901: the idea that species originate by major changes or "mistakes" in the replication process. It was De Vries, too, who a year earlier had been one of the three workers who independently rediscovered Mendel's paper of 1866 and brought it to the attention of the world.

The person who is usually credited with bringing Mendel's work to the English-speaking world is William Bateson (1861–1926). Already in 1894, Bateson had challenged the idea of natural selection working on small, continuous variations (an idea attributed to Darwin and his successors) by saying that new species were formed when large changes occurred. It was Bateson who first proposed the word *genetics* (in 1906) and who also proposed several other basic terms used in that science. Throughout his life Bateson remained an advocate of evolution by discontinuous variations, or saltations, and he was one of the leaders in supporting Mendelism as opposed to Darwinism, a dispute unresolved until 1930, when R. A. Fisher published *The Genetical Theory of Natural Selection* and ushered in the modern synthetic theory of evolution.

While genetics was being founded on the basis of breeding experiments, the embryologists were at work on problems of inheritance from another point of view. Wilhelm Roux (1850–1924), an important essay of whose was translated by Wheeler in 1895, was much concerned with "developmental mechanics," that is, the processes by which the inherited features become expressed in the organism. Roux's pioneering work, which gave him the title of founder of experimental embryology, was done with frog eggs. He separated the two halves of a newly divided, fertile egg, destroyed one of the halves, and allowed the other to develop. The result was half of a normal-sized embryo. In 1891, however, Roux's work was challenged by Hans Driesch (1867–1941). Driesch, working with the eggs of sea urchins, performed experiments similar to Roux's, but with different results. Instead of half of a normal-sized individual, he produced a complete animal which was only half of the normal size. What had happened? Why had these two similar experiments produced such dissimilar results? The Roux-Driesch controversy was an interesting one, but belongs more to the history of embryology than to our story. Perhaps more pertinent here is the fact that Roux, along with Loeb, held a mechanistic conception of life, while Driesch was a strong vitalist and eventually, late in his life, an out-and-out mystic. But in the 1890s, the work of the two men aroused much interest in experimental biology, and one of those caught up in this excitement was a young zoologist teaching at Bryn Mawr, Thomas Hunt Morgan (1866–1945). Morgan had already done some work on the descriptive embryology of fish, starfish, and sea urchins, but his year in Europe (1894–95) redirected his interests. In his *Biographical Memoir* of Morgan, A. H. Sturtevant wrote: "Ten months in 1894 and 1895, and also the summer of 1900, were spent at the Zoological Station at Naples, which he had visited in 1890. Here he collaborated with Hans Driesch in the use of experimental methods in the study of embryology. This association was important in influencing the course of his later work. He found Driesch congenial and stimulating, and remained on close friendly terms with him—even though he found himself wholly out of sympathy with Driesch's later vitalistic views and preoccupation with philosophy."[3]

Morgan proceeded to experiment with developmental problems, and gradually he became intrigued with the probable causes of sex determination. Many others were interested in this, too, including E. B. Wilson, C. E. McClung (who had studied briefly under Wheeler in Chicago), T. H. Montgomery (who replaced Wheeler in Texas), Boveri, W. E. Castle, and others. But it was not until he began to use the fruit fly, *Drosophila,* whose convenience as a laboratory animal was apparently first pointed out by Castle sometime shortly after 1900, that Morgan came into his own. The science of genetics was thereafter assured a permanent place in biology.

But what did this all have to do with William Morton Wheeler? In fact, Wheeler had been excited by the work of all of these men and was personally acquainted with most of them. Why was he uninterested in the developmental sciences, or worse, why did he even seem to be antagonistic to them? Under Whitman and Patten, Wheeler had become an excellent descriptive embryologist, and his graduate work plus his year's work in Europe with Boveri and van Beneden, and at the Naples zoological station, added not only to his knowledge and experience, but also to his rapidly expanding reputation as an embryologist. But through all those years he retained his interest in taxonomy (chiefly of certain flies). Nor was his one paper on tropisms of an experimental nature! Why had Wheeler, unlike Morgan and E. B. Wilson, not succumbed to the lure of experimental embryology—especially as he was seemingly subjected to almost identical influences as were those men? It becomes more and more apparent that his deepest interest was in whole organisms and their natural relationships. He had always been interested in insects, and when the ants fell under his eye, they were a natural subject for his inquiring mind. Once again the influence of Rauterberg, Peckham, and Ward asserted itself, and he became a staunch defender of natural history and a scoffer at the genetics of his time, partly in self-justification, perhaps, and partly because he enjoyed pricking the egos of his colleagues. Wheeler was not to join the crowd meeting every summer at Woods Hole—at least not ideologically. He would become his own man, a pioneer in the study of social insects and ants in particular and all their complex interrelationships. Here his abilities as an observer, interpreter, and philosopher could best be utilized.

But this did not exclude him physically from Woods Hole or from the world of the experimental biologist and geneticist. Wheeler had known the Marine Biological Laboratory from its beginnings, and though he skipped a season there now and then when other parts of the world demanded his attention, he always kept his ties with the institution and the men working there. He was a trustee from 1919 continuously until his death, and through the years gave many lectures there. In one of these, his lecture on Casper Friedrich Wolff (1898), he stressed that "grossly underestimating the complexity of the problem . . . has always been a great pitfall in attempting an explanation of life." He noted that there was divided opinion as to the relative importance of heredity versus environment in the story of development and differentiation, and that mathematical proof was needed before the truth could finally be determined. Perhaps Wheeler thought that insects, or certain insects, might be good experimental animals for obtaining such data. At any rate, in 1907 he published a paper called "Pink Insect Mutants," in which he described the pink, brown, and green forms of a common katydid and surmised that the brown and pink forms represented "sports, or mutants, as we should now call them." Then he continued: "Conclusive proof of the correctness of this view can be obtained only by experimental breeding. On the sport or mutation hypothesis we should expect pink individuals mated *inter se* to produce only pink individuals, and the same should result *mutatis mutandis* in the case of the brown forms. Pink or brown individuals crossed with the common green form may be expected to give offspring in the Mendelian proportion, with the pink and brown characters acting as recessives. Perhaps some student at the Marine Biological Laboratory at Woods Hole, where pink individuals . . . seem to be less rare than in other localities, may find it worth while to perform these and other experiments for the purpose of determining the inheritance value of the characters above discussed."

Wheeler's previously mentioned paper on insect parasitism was given first as a lecture at Woods Hole (delivered in August 1911), and the following quotation from that paper states succinctly his position in biology in general, as well as his opinion of the value of studies of parasites:

Having made the pilgrimage to the American Mecca of experimental zoologists, I could hardly hope for salvation if I departed without at least saluting the Kaaba. This I can do most effectively, perhaps, by calling attention to the great need of experimental work in animal and especially in insect parasitology. Biologists, during the romantic period of Darwinism, made much of the parasites. These organisms, in fact, supplied them with no end of ammunition in defence of natural selection, the influence of the environment and the biogenetic law. Then came the period of morphological minutiae with its tacit assumption that particles of a dead organism are vastly more interesting and illuminating than the whole of a living one. During this period the parasites were, of course, sectioned and studied in the same manner as other organisms, but since it is impossible to explain a living whole by pulling it to pieces and sticking the inert fragments together again, parasitism, which is a process and not a thing, retained its ethological interest mainly for biologists who were engaged in the practical applications of their science.

Now that we have reached the third period, or that of emphasis on experiment with the living organism as the best means of elucidating the life-processes, those of us who had the misfortune to live and exhaust our greatest enthusiasm during the romantic and morphological periods, can, I suppose, do nothing better with the meager remnant of our vitality than pray for breadth of sympathetic vision on the part of our younger, more numerous and more vigorous contemporaries. The splendid achievements of the investigators who assemble here every summer certainly whet one's desire to see experimental work of the same character accomplished in parasitology. A certain amount of simple experimental work on social parasitism in ants has been inaugurated by Wasmann and myself . . .

Experimental zoologists, including the students of animal behavior, are most keenly interested in the modifiability of the organism, and their experiments are usually devised for the special purpose of determining the amplitude and peculiarities of this modifiability. The entomologist, however, who is attempting to use parasitic insects as tools or implements in controlling the depredations of other insects, is primarily interested in the stability of structure and constancy of behavior. This follows from the very nature of his work. As the essential excellence of a tool consists in its remaining the same as it was when it left the hands of the manufacturer, so a parasitic species can be used as an efficient tool only if it behaves generation after generation with uniform constancy. Hence in combating pests, only those parasitic insects can be utilized to advantage that are not only prolific and will endure the climatic conditions into which they have been artificially introduced, but will maintain very definite relations only with individuals of a single or a very few host species and destroy them in their earliest possible ontogenetic stage before they can do extensive damage.

For a subject which tended to be confined pretty much to Farmers' Bulletins, parasitism and its use in biological control reached new heights under the talented tongue and pen of Professor Wheeler, who made it both interesting and informative to a group of the nation's foremost biologists.

Wheeler's friendship with the "Woods Hole group" was a matter of much importance to him throughout his life, and the members of this group often visited in each other's homes. The Wheelers (for Mrs. Wheeler often accompanied her husband) enjoyed the brief informal visits and overnight stays at Woods Hole with the E. B. Wilsons, E. G. Conklins, the Jacques Loebs and the Morgans. Many of these same people belonged to the National Academy of Sciences, and on the occasions of the academy meetings in New York, homes were always open to out of town visitors. Conversely, when visitors were coming to Boston and Cambridge, the Tom Barbours, the G. H. Parkers, and the Wheelers vied for the privilege of entertaining.

Sometimes the relationship between these men was stormy, though seldom were the arguments of long duration. For example, in Dr. Wheeler's study at home is a carefully framed letter, dated

July 28, 1900, which says: "My dear Wheeler: I think I ought to let you know that I regret having lost my temper at the end of our last conversation. Yours, Jacques Loeb."

Another interesting exchange of letters occurred between Loeb and Wheeler in 1918, when, in a letter dated May 20, Loeb wrote saying that his receipt of the Walker Prize (given by the Boston Society of Natural History for meritorious scientific writing) was very unexpected and gratifying. He knew that Wheeler had been one of the councilors and suspected that he had had something to do with the verdict. He thanked him, and said he hoped to see him at Woods Hole in the summer. He continued: "We are getting along in years—at least I am—and it would be nice to see a little more of each other."

In December of that same year, Wheeler wrote to Loeb to thank him for a book, and then said: "[I] agree with you in the main that the activities of insects are highly mechanized . . . I do not believe, however, after thirty years of working with insects that their behavior can be completely resolved into tropisms."[4]

Correspondence between T. H. Morgan and Wheeler was always enjoyable (if one could decipher Morgan's not-too-legible handwriting, and many of his letters were handwritten). "By the way [wrote Morgan to Wheeler in 1922], we have heard a rumor to the effect that the Bussey is about to announce a very important discovery, one that will put the Columbia laboratory entirely out of commission. We are told that it is a great secret and is not to be released for about two weeks, and then good-bye Drosophila. You can well picture to yourself our trepidation. We can hardly drag ourselves through the ordinary daily work of emptying fly bottles, for alas we know our days are numbered." "In regard to the important discovery you mention," replied Wheeler, "I would say that it must be kept with such secrecy that I do not know anything about it, but probably even the Faculty believe that I am so communicative that they would not let me in on it for fear of its leaking out."

In 1923, Wheeler wrote that he had received no accounting from the Columbia University Press for four years on his ant book, and, as he thought it had been selling well, he wanted Morgan to go over and check for him. Morgan replied that he also was on the "outs" with the Press, but that he had asked Wilson to go over and check. And then a postscript from Morgan: "[I] became interested in some of the egg rafts of cockroaches before I found out that you had discovered all that. I wanted to know. How did you manage to do such a fine piece of work when you were so young!"

Another 1923 letter from Wheeler to Morgan, then at Woods Hole, was an invitation to come and stay with Wheeler for a visit. "You need a change and to get some of that drosophiline out of your system."

In November of 1927, Wheeler thanked Morgan for the beautiful book [perhaps *Experimental Embryology?*], which he said covered matters which always interested him, but about which he was ignorant because he had not kept up. "You should not feel that I am in any way unsympathetic with your ideas, I simply happen to be intensely interested in the historical aspects of biology and I realize that most investigators have no interest in history."

In 1928 Morgan left Columbia University and moved to the California Institute of Technology in Pasadena, and he had apparently suggested to Wheeler that he move west with him. In the same letter of November 1927, Wheeler commented: "I have been thinking much of the California situation. Probably Milliken would prefer me to keep out of that state. He certainly will if he ever reads the introduction to my 'Foibles of Insects and Men.' Mrs. Wheeler, rather to my surprise, thinks favorably of our moving." He also added that California would probably be good for his children, Ralph and Adaline. However, he said, "I am thinking very seriously of retiring next September [and] it will take me another year, I fear, to liquidate all the matters connected with my collections, library, etc." But in May 1928, Morgan wrote saying that he was sorry to hear that President Lowell had forbidden Wheeler to retire, and that "this is the first time I ever heard of your being forbidden to do anything that you did not promptly go ahead and do, so I hope he will continue to forbid you, and then I shall have the pleasure of seeing you kick over the traces."[5]

It is interesting to note that T. H. Morgan, whom geneticists claim as their own, in his latter years left the field of genetics and returned to his first interest, experimental embryology.

A. H. Sturtevant, in a letter written to the authors in 1967, summarized what he considered to be the bond between these two men:

> I do not think that either W. M. Wheeler or T. H. Morgan had much sympathy with or appreciation of the scientific work or philosophy of the other. But they were nevertheless close and congenial friends. They

shared a large group of common friends, both European and American, and a strong interest in natural history. I don't think that one could appreciate Morgan's enthusiasm for field work unless he had been on a field trip with him. I shall never forget one field trip, to the Elizabeth Islands (off Woods Hole), with Morgan and Wheeler.

Their strongest bond was their common ability as witty conversationalists. Each had a very quick mind and a highly developed sense of humor, and they thoroughly enjoyed mental sparring with each other. Those of a younger generation who had the privilege of listening to this display of fireworks will never forget it.[6]

One of the many reasons why Dr. Wheeler could not let the new science of genetics pass him by was that he was surrounded by it on every working day at Bussey. And after he became dean, he had to handle the affairs of the staff and students specializing in genetical studies just as he did those in all the other departments of the school. His annual reports dealt fairly with all, providing details of the student and staff research in genetics (and plant morphology, taxonomy, and so forth) as well as in entomology. Furthermore, when Morgan wrote to Wheeler in November 1921 saying that Bateson was coming to the Christmas scientific meetings and wanted to stop and see the men at Bussey, Wheeler replied saying he would be delighted to see him. Professor Wheeler not only received Bateson warmly, but he arranged for an informal talk to Bussey students and even took him to see some Chinese works of art in Salem, which apparently were of great interest to Bateson. Two years later, Wheeler did the same for Hans Driesch: arranged an address, had Mrs. Wheeler arrange a dinner and invite some philosophers and zoologists who wanted to meet Driesch and his wife, and took them on a guided tour around Cambridge and Boston, including a visit to the art museum which Driesch had specifically mentioned as wanting to see.

The exact relationship between Wheeler and his own staff is not easy to discern. Obviously there was not much written communication, and much of what we think we know is surmised from the meager reports and the recollections of the people who were at Bussey at the time. Professor W. E. Castle was a quiet man, dedicated to his science and often stubborn in his convictions, although he would change his mind if experimental results indicated clearly that he was wrong. He had been introduced to experimental morphology in 1893 in a course at Harvard taught by C. B. Davenport, who had later replaced Wheeler at Chicago, and in 1901 he became the first American to publish on Mendelism. Castle's first scientific papers were concerned with problems of development and differentiation, but once he knew of the rediscovery of Mendel's work, he never again left the field of genetics. Although he is credited with the introduction of *Drosophila* as an experimental animal, his major interest was in mammalian genetics, and the laboratories of Bussey were filled with rats, guinea pigs, rabbits, and the like. The annual report to the president of Harvard, written by Dean Wheeler for the Bussey Institution for the year 1915–16, contained the following summary of Professor Castle's work:

> In addition to the work of formal instruction at Cambridge, Professor Castle has as usual devoted himself to investigation at the Institution during the entire year. His work has been carried on with substantial and indispensable assistance from the Carnegie Institution with general results as follows: The central problem has been the inherited characteristics in guinea-pigs, rats, and mice, with special reference to the constancy and inter-relations of Mendelizing characters. In the guinea-pig studies attention has been centered on size inheritance in species crosses and on the influence of inbreeding on size. In the rabbit work attention has been given principally to quantitative studies of two color-patterns involving white-spotting, the so-called Dutch and English varieties. Among the rats the selection experiments for modification of the hooded pattern, a Mendelizing character, are being continued, now in their nineteenth generation, and the linkage-relations of two yellow varieties of recent origin are being studied intensively. The mouse work centers upon the study of the several allelomorphs of yellow, an unfixable because always heterozygous character.

At the same time that Castle was working in animal genetics, Bussey was supporting active work in plant genetics under the guidance of E. M. East (1879–1938). Trained as a chemist, Dr. East became involved in the analysis of corn for percentages of protein, carbohydrates, and fats. Noting much variation in different strains, East and his co-workers thought of the possibility of improving nutritive value by selective inbreeding. Hybrid corn was an astounding success, and subsequently East turned to the breeding of many other economically important plants, such as potatoes and tobacco. This work was of immediate value to farmers, but it was creating some problems in theoretical genetics. Dunn, in *A Short History of*

Genetics, remarks that prior to 1911 genetics was not penetrating into other fields of biology because no one believed in the stability of the gene.[7] Even Morgan had said, in 1905, "once crossed, always contaminated." Only after a decade or more of arduous experimenting (Castle held out longer than Morgan) were the geneticists convinced that the genes were stable, that changes could occur only through mutations, and that most crosses resulted in predictable offspring ratios. Plants, which were bred for many quantitative characters, such as size, yield, height, color, and so forth, had seemed to be exceptions to Mendelism. East and his students and contemporaries worked with these multiple-factor characters, and soon proved that they also operated according to Mendel's laws, producing varying offspring as a result of interactions among several genes and of conditions of the environment. Thus, East and the others brought population genetics into line with the general development of genetics and helped create a whole new basis for understanding evolution.

Something of the nature of East's research is revealed in Dean Wheeler's annual report for 1919–1920:

Professor East made a study of several plants with the idea of finding out their desirability for use in certain genetic problems. Material was collected for investigating the problem of so-called "false-hybrids" in strawberries. Studies on the inheritance in trimorphic flowers were begun with *Lythrum* and *Oxalis* and materials were gathered for studying the origin of certain teratological organs in maize and for the purpose of initiating some physiological and cytological studies on the development of seedless fruits. The final results of researches on the inheritance of protein in maize were brought together for publication. This investigation was begun 12 years ago at the Connecticut Agricultural Experiment Station and since that time has been carried on continuously with the cooperation of that institution. A number of theoretical points in genetics were cleared up, a method of practical breeding devised and strains of maize produced yielding more than 70 bushels per acre and containing about 18 per cent protein, 5 per cent more than the average for wheat. As the result of a second cooperative project with the Connecticut Agricultural Experiment Station, a new cigar wrapper tobacco of high quality has been given to the growers. Their reports on it are highly favorable. The interesting thing about this tobacco is that it was made to order. The desirable qualities were listed before hand, existing varieties selected from combinations of which the desired type might be expected to be produced, and the strain built up by the application of the methods of theoretical plant genetics. In collaboration with Dr. D. F. Jones, a former student at the Bussey Institution, Dr. East has published an important volume entitled "Inbreeding and Outbreeding," based on his extensive series of experiments carried on during the past fifteen years.

Each of these three men, East, Castle and Wheeler, was well placed in his professional field and possessed a strong, individualistic personality. The interaction of the three on each other and on their students must have been interesting, to say the least. Castle was probably the best teacher and hence closest to the students. East was more difficult to know and to like, especially in his later years when he was often ill. Wheeler tended to be aloof, except with a few of his favorite graduate students, found undergraduate teaching less than inspiring at times, and may have felt a lack of interest and understanding of his ethological studies on the part of the geneticists.

One subject in which these men did all share an interest, however, was eugenics. In 1916 Castle published a textbook called *Genetics and Eugenics,* which went through four editions, the last in 1930. East was also concerned about the human aspects of genetics, even though he worked on plants. In 1916, for example, he wrote a note to Wheeler about a young sociologist from Columbia who was framed for passing out birth control information and sentenced to three years in jail. East was upset, and Wheeler agreed with him. In 1931 Wheeler was pessimistic about the future of eugenics, saying, "But those which deal with eugenics, sex-hygiene and voluntary limitation of the population, encounter such a resistant barrier of emotions, prejudices and ancient mores that their general acceptance will probably be long delayed. So far as these matters are concerned, therefore, the biologist will have to possess his soul in patience."[8] Meanwhile, in 1923, Wheeler became one of the charter members of the Eugenics Commission of the United States of America, and served as a member of the Advisory Council. In the annual report from Bussey for the year 1927–28, he wrote: "Dr. East, in addition to several technical papers, published a volume on human heredity entitled 'Heredity and Human Affairs,' under the imprint of Charles Scribner's Sons, and several magazine articles. At the meeting of the International Congress on Population at Geneva in 1927, Dr. East was appointed a member of the International Union for the Scientific Study of Population Problems. He attended a meeting of the Committee in Paris during the summer of 1928, at which this international scientific project was formally launched."

While Castle tended to look at eugenics from the point of view of an animal geneticist, that is, how the human mechanisms of inheritance might be manipulated, East looked at the subject from the point of view of feeding an ever-expanding population. Improved agricultural practices, including the use of hybrid seeds to increase yield, could not possibly take care of all the problems which would arise when there were too many people in the world. East was a man with remarkable foresight, as Professor Wheeler undoubtedly sensed.

Perhaps partly as a result of East's influence, Wheeler encouraged his wife's work on making better food available for all at reasonable prices. She belonged to the Women's Municipal League in Boston and was chairman of its Department of Food Sanitation and Distribution just before and during the First World War, when there appeared to be a scarcity of food. For example, in May 1917, one of Mrs. Wheeler's duties was the supervision in Jordan Marsh's department store in downtown Boston of a daily demonstration of how to preserve and prepare inexpensive foods. When the war was over, Mrs. Wheeler continued her activities in public affairs. She became one of three women members of the Massachusetts Board of Food Administration, and was especially concerned with the packaging of food in such a way as to protect it from flies. In 1921 she was elected Boston city chairman of the Women's Division of the State Republican Committee. She began her political career, perhaps without political intent because of her concern with food problems, but she eventually became a champion of good government and an opponent of the Boston Democratic machine and its "boss," Mayor James M. Curley. One story has it that after one of her many defeats in attempting to eliminate the "tombstone voters" on the Democratic registration lists in Boston, the newly reelected Mayor Curley sent her a large bouquet of beautiful red roses. Mrs. Wheeler is said to have been delighted with the flowers, but undeterred in her efforts to defeat him in the next election.

When Herbert Hoover left his food administration work to run for president, Mrs. Wheeler supported him vigorously and served as secretary of the Boston Volunteer Campaign Committee for the Republican ticket. The Wheelers evidently enjoyed cordial relationships with the Hoovers, and Professor Wheeler sent Hoover copies of his books, which the latter acknowledged and professed to have enjoyed.

Food production and marketing, overpopulation, eugenics, genetics, animal behavior—these were subjects of consuming interest to the Bussey group, and all fell within the realm of interest of another versatile and energetic biologist of those years, Dr. Raymond Pearl (1879–1940) of Johns Hopkins University and almost of Bussey Institution at Harvard University. In their later years, Wheeler and Pearl were very close friends and mutually interested in philosophical biology. We are not sure when they first became acquainted, but it seems to have been during World War I, when Pearl, as chairman of the Agricultural Commission of the Natural Resource Council located in Washington, D.C., wrote to Wheeler for advice on ways in which economic entomology might help improve food production, reduce losses in agriculture, and hence help in the war effort. We do not know what Wheeler's reply was, but Pearl must have liked it, because he asked Wheeler (and East, too) to become members of the commission, and both accepted. But the friendship apparently did not ripen until after Pearl moved to Johns Hopkins shortly after the end of the war.

Although Pearl was a man of very broad interests, his first studies were on the behavior of lower invertebrate animals, and these led him to an interest in the genetics of these organisms. He was soon drawn into the application of statistics to genetics and published *Modes of Research in Genetics* in 1915. Then, as head of the Maine Agricultural Experiment Station, he used statistics in his work on the studies of genetics and biology of poultry and other domestic animals. As a result of his work, he was asked by President Wilson to come to Washington to be chief of the Statistical Division of the new U.S. Food Administration. This led logically to his appointment as professor of biometry and vital statistics in the newly formed School of Hygiene and Public Health of Johns Hopkins in 1919. In a biographical memoir, H. S. Jennings wrote:

Pearl remained at Johns Hopkins for the remainder of his life, though with some changes of work and title. It will be well at this point to note certain characteristics of the man and his work. He was a man of unusual height and weight, physically an impressive figure. His was a masterful personality, of extraordinary resourcefulness and initiative, of wide knowledge, astonishing power of work, remarkable versatility and scope, and strong ambitions. His interest in biology was encyclopedic. In his contributions he touched upon most aspects of the subject. This was not a matter of merely the extent of scattered interests, but rather of the kind of interest and of the kind of man he was.

With the transfer to Johns Hopkins (1919) came gradually a centralizing of all his interests in the biology of man. A break with past interests was reinforced by a fire which in 1919 destroyed his notes on past work, as well as his large library of reprints. During the organization of the statistical laboratory and of courses in the statistical treatment of biology and medicine there was naturally an accentuation of interest in the problems of method, evidenced by the publication in 1923 of his well known textbook "Introduction to Medical Biometry and Statistics." As statistician of the Johns Hopkins Hospital (1919–1935) he systematized autopsy records and published at intervals data and conclusions based on study of these.

Soon his studies took a more experimental and broadly biological turn. Though they were henceforth directed mainly toward the biology of man, he employed other organisms for experimental purposes. He carried through extensive breeding and experimental studies on Drosophila, with relation to duration of life and its inheritance, mortality, and growth of populations; and on the factors, genetic and environmental, that influence these. They were accompanied by statistical investigations on the same kinds of problems in man.[9]

Pearl's publications, such as *The Biology of Death, The Biology of Population Growth,* and *Alcohol and Longevity,* attracted attention in the press and aroused much interest as well as controversy. The thesis of *Alcohol and Longevity*—that alcohol in moderation is not harmful to health—was not popular with some segments of the populace (especially during prohibition), nor was his later claim that tobacco *is* harmful, even in small quantities. But Pearl enjoyed keeping the public as well as the scientific world informed on his studies and beliefs, and as a result he lectured often, wrote prodigiously for innumerable journals, both popular and scientific, and for the newspapers. He also founded two journals of his own: *Quarterly Review of Biology* (1926) and *Human Biology* (1929). The former has been especially appreciated and successful in the scientific world.

In October 1923, Pearl wrote to Wheeler about another publishing project, and about his friend, H. L. Mencken:

I have agreed to be Mencken's biological adviser and sort of *agent provocateur* in his new enterprise "The American Mercury." It should be a first rate thing, a magazine which an intelligent person can read without being bored to death. I have become so fed up with that unctuous dirty brown purvayer [*sic*] of near culture which is distributed from your town that I have refused to pay them any more money and stopped my subscription. Mencken is an extremely good fellow of a realistic turn of mind who, as you know, can really write. This enterprise he has been working on for a long time and has his heart very much in it. I am anxious to help him out not only because he is a good fellow, but because it seems to me that those of us who perceive and sometimes comment upon the sad tendencies of our times ought to be willing, when an occasion offers, to help along any enterprise which is calculated to shoot a ray of real light into the prevailing duskiness. Mencken is very kind and considerate about asking my help, and is anxious that I shall not do any more in connection with it than I feel that I want to do and can spare the time from research work to do. All that I have agreed is occasionally to write an article for him myself and to scare up for him occasionally some articles by biologists who can write and who have something to say.

Pearl was, of course, trying to encourage Wheeler to write for the new magazine, and he soon invited him to attend the famous Saturday Night Club of Baltimore. In March 1924 he wrote Wheeler:

It was a great pleasure to hear from you once more. I need not say I am looking forward to seeing you. In point of fact, I was just about to write to you regarding a scheme.

As you will have noticed, the Philadelphians are, like everybody else, becoming tight-wads, and if we get a dinner this year we have to pay for it. [Reference, apparently, to the coming meeting of the Philadelphia Academy of Sciences.] The speeches are sure to be dull (unless you are to make one), and therefore I suggest that we leave Saturday afternoon, come down here and go to the Saturday Night Club instead. The Saturday Night Club is an institution, fathered by Mencken, which has been in existence about twenty years. We meet together at eight o'clock each Saturday night and play symphonies until ten. Then we drink beer, eat and converse upon sundry subjects until midnight or later. I think you will not find anywhere now in existence a more highly "he" group than this, nor one in which Rabelaisian conversation reaches such genuinely high flights. I mentioned the possibility last Saturday night to the group that I might persuade you to come down for a meeting. If you do, you will be received in a manner which sincerely and genuinely befits your eminence as a gentleman and a scholar. You will be assured of good food, good beer, good conversation, and not too bad music. Mencken is extremely keen to meet you, and altogether I can assure you that it renews your youth as nothing else I know of. The association has done wonders for me. What better recommendation could one give than that?

Over the years Wheeler attended a number of meetings of the Saturday Night Club, for Wheeler greatly admired Mencken and may have been influenced by Mencken's free-swinging style and his iconoclastic views. Why Wheeler never published in *The American Mercury* we do not know; perhaps he felt that his essays were a bit too technical even for Mencken's audience of intellectuals. Through their mutual friend, Pearl, however, Wheeler and Mencken enjoyed each other's keen mind and wit. Mencken's belief that scientific knowledge was the only acceptable philosophy and that religions were essentially absurd appealed to Wheeler as well as to Pearl, and could only have drawn the three closer together. After the publication of *Treatise on the Gods* in 1930, Mencken sent a copy to Wheeler with the following inscription: "For William Morton Wheeler with the devotion of the undersigned theologian, H. L. Mencken." In a letter dated March 31, 1930, Wheeler replied:

I appreciate your great kindness in sending me a copy of your splendid "Treatise on the Gods," which I devoured at once with great glee and profit. There is in it not a word with which I do not heartily agree. Its perfect sincerity and integrity makes it seem to me the best book I have ever read on the subject and I feel that we all owe you not only congratulations on the completion of the work, but thanks for giving our poor religion-ridden population an opportunity to read something helpful and enlightening. Yours is educative work of the right kind. I was glad to hear from one of Lauriatt's salesmen that the book is having a wide sale. This is most gratifying, both on your account and on account of the pleasure it must give our friend Mr. Knopf.

On April 3, Mencken wrote:

Thanks very much for your friendly letter. I am delighted to hear that you found the book interesting. Knopf reports that it is doing very well—a new experience for me, for my books have never sold in the past.

I surely hope that you and Mrs. Wheeler will come back from the West by way of Baltimore. It would be an immense pleasure to see you.

Pearl, after his illness, is now flourishing, and his thirst seems to be as heroic as ever.

Mencken's last remark must have brought considerable relief to Wheeler, for the two men, Pearl and Wheeler, had just been through a most unfortunate and regrettable experience. Wheeler, as previously mentioned, had seriously considered resigning or retiring from Bussey and Harvard and joining his old friend T. H. Morgan in California. Wheeler apparently had even gone as far as discussing this with Harvard's President Lowell, or so one would gather after reading the following paragraph from one of Pearl's letters to Wheeler (Feb. 20, 1928): "I had a little suspicion that perhaps Lowell would not fall so easily into your scheme of leaving Harvard. After Cattell's article in *Science* the old boy will be so puffed up that he will never let any of you elder statesmen in the University resign or die. I suspect that what he will do ultimately is what the Russians did with Lenin—have you all embalmed and arranged in a row in Memorial Hall as evidence of your continued presence and watchful care over his high-toned institution."[10]

In early 1928 George R. Agassiz, great-grandson of Louis Agassiz and president of the Harvard Board of Overseers, wrote to Professor Wheeler, mainly to discuss the various possibilities for his replacement as dean, but he began his letter with the following statement: "I am extremely sorry that you have decided to pull out; for I appreciate fully that you are one of the few distinguished men now left in the University.[11] Then he went on to discuss Wheeler's suggestion of Raymond Pearl as dean. Dr. Pearl, knowledgeable in genetics and behavioral biology and an indefatigable worker, seemed a good choice. The Harvard biologists thought highly of Pearl, as apparently did President Lowell, the trustees, and the Bussey visiting committee, headed by Mr. Henry James, son of William James and namesake of his literary uncle. Pearl was interested in moving to Harvard, and all that remained was for the Board of Overseers to pass officially on the appointment. Only Mencken seemed to object, for in January 1929, he wrote to Wheeler that it was "bad news, indeed, that you are contemplating seducing Pearl," and then he added that he would be "tempted to quit [Baltimore] myself if he moved out."

But Pearl never moved out, because on the first of May a loud cry of protest was issued from the Harvard School of Public Health by a professor of vital statistics named Edwin Bidwell Wilson (not to be confused with the cytologist Edmund Beecher Wilson of Columbia and Woods Hole). Wilson apparently wrote first to Henry James and then to President Lowell, saying that he had just returned from sabbatical leave and heard the news of Pearl's succession to Wheeler. He felt sorry for poor old Bussey because it was going to be Menckenized! He added that he did not think Pearl's fallacies, and particularly his statistical papers on cancer and tuberculosis and similar studies, should be fostered at Harvard. He continued actively to stir up doubts and gradually outright opposition to Pearl's appointment at Harvard. One of those so persuaded was East. In a letter dated March 31, 1929, East had written to Pearl as follows:

Not being in the confidence of the powers, I did not hear the great news until yesterday. No matter! I am writing immediately after. In recent years I had about lost hope the old B.I. would grow into a sizable youngster. It was good for its size, but too much of a bantam. With prospects of you at the helm the hopes revive. If anyone in the world can make it great that man is you. And if the President and Fellows keep their lately aroused interest in biology there ought to be a wonderful future. Naturally the Deanship at Bussey will be only a stepping stone, and the logical thing will be to make you Director of the whole biological institute.

I am sure that you will find Boston and Cambridge congenial. I have found that the place grows on one with the years. There is only one drawback; good liquor is terribly high—too high for me.

If in any way I can be of service before, during and after the change count on me. I shall be ready and willing at least, perhaps in some cases able.

Mary was telling me right after Christmas that she wanted to invite Mrs. Pearl and yourself to stay with us during Ruth's commencement. She had not gotten around to it sooner. But now with this turn of affairs you [had] better consider it. You can use our car so it would be convenient enough. We'll talk it over at W.

Well! Well! It sure will be nice to grow old together at deah ol Hahvad!

But on May 17, 1929, East wrote to George Agassiz about the Pearl appointment, concluding his letter as follows: "There is certainly this to be said; a great deal of opposition has developed in the Medical School, the School of Public Health, and the Department of Economics, where the professors are well acquainted with the type of statistical, sociological, and medical work which Professor Pearl has been doing during the past ten years. Do you not think it would be wise to find out whether this opposition is serious or not? I have been friendly with Dr. Pearl for a number of years; and in trying to view matters as I think he would view them I have wondered whether he would really want to come to Harvard if there is any considerable opposition."[12]

Meanwhile (May 13), Wheeler had written to James:

I very greatly appreciate your kindness in writing me about Pearl. After I had been informed of Wilson's opposition to his appointment I have again looked into the matter and, so far as I can see, Wilson's opposition is largely due to disagreement with Pearl in matters of statistics and fear that Pearl will invade his department in the Medical School. Personally I feel very badly about the matter because Wilson is one of my intimate friends and I feel that he should have discussed the matter with me before writing to you, President Lowell, and others, and creating a most unfortunate situation for Pearl if he is appointed. I have known Pearl for many years very intimately and am sure that many of Wilson's criticisms are quite unfounded. Of all the younger biologists who would be available to expand the Bussey interests and represent them adequately to the public Pearl is certainly the best, precisely because he has sympathy with all lines of biological work. In my long intimacy with him I have never heard him speak disparagingly of any branch of biology. Moreover, he is not a person who enslaves his environment as Wilson maintains, although he very naturally gets a great deal of work out of his technical assistants. In order to justify the expenditure of a considerable amount of money for his institute at Johns Hopkins he has had to do a large amount of publishing. Some of his work I will admit is mediocre, but much of it is very valuable biologically and is in fields which applied biology will cultivate increasingly in the future. No one who is very productive, like Pearl, can fail to make mistakes or publish a certain amount of work that does not come up to the highest standards. One of the greatest advantages in having Pearl with us would be his sympathetic attitude towards the great work which Barbour is carrying on in the Agassiz Museum. I feel that with the aid of these two physically magnificent Gargantuan human individuals we may get biology at Harvard University out of the rather narrow academic rut in which it has been vegetating for so many years. I do not believe that

Pearl's mind would be obfuscated by the glamor of a Harvard position or fail to realize at once that every member of his department was to be helped and sympathized with in his research work as an individual investigator.[13]

But in spite of many excellent recommendations, such as those from David L. Edsall, dean of the Harvard Medical School, who said a man should not be condemned for originality, and Lawrence J. Henderson, professor of biochemistry in both the Medical School and School of Public Health, in spite of the valiant efforts of Dr. Thomas Barbour, director of the Museum of Comparative Zoology, in obtaining favorable reports from many scientific men from many places—in spite of everything Wheeler and others could do, the Board of Overseers did not give Pearl the official appointment. The feelings of Pearl and Wheeler and their friends and sympathizers can only be imagined. Wheeler immediately moved out of Bussey and stored his collection and library in the museum where Barbour had made room for him until such time as his office in the new biology building would be ready. Pearl collapsed and was quite ill through most of the fall of 1929. But on December 28 he answered Wheeler's Christmas greeting, saying how happy he was to get his letter, that the Lamarck manuscript affair sounded most exciting, that he was glad to know that things were pleasant at the museum, and that he should forget the whole Bussey mess. "Professors are always sore in general [Pearl went on] and can be induced to be sore against a specific person with the greatest ease in the world, as you know. So I do not see why we should not forget the whole business, and charge it up to professional behaviorism, in which category it is by no means unique. I have been reading lately a good deal of the contemporary literature at the time of the appearance of *The Origin of Species* and immediately following, and I find, to my great interest, that the things said about Darwin, his character, and his work at that time were *precisely identical* with what Wilson and his gang have been saying about me and my work. So, after all, on a long time base why worry?"

Why indeed! For no dean was ever appointed again to Bussey, and the various faculty and students were gradually moved out of Jamaica Plain as room was found for them in the new biological laboratories or museums of Harvard in Cambridge. And Pearl and Wheeler, now closer at least in spirit than ever, continued to make their many contributions to science.

12 / Student of Insect Sociality

In 1921 Wheeler was invited to present a series of six lectures at the Lowell Institute in Boston. His lectures, entitled "Social Life Among the Insects," were given in February and March 1922 and appeared in several issues of *Scientific Monthly* beginning later that year. In 1923 they appeared as a book of the same title, published by Harcourt, Brace and Company and dedicated to his son, Ralph Emerson Wheeler. Two years later Wheeler held an exchange professorship at the University of Paris. His lectures at that institution were given in French and were later published in Paris under the title *Les Sociétés d'Insectes, leur Origine, leur Evolution,* a publication that was awarded the prix Dollfus by the Societé Entomologique de France. These lectures appeared in English as the book *The Social Insects, their Origin and Evolution,* published by Harcourt, Brace and Company in 1928. This book is by no means a mere updating of the 1923 *Social Life Among the Insects.* Although the subject matter is similar and some of the same illustrations appear in each, the organization and to some extent the content is different. The two books complement one another, and the two together represent Wheeler's major contribution to the field of animal sociology.

The social insects proved a popular subject in the years following World War I. People were satiated with war and alarmed by the spread of social Darwinism, the attempt to apply the doctrine of survival of the fittest to human affairs. They were receptive to Wheeler's statement in his initial Lowell lecture that "an equally pervasive and fundamental innate peculiarity of organisms is their tendency to cooperation . . . Living beings not only struggle and compete with one another for food, mates and safety, but they also work together to insure to one another these same indispensable conditions for development and survival." Wheeler continued: "If asked why it seems advisable to devote six lectures to social life among the insects, I might say that these creatures exhibit many of the most extraordinary manifestations of that general organic co-operativeness which I have just mentioned, and that these manifestations have not only an academic but also a practical interest at the present time. For if there is a world-wide impulse that more than any other is animating and shaping all our individual lives since the World War, it is that towards ever greater solidarity, of general disarmament, of a drawing together not only of men to men but of nations to nations throughout the world, of a recasting and refinement of all our economic, political, social, educational and religious activities for the purposes of greater mutual helpfulness."

Thus Wheeler was once again able to visualize the ants and other social insects as a measure of all things biological and sociological. He was nevertheless quick to point out that because of his many different features, "man can from the social insects learn nothing that might be applied with advantage to the solution of his intricate social problems." However, "it must be remembered that he will never cease to be an animal and that his activities are broadly and irrevocably grounded in the nutritional, reproductive, relational, and appetitive functions common to all other living organisms."

Although Wheeler returned to man in the final chapter of *The Social Insects,* both of these volumes are largely devoted to descriptions of various groups of social insects, with a special effort to understand the cohesive forces in their societies and the origin of these forces. Wheeler reviewed a very large amount of published information, much of it poorly known to the English-speaking world. He had, of course, discussed many of these topics in his own earlier publications, not only in *Ants* (1910), but in several important papers appearing between 1910 and 1923. Some of these papers will be discussed before we consider the 1923 and 1928 books further. Wheeler's interest in social insects other than ants was revealed well before 1910, for example, in a paper on the phylogeny of termites published in *Biological Bulletin* in 1904, which was based on a lecture given at Woods Hole the previous summer. It will also be recalled that the Peckhams had achieved renown for their book *Wasps, Social and Solitary* and that Carl Hartman had published on the solitary wasps of the vicinity of Austin, Texas, as early as 1905. Wheeler made much use of the findings of these close acquaintances as well as of those of various students, including C. T. Brues (to whom *The Social Insects* is dedicated), W. M. Mann, F. X. Williams, and others.

One of Wheeler's most important papers bearing on insect sociality appeared in 1911 in the *Journal of Morphology*. This, too, had been presented as a lecture at Woods Hole the preceding summer. It was titled "The Ant-Colony as an Organism" and developed the thesis that the colonies of ants and other social insects are "true organisms and not merely conceptual constructions or analogies," since each colony was coordinated and individualized for the business of nutrition, reproduction, and protection. He reviewed the ways in which colonies maintain their identity and defend themselves against intrusion; he pointed out that each has a "germ plasm," consisting of the reproductive castes, and a "somatic tissue," consisting of the sterile workers; furthermore, each colony has its life cycle and its own parasites and diseases. "If it be granted that the ant-colony and those of the other social insects are organisms," Wheeler remarked, "we are still confronted with the formidable question as to what regulates the anticipatory cooperation . . . and determines its unitary and individualized course." Maeterlinck's "spirit of the hive" he found charming and poetical, but mystical and nonscientific. As for Hans Driesch's "entelechy," "his angel-child . . . comes, to be sure, of most distinguished antecedents, having been mothered by the Platonic idea, fathered by the Kantian Ding-an-sich, suckled at the breast of the scholastic *forma substantialis* . . . but nevertheless, I believe that we ought not to let it play about in our laboratories, not because it would occupy any space or interfere with our apparatus, but because it might distract us from the serious work in hand. I am quite willing to see it spanked and sent back to the metaphysical house-hold."

Wheeler was left with no theory as to the factors controlling "the correlation and cooperation of parts," "an old and very knotty problem." "Every organism manifests a strong predilection for seeking out other organisms and either assimilating them or cooperating with them to form a more comprehensive and efficient individual." He suggested an ontogenetic approach, that is, an endeavor to understand the stages in the "process of consociation," not only in colony founding but also in parasitism, symbiosis, and sexual and parthenogenetic reproduction. In all of these phenomena, he felt "our attention is arrested not so much by the struggle for existence, which used to be painted in lurid colors, as by the ability of the organism to temporize and compromise with other organisms, to inhibit certain activities . . . in the interests of the unit itself and of other organisms; in a word, to secure survival through a kind of egoistic altruism."

Seven years later Wheeler felt that he had at least partially solved the knotty problem and found the basis for the process of consociation. His ideas were presented in an article in the *Proceedings of the American Philosophical Society* entitled "A Study of Some Ant Larvae, with a Consideration of the Origin and Meaning of the Social Habit among Insects." His major insight was simply that care of the brood by the adults is not a one-way affair, for the larvae produce substances, either from salivary glands or from "exudatoria" on various parts of the body, which are eagerly fed upon by the workers. Wheeler described the exudatoria of several different ant larvae and supposed that these contained nutrients which diffused through the cuticle to the outside of the body. Furthermore, Wasmann and others had shown that many of the guests of ants and termites are physogastric, that is, their abdomens are greatly distended with fluids which evidently escape through pores in the integument and are fed upon by their hosts. The French worker Roubaud had shown that the larvae of several kinds of social wasps produce an abundance of salivary fluid which is imbibed by the adults soon after they have fed the larvae; for this mutual exchange of nutriment between larvae and adults he had coined the word *oecotrophobiosis*. Wheeler agreed with Roubaud as to the importance of this phenomenon, but chose to call it *trophallaxis,* formed from Greek words meaning the exchange of nourishment. Trophallaxis became a major theme in subsequent discussions of insect sociality, not only by Wheeler but by many others. Wheeler used the word to denote not only interchanges between adults and larvae but also between the adults of a colony and between social insects and their various guests; he considered the trophic relationships of social insects outside the colony an extension of trophallaxis (for example, the relation of ants to aphids). In the final pages of his paper he even suggested that the concept of mutual stimulation might apply to humans, citing LeDantec's remark that maternal love may originate in the female's desire to rid herself of her milk.

Needless to say, the elevation of trophallaxis to a major role in insect sociality did not please Father Wasmann, whose theories of "amical selection" and "symphilic instincts" had been proposed some years earlier. In 1920 Wasmann published a 176-page paper attacking Wheeler's 50-page paper; but the latter retaliated by devoting a large section of *The Social Insects* to a further discussion of trophallaxis, spicing it with amusing quips about the "reverend fathers" and their "pet lucubrations." For example, Wasmann had pointed out that larvae or ant guests could scarcely produce exudates in proportion to the food they received from their hosts, or they would fail to survive. Wheeler replied that, as he had already made clear, many exudates, "though produced in very small amounts, may nevertheless elicit intense reactions! One would suppose," he went on, "that the food served in a Jesuit refectory must be even more tasteless than that served in a fifth-rate New York restaurant!"

We shall take another look at Wheeler's concept of the colony as an organism and his appraisal of trophallaxis as the lifeblood of that organism later in this chapter. For the moment we wish to call attention to another major paper which he published in the *Proceedings of the American Philosophical Society* in 1919: "The Parasitic Aculeata: A Study of Evolution." In this paper he reviewed the available information on those ants, wasps, and bees (both solitary and social) that have become parasitic on other members of their groups. One of his major conclusions was that the majority of parasitic forms appear to have evolved from their hosts; for example, the parasitic leafcutter bee genus *Coelioxys* from *Megachile,* the parasitic bumblebee genus *Psithyrus* from *Bombus,* the true bumblebees, and so forth. (Actually Carlo Emery had drawn the same conclusion with respect to parasitic ants some years earlier.) Parasitism has arisen many times independently among the Aculeata, its origin being attributable "to the urgency of oviposition and temporary or local dearth of the supply of provisions for the offspring." The specialized forms of parasitism found among certain ants were seen as the extremes of evolutionary trends beginning among the solitary wasps and bees. It is true that parasitic solitary wasps and bees destroy the egg or larva of the host, whereas social parasites foster the host brood in order to insure the rearing of their own young; this change is attributable to "the formation of trophallactic relations between the parasite and the host brood." So much has been learned in this field since 1919 that parts of Wheeler's paper are now obsolete, but it should be noted that no one since his day has had the courage to review this subject in a comprehensive manner. Half a century later, this paper is still a basic reference regarding parasitism among these insects.

Wheeler spent the summer of 1920 at the Tropical Laboratory of the New York Zoological Society at Kartabo, British Guiana, a locality made famous by William Beebe's book *Edge of the Jungle* (1920). Here he made an intensive study of the ants associated with various "ant plants," while his colleague Irving W. Bailey studied the plants themselves. This was a subject that interested Wheeler for many years, his first original contributions having been made in 1913 following a trip to Central America. It was Wheeler's contention that certain ants merely exploited cavities, extra-floral nectaries, and other features of the so-called myrmecophytes (ant-plants), that the bull's-horn and other acacias, for example, "have no more need of their ants than dogs have of their fleas." Father Wasmann took exception to this view, remarking in a paper in 1915 that the plants surely profited from the association and that this was an instance of true symbiosis. Never one to let Wasmann have the last word, Wheeler pointed out that Wasmann had never seen these acacias and their ants alive. But in this case Wasmann was evidently right. In a series of experiments, the results of which were published in 1967, Daniel H. Janzen, now of the University of Chicago, showed that artificial removal of *Pseudomyrmex* ants from bull's-horn acacias resulted in extensive damage by phytophagous insects, damage that was minimal when the ants were present. It now seems evident that at least in this case, the plants are as dependent upon the ants as the ants upon the plants. In his final paper on this subject (which included the results of his 1920 expedition and several later trips), Wheeler persisted in his exploitationist theory, even though the majority of observers had concluded that the plants indeed profited from the association. In other respects this paper, "Studies of Neotropical Ant-Plants and their Ants," was a valuable review of the subject, 262 pages in length with fifty-seven plates. It remained unpublished at his death and was edited and prepared for publication in 1942 by Joseph Bequaert.

A more immediate result of Wheeler's 1920 trip to Kartabo was a lengthy paper in *Zoologica* on the biology of certain previously unknown social beetles found in association with the tree *Tachygalia paniculata.* These beetles live in hollow petioles which they share with certain mealybugs; the bugs suck the sap of the plant and produce droplets of honeydew, which are fed upon by the beetles. The beetles and their larvae stroke the mealybugs to obtain this fluid, several of them sometimes standing around a bug "like so many pigs around a trough." The beetles lay eggs in the petioles and the young remain with their parents, so that the colony soon comes to consist of beetles of all stages. When the petiole becomes too crowded, pairs of beetles leave and start colonies in other petioles.

Wheeler's study of the *Tachygalia* beetles made up part of the first chapter of *Social Life Among the Insects,* along with accounts of other social beetles: dung beetles, ambrosia beetles, the horned Passalus, and others. Wheeler recognized six stocks of beetles as social, and among all the insects he concluded that "social habits have arisen no less than 24 different times in as many different groups of solitary insects." It should be explained that he here used the word *social* rather loosely. In fact, Wheeler explained that he considered fourteen of the groups "incipiently social or subsocial" and only ten as "definitely social." Since his purpose was to point out the multiple origin of cooperative associations among insects, it was important to treat associations of parents and offspring among such groups as beetles and earwigs, even though no members of these groups ever became "definitely social." Quite clearly the ants, termites, and social wasps and bees must have passed through various stages of incipient sociality.

After discussing the social beetles and general conditions of sociality in the first of his Lowell lectures (and the first chapter of his 1923 book), Wheeler went on to devote one lecture each to the wasps, bees, ants, and termites, reserving one lecture for parasitic ants and ant guests. The lectures on ants and their guests were largely a condensed updating of the review he had presented in 1910, but with special emphasis on trophallaxis, a matter that had not concerned him until 1918. The lectures on wasps and on bees provided an excellent review of the literature of that time, and through them the work of such persons as Roubaud, Ducke, Brauns, and von Buttel-Reepen was brought to the attention of American biologists. Both the wasps and bees are of special interest in that the majority are solitary, but presociality and true sociality have evolved several times in both groups, and a number of existing genera exhibit behavior which must have characterized the fully social forms at one stage in their evolution. The social wasps,

Wheeler pointed out, exhibited trophallaxis more clearly than any other group, although the bees provide no good examples of trophallaxis at all. Wheeler supposed that the ancestors of bees may have gone through a stage when trophallaxis was prevalent, but when they switched from insect food to nectar and pollen, which are available in great abundance and which are often stored in the nests in quantities, "the exploiting of larval secretions [became] unnecessary." Wheeler's discussion of bees now seems rather obsolete, since research on these insects has flourished in the past few decades; in 1923 Karl von Frisch had barely begun his work on the senses and language of honeybees, and Charles Michener, who has recently contributed so much to this field, was still a youth. On the other hand, a review of the evolution of sociality in wasps by Howard Evans as late as 1958 still made much use of Wheeler's lecture of thirty-five years earlier. Fortunately the study of wasps is now proceeding more rapidly than it was a few years ago.

Wheeler's final Lowell lecture considered the termites, or "white ants," and also treated two other remotely related groups of subsocial insects, the earwigs and embiids. These are insects without complete metamorphosis, none of them more than very remotely related to the Hymenoptera. Yet the termites live in highly developed societies and show many resemblances to ants: they have well defined worker, soldier, and reproductive castes; their nests often resemble those of ants; they have various beetles, flies, and other insects living in their nests which are modified in much the same ways as the guests of ants; and so forth. Yet there are many striking differences, not the least of which is the fact that the termites are the only fully social insects which have succeeded in "domesticating the male sex," as Wheeler often pointed out, for in these insects (in contrast to ants, bees, and wasps) the worker caste consists of sterile males as well as females, and a male of the reproductive caste (the "king") remains as a consort of the queen. When such utterly different insects as ants and termites have evolved so many similarities, Wheeler remarked, it should not surprise anthropologists that human societies that have developed independently have sometimes "hit upon the same inventions."

Wheeler's lectures at the University of Paris were prepared for a more learned audience than his Lowell lectures, and consequently *The Social Insects* of 1928 is a good deal more technical than its forerunner. His chapter "The Evolution of Wasps" discusses the ancestry of the true wasps among the parasitoids and presents much interesting information on various poorly known and primitive families before proceeding to a brief survey of the social wasps. The chapter on the evolution of bees relies heavily upon the work done a few years earlier by Hans Brauns on South African members of the genus *Allodape.* In this genus one finds a series of stages leading from strictly solitary forms to others in which there is an association of the mother with her offspring. Chapters on the evolution of ants and of termites explore the fossil record and probable ancestral stock of these very different groups of insects. In the first of two chapters on polymorphism, Wheeler discusses the origin of marked caste polymorphism in ants and the progressive loss of certain castes, finally all but the reproductives, in the social parasites. In the second chapter he reviews various theories of caste determination and concludes that while doubtless in the wasps and bees "the feebly differentiated queen and worker castes develop from eggs of the same kind"—that is, are "trophogenic," or the result of differential feeding—"it is more probable that in the two more ancient groups, the ants and termites, the pronounced fertile and sterile castes may be blastogenic [that is, predetermined in the egg]." Wheeler felt that the sterile castes reproduce more frequently than had been appreciated and that they transmit their special features through their offspring. The chapter ends with a typical blast of skepticism regarding the efficacy of genetic and evolutionary theories:

No doubt, we have hitherto laid too much stress on the static, morphological aspects of polymorphism and too little on its dynamic, physiological and behavioristic aspects . . . It has been suggested that the production of workers and soldiers in the colony is a kind of experimental teratogeny carried on by the worker nurses, but it is certainly strange that the monsters produced, e.g., the janitor soldiers of Colobopsis among the ants and the nasuti among termites, should be structurally and functionally so exquisitely adapted to their particular professions. And it strains our credulity to be told that such forms arise either from peculiar genes popping out of nowhere into the germ plasm or develop gradually under the guidance of natural selection from forms which, so far as we can see, must have an equal or even greater survival value. When we encounter such *impasses* as the foregoing, instead of embracing the Aristotelian Entelecheia . . . or joining the apostles of the survival of the fittest and forever croaking "natural selection!", it is surely more commendable to sit down in the laboratory or in the field and say nothing but *"ignoramus"* till we have made a much more exhaustive behaviouristic and physiological investigation of the phenomena.

The heart of *The Social Insects* is the chapter entitled "The Social Medium and Trophallaxis," a more thorough treatment of a subject he had first discussed in 1918 and an attempt to counter some of the criticisms raised against his embracing of trophallaxis as the essence of "the social medium." Wheeler believed that he had overemphasized the exchange of food among members of a colony and said too little about the exchange of chemical and other messages. Von Frisch had recently demonstrated that nectar-laden honeybees "dance" on the combs and call attention to the odors of the flowers they have visited adhering to their bodies, and evidence was accumulating that various insects produce glandular secretions which have specific effects on the behavior of other members of the species. Wheeler felt that

extensive as is the scope of trophallaxis, or exchange of food in the various social insects, the principle is sufficiently elastic to cover an even greater number of phenomena, if we include besides the substances that are socially excitatory through the taste receptors also those that affect the other chemical sense, namely olfaction . . . It would seem . . . that the question as to whether an ant or bee smells or tastes its food, a larva, pupa or another ant or bee with its antennae, is largely academic . . . And since, moreover, the food stimuli are necessarily chemical, I can see no reason to change the term "trophallaxis" because it happens that much of the behaviour of social insects is what we have been calling "olfactory." Nor is the fact that some of this behavior depends on stimuli other than food a valid argument against trophallaxis, since I have never asserted that it includes all the social activities. I believe, nevertheless, that it constitutes the most essential characteristic of the social medium. If we compare the distribution of food in the colony regarded as a superorganism with the circulating blood current ("internal medium") in the individual insect or Vertebrate, trophallaxis, as the reciprocal exchange of food between the individuals of the colony, may be compared with the chemical exchanges between the various cells themselves . . . This process necessarily involves not only the transportation and distribution of nutritive materials but also the transmission of stimuli, or excitations among the living elements.

As to the colony as a superorganism: this was seen as not a mere summation of individuals but "a different and at present inexplicable 'emergent level,' " to use a term proposed by C. Lloyd Morgan and then very much in vogue. Wheeler himself had discussed "Emergent Evolution and the Social" before the Sixth International Congress of Philosophy in Cambridge in 1926. His talk was published in *Science* and later expanded into a small book published in London in 1927 and by W. W. Norton in New York in 1928. Wheeler's name became widely associated with the theory of emergent evolution as a result of these publications, and since his thoughts in this field were very much the result of his concern with the social medium, it seems appropriate to say more about emergent evolution at this point. Several of Wheeler's close friends and associates had also become concerned with this subject and had published books or articles relating to it: H. S. Jennings, E. B. Holt, C. K. Ogden, and G. H. Parker, among others. Emergent evolution, in Wheeler's words, is the concept "that the unique qualitative character of organic wholes is due to the peculiar non-additive relations or interactions among their parts. In other words, the whole is not merely a sum, or resultant, but also an emergent novelty, or creative synthesis."

While in a sense every organism is unique, novel, and emergent, it was fashionable to distinguish certain major levels of emergence: the cell, the organism, the mind, and deity, for example. Wheeler felt that the concept of "mind" as an emergent was not very illuminating because of the complexity of the subject. As for "deity:" "to the observer who contemplates the profuse and unabated emergence of idiots, morons, lunatics, criminals and parasites in our midst, [the] prospect of the emergence of deity is about as imminent as the Greek Kalends."

On the other hand, it seemed obvious to Wheeler that animal societies were true emergents and far more amenable to study than others that had been proposed. In societies, "whole organisms function as the interacting determining parts . . . Experiments in subdividing, compounding, castrating and grafting, and in introducing foreign elements with a view to observing their effects on animal and plant societies as emergent wholes, can be carried far beyond the limits of such experiments on the single living organism." After reviewing briefly several types of animal societies, Wheeler remarked that in every case "we are, I believe, bound to assume that the organization *is entirely the work of the components themselves* and that it is not initiated and directed by extraspatial and extratemporal 'entelechies' (Driesch) . . . or 'élan vital' (Bergson) . . . The tender-minded may still delight in assuming their intervention in the development and maintenance of unicellular and multicellular organisms, whose integration is so exceedingly complicated and opaque that we are probably still centuries removed from any adequate understanding of their functional composition, but on the next level, that of the very loosely organized social, or superorganisms, in which the actual play of the components is open to inspection, it is not easy to tolerate these ghostly presences."

Wheeler's talk on emergent evolution was well received and the resulting book had a considerable circulation. Raymond Pearl requested several copies of the paper and reviewed the book for the *Baltimore Sun,* and David Starr Jordan, then seventy-seven, commented that "there is much in this subject that interests me, [but] the younger fellows will have to settle these problems."[1] H. S. Jennings wrote that it was a "great pleasure to see you back in your old form . . . I had great joy in what you said about entelechies and similar creatures." In replying, Wheeler remarked that Driesch was bound for the University of Wisconsin, which "I hope he will not infect . . . with his crazy notions."[2]

The concept of emergent evolution underwent a period of considerable popularity during the 1920s, but the phrase is now rarely encountered. However, the physiologist C. Ladd Prosser has recently reviewed some aspects of the subject and concluded that "biologists are so occupied with seeking mechanistic explanations of life at all levels that they have partially overlooked certain important applications of the principles of emergent evolution." He points out that natural selection acts on whole protein molecules, for example, and this may account for the multiple forms of some enzymes and for the "dispensable" sections of some proteins. In the final analysis, however, it is the whole organism which is molded by natural selection. Prosser feels that "the theory of emergence of qualitatively different properties in integrated whole organisms is an important philosophical bridge between biology and the humanities."[3]

The idea of levels of emergence has remained popular with certain philosophically inclined writers such as Teilhard de Chardin and Michael Polanyi. Theodosius Dobzhansky, in his book *The*

Biology of Ultimate Concern (1967), speaks of radical innovations such as the origin of life and of man as emergences, but he prefers the term *transcendences*. "Evolution," Dobzhansky remarks, "is not simply an unpacking of what was there in a hidden state from the beginning. It is a source of novelty, of forms of being which did not occur at all in the ancestral states . . . The word 'transcendence' is obviously not used here in the sense of philosophical transcendentalism; to transcend is to go beyond the limits of, or to surpass the ordinary, accustomed, previously utilized or well-trodden possibilities of a system."[4]

What of Wheeler's pet emergence, the colony as a superorganism? For many years Alfred Emerson, of the University of Chicago, has been the leading exponent of this concept, although approaching the subject from a somewhat different point of view than Wheeler. To Emerson, the colony serves as a unit of natural selection and in the process of evolving homeostatic mechanisms has developed adaptations closely analogous to those seen in the behavior and physiology of individual organisms. Many current students of social insects have little use even for this broadened concept of the superorganism. In a recent review of the subject, Edward O. Wilson remarks that "seldom has so ambitious a scientific concept been so quickly and totally discarded." Wilson continues as follows: "The superorganism concept faded not because it was wrong but because it no longer seemed relevant . . . The concept offers no techniques, measurements, or even definitions by which the intricate phenomena in genetics, behavior, and physiology can be unravelled. It is difficult even to cite examples where the conscious use of the idea led to a new discovery in animal sociology . . . [Yet] the idea was inspirational and . . . as originally formulated by Wheeler had just the right amount of fact and fancy to generate a mystique . . . It would be wrong to overlook the significant, albeit semiconscious role this idea has played in the history of the subject."[5]

As Wilson points out, modern students of animal sociality are inclined toward the experimental analysis of individual problems, using techniques wholly unknown to Wheeler, and they are discovering a far greater diversity of integrative mechanisms than could have been imagined in Wheeler's time. For this reason, even the word *trophallaxis* has fallen into disfavor in some quarters. As we have seen, Wheeler in 1928 broadened his earlier definition to include olfactory stimuli, and later other workers broadened it still further to include tactile stimuli. The fact is that so many different phenomena were labeled trophallaxis that the term became meaningless. Also, some have argued that the secretions of the larvae of social wasps were simply a means of ridding the body of excess water and wastes. More recent work, however, has suggested that these secretions may contain carbohydrates and proteinaceous materials required by the adults, and the concept of trophallaxis, in its original sense of exchange of nutriments, may yet be found to be useful at least in the social wasps, where it was first promulgated. A recent paper on the Oriental hornet by Ishay and Ikan (1968) concludes with the statement that "recent observations confirm Wheeler's (1928) opinion as to the importance of trophallaxis in forming a close bond between the various members of insect families, particularly between the larvae and the nursing workers."[6]

The chemical senses are now known to play a far more important role than was apparent in Wheeler's day. Ants and termites, in particular, communicate extensively by specific chemical substances, called pheromones, which are produced by a variety of special glands. Visual, auditory, and tactile cues are also important in some social groups, and generalization from one group to another is often difficult. Such cues, particularly pheromones, play important roles in caste determination which were unsuspected in Wheeler's time. For example, queen honeybees secrete a substance which is licked from their bodies by the workers and passed about from worker to worker; workers receiving this substance do not construct "queen cells." The active substance in this case is ketodecenoic acid. Ants are now known to produce a great variety of pheromones, many of which still await behavioral or chemical analysis. Not all of these are passed around from mouth-to-mouth by any means, but at least we can say that trophallaxis in the sense that Wheeler used it in 1928 was a good deal more meaningful than Wasmann's amical instincts. Wheeler was genuinely seeking a mechanistic explanation in an age when the spirit of the hive, élan vital, and other "ghostly presences" were popular.

Wheeler's final views on the problem of caste determination were summarized in his book *Anomalies Among Ants,* the manuscript of which he delivered to the Harvard University Press a few days before his death in 1937. This book was based largely upon a series of anomalous individuals from a colony of fungus-growing ants collected by Neal A. Weber in Trinidad. It was originally planned as a joint paper with Dr. Weber, but the latter withdrew his name as coauthor as a gesture of respect for his departed mentor. This colony contained 164 anomalous individuals out of a total population of 8,174. Some of these were "of unusual interest both because they are quite unlike any previously observed among ants or indeed among any other social insects and because they enable us to decide between two theories of caste determination which have baffled and divided the students of ants for more than half a century."

Most of the aberrant individuals were either mutant workers or gynergates (mosaics between workers and queens), but there were also ten gynandromorphs (male-female mosaics) and a few mutant females. Since gynandromorphs are regarded as an indication of genetic differences between male and female, Wheeler took the presence of so many worker-queen mosaics to mean that these castes are also genetically determined. Thus the weight of evidence, he felt, now supported the blastogenic theory of castes as opposed to the trophogenic.

However, the anomalies were also capable of other explanations, as Wheeler's former student P. W. Whiting was quick to point out in a review of the book in *The Journal of Heredity*. Nearly all the unusual features of these insects were confined to the head, suggesting that these were not true mosaics but intergrades resulting from an alteration in physiology occurring at a time in development when the major part of the body had passed the critical threshold for queen or worker determination. That is, these gynergates may have begun life as workers but as a result of nutritional changes at an inappropriate time underwent a partial modification of the head toward that of the queen. It is well known that the head of ants is the most labile part of the body with respect to the allometric growth associated with the female castes.

Clearly one could not be sure what had happened in this particular ant colony, but just as clearly Whiting was correct in concluding that "the moot question of caste-determination seems not yet to have been solved . . . Differences of size and correlated characters between [the sterile castes] is readily explained by nutrition. The basis for the wide differences between workers, soldiers and sexual females may not be altogether incomprehensible to the developmental geneticist aware of the far-reaching influence of slight stimuli if applied at critical temperature-effective or nutritive-effective periods. Thus the distended abdomen of the sexual female or the enormous head and jaws of the soldier may be but trophogenically different from the modest proportions of the worker. This trophogenic influence may have taken place before the hatching of the egg."[7]

The problems of caste determination have by no means been finally solved, even now, but the weight of evidence is against genetic determination (except, perhaps, in one genus of stingless bees). We now know that to say that caste is predetermined in the egg is not to say that it is necessarily genetic. As Whiting implied, there may be variation in the amount or quality of yolk deposited in the egg, and this in turn may be caused by certain influences upon the queen. Contemporary workers often speak of "egg bias," meaning that queens are at times capable of producing eggs in which the trophic substances within the egg are such that the resulting larvae are more likely to develop in the direction of one caste or another. But the food and pheromones received by the larvae themselves remain of major importance even here, as do temperature and other external influences. Nevin Weaver, of the University of Massachusetts in Boston, fittingly concludes a recent review of this subject with a quotation from Shakespeare's *Hamlet:* "There are more things in heaven and earth, Horatio, than are dreamt of in your philosophy."[8]

During Wheeler's later years, he taught a course in animal sociology jointly with Professor Pitirim A. Sorokin. Sorokin, who died in 1968 in his eightieth year, was a well-known scholar and influential teacher in Harvard's sociology department. He was a Russian exile, and a lively and controversial individual—perhaps a bit reminiscent of Loeb to Wheeler. The course they gave was described by the *Boston Post* as "several degrees more radical than the internationally famous controversy upon which the Scopes trial hinged," concerned as it was with the evolution of societies "from fish and insect groups to modern human societies."[9] In 1933 Wheeler lectured on "Animal Societies" to a meeting of the American Society of Naturalists, publishing his paper in *Scientific Monthly* a few months later. This lecture also received wide publicity, and the *Brooklyn Eagle* devoted the first page of its Sunday *Magazine,* January 7, 1934, to a large picture of Wheeler and commentary on his talk. "Women Resent Charge They Would Stagnate the World," the headline read. "If this were a women's world," the article began, "according to Prof. William Morton Wheeler of Harvard, peace and order would reign, but society would remain stationary and make no advances." Wheeler was quoted as follows:

> Throughout the ages, the aggressiveness, curiosity, unstable intelligence, contentiousness and other anti-social tendencies which the male inherited from his anthropoid ancestors have kept society in a constant turmoil.
>
> But the restlessly questing intellect, driven by the dominance impulses of the mammalian male, furnishes the necessary stimulus to progress in human societies. Female societies are harmonious, but stationary and incapable of further development.

These comments, according to the *Brooklyn Eagle,* were met by "Bronx cheers of a withering, ladylike sort. Said Fannie Hurst, the militant: 'Such arbitrary statements are ridiculous. They are the epitome of the very complacency which the professor attributes to the opposite sex.' " And Miss Lena Phillips, lawyer and president of the National Council of Women, was quoted as saying: "Why, the first curious person in the world was Mother Eve. Adam was contented enough."

The *Eagle* managed to fill a whole page with similar remarks, undoubtedly to the amusement of many of its readers, including, one supposes, Wheeler and his colleagues. The excerpt from his talk (which differs slightly from the published version) was, of course, taken wholly out of context. The social insects have "solved the problem of the male" by completely socializing him (termites) or reducing him to a "stored convolute of sperm" (Hymenoptera), and there can be little question of the greater stability of insect societies as compared to human societies. Wheeler's point was that in human societies the female sex also tends to promote stability, along with a majority of the male sex. A small minority of "less socialized" males provide most of the social turmoil, so that "human history is little more than an interminable record by sober and impressionable males of the abominable behavior of other males." Another small minority, of course, provide us with most of our new ideas, great works of art, and "great cultural illusions." Human social ills, Wheeler concluded, might best be solved by education, by research into the biology and psychology of the sexes, and perhaps ultimately by "the youthful sciences of endocrinology, genetics, eugenics, penology, and psychiatry." Wheeler's innumerable comparisons between humans and insects were forever in danger of being taken for more than intended. As he wrote to John H. Vincent, an amateur biologist of New York City, in 1932: "while I have in my various books pointed out certain analogies between human and insect societies, I do not believe that they can be taken very seriously because vertebrate and insect societies are built on quite different plans. Howver, I believe that social life always implies constraint of the individual."[10]

To what extent one can draw conclusions about human behavior from a study of animal behavior is still very much a controversial subject. It is to Wheeler's credit that he at least stimulated thought along those lines—and that he provided the basis for a much more complete understanding of insect societies than had previously been available. That some of his favorite concepts—trophallaxis, the superorganism, and emergence—no longer seem so important as they once did may be in part a consequence of the pervasiveness of his spirit of inquiry. His books and articles have for many years been the starting point for investigations of social insects, and few readers fail to become infected with his skeptical eye and his readiness to plead "ignoramus" when the situation requires it. Animal sociality now seems much more complex and less amenable to generalization than it did in Wheeler's time, and contemporary students are committed to a reductionist approach in the hope of providing the material for a fuller synthesis at some future date. The study of social insects is a very active field at the present time, and much of the progress is attributable to the sound descriptive groundwork provided by Wheeler and to the stimulating argument of his publications.

13 / Distant Places and Home Port

After more than thirty years devoted to the study of ants I find myself wondering why so very few of our nature lovers have engaged in this pursuit. I find that observing and collecting ants in many lands, quite apart from the benefit to my health, have yielded a keen delight which has remained with me and seems to have colored my recollections, so that I have acquired the habit, when regrets and unpleasant memories assail me, of effacing them with the memories of excursions in mountains, forests, and deserts peopled by colonies of thrillingly interesting ants. Certainly the pursuit of any branch of natural history may be recommended as an avocation to our youth, to convalescents, to our tired business men, or in fact to any one who craves a hobby, a surcease from the nerve-racking routine of our city life, or a valid excuse for remaining as many hours as possible in the open air of the woods and fields.

Although this was not published until 1932, William Morton Wheeler, field biologist *extraordinaire,* had been saying this in essence since his school days in Milwaukee. Although he enjoyed writing and lecturing, and spent many long hours at his desk in his taxonomic studies, Wheeler loved his days in the field, and it was there that his students found him most relaxed and companionable. In his Texas days, he had studied the surrounding countryside because the ants were all so new to him, but with his move to New York, Wheeler began to widen his field excursions, as we have shown. After he came to Harvard, he continued to travel to ever more distant places, and he encouraged his students to do likewise—as William Mann explained so well in his *Ant Hill Odyssey*.

Wheeler's trips to Europe in the summers of 1907 and 1909 were not so much to collect ants in the field (that had already been done fairly well), but to become acquainted with the European myrmecologists and their collections. This was to give him a better basic knowledge of what had been done and perhaps what still needed to be done, particularly to advance knowledge in and of the western hemisphere. Probably he also wanted to check on some last minute details before his *Ants* was finally published in 1910.

After coming to Harvard, Wheeler's first real collecting trip was back to familiar countryside, namely the southwestern United States. From November into January (1910–11), he and Mrs. Wheeler collected and observed in Arizona and southern California. No correspondence remains from these years, so we do not know exactly where they went, who went with them, and precisely what they found. But we know from the resulting papers that in November they found some interesting fungus-growing ants in Miller and Hunter canyons in the Huachuca Mountains. In fact, one of the largest nests "was situated in front of Mr. Joseph Palmerlee's ranch in Miller Canyon at an altitude of 5,500 feet. It was in such hard, stony soil that I was unable to reach its lower-most galleries even when Mr. Palmerlee came to my assistance with a large pick and a pair of powerful arms." Later in November, Wheeler found another species of this genus of ants in a dry arroyo near the Carnegie Desert Botanical Laboratory near Tucson, Arizona.

Another interesting group of ants was found in the Huachucas, but not in tremendous nests in the ground. Instead the ants lived in the hollow stems of mistletoe growing on live oaks in one of the canyons and were busily attending red scale insects. The remarkable interrelationships there did not go unobserved for Wheeler described "the existence of a peculiar coenobiotic association, in which at least five or six different organisms regularly cooperate: a live oak, a mistletoe, a weevil larva, one or two scale insects and an ant. The mistletoe is a parasite on the oak, the weevil and the scales are parasites on the mistletoe and the ant is, in a sense, a parasite on the beetle-larva and the scales, since it owes its dwelling to the former and derives its food-supply from the latter." And then Wheeler added: "Some resident entomologist in Arizona will probably find that the exhaustive study of the coenobiotic association here briefly outlined has not only a theoretical but also a practical interest, for [the mistletoe ant] is to be regarded as a useful forest insect, since it cultivates scales that are injurious to a serious plant parasite of the live oaks and other trees."

In California, Wheeler stopped at Pomona College in Claremont, California, to collect and to visit with the zoologists there, and especially with Professor C. F. Baker (1872–1927) who was himself an outstanding collector and quite knowledgeable of both insects and fungi. Wheeler caught Baker just in time, for Baker was about to leave for the Philippines, where he would become dean of the College of Agriculture at Los Baños.

Wheeler also went to Pasadena to collect and to visit with Mr. H. C. Fall (1862–1939). Fall was a native New Englander and a graduate of Dartmouth College, but he had moved to California because of poor health. There he had taught physics and chemistry in a high school in Pasadena to support himself, and at the same time had become one of the country's best coleopterists. Wheeler enjoyed Fall's company as well as a chance to have him show him his favorite collecting places. In fact, unknown to Fall, Wheeler tried in 1909, through his old friend William Patten who was teaching at Dartmouth, to have that college give Fall a doctorate in recognition of his outstanding contributions to entomology. This honor, however, was not conferred until 1929.

In the summer of 1915 Wheeler was again in California to attend the scientific meetings which were held in August in San Francisco during the Panama-Pacific International Exposition. According to Wheeler in a letter to his former student A. L. Melander, he would rather have stayed home and worked, but Mrs. Wheeler felt he should take the family to the Exposition because the children were old enough to appreciate it, and they had worked hard during the year.[1] In fact, the children did enjoy the trip immensely, and Adaline still remembers her first view of the Rockies as well as many of the foreign exhibits in San Francisco. They "took a house" on a hill near the Presidio, where they could walk to the Exposition. Professor Wheeler also attended an "entomology camp" for about ten days at Fallen Leaf Lake near Lake Tahoe in the Sierras. His report of his findings there was published about a year later, along with a description of how he had almost the whole camp of twenty or so out watching and taking notes on slave-raids of the local amazon ants. En route home he also had a chance to collect in the Canadian Rockies in Alberta and British Columbia.

In March of 1917, Wheeler went to Ithaca, New York, at the invitation of Professor John Henry Comstock, to give a lecture to the Cornell chapter of Sigma Xi. While he was there he heard about a collecting expedition that a number of Cornell biologists were planning for the following summer across the United States. They expected to camp each night, and to collect plants and animals, both vertebrates and invertebrates, all along the way. Would Wheeler like to go, too? He certainly would, but because of his duties at Harvard as dean he could not get away until the end of June. Could he come then? Of course, and they would be delighted. The expedition finally had a total of thirteen people after Dr. Wheeler and his son, Ralph, joined it, and an abundance of hymenopterists, for also along were Joseph Bequaert from the American Museum in New York and J. C. Bradley from Cornell. The party had three cars, named appropriately "Ezra Cornell," "John Harvard," and "Simon Henry" (after a zoology professor at Cornell, Simon Henry Gage) plus a trailer to carry all their equipment. Some of the highlights of this remarkable trip were written up by Professor Bradley in *Scientific Monthly* and published in 1919.

Just before the Wheelers met the group in El Paso on July 9, the Expedition had spent some very profitable days in the vicinity of New Braunfels, thanks to some prior arrangements made by Wheeler with his old friends, Lisa Grob Dittlinger and her husband, Hippolyt. On May 28, 1917, Wheeler had written to Mr. Dittlinger saying: "Please remember me to Elise and Amalie and tell them that I expected to see them this summer, but I find that I cannot get away until the end of June." Then he explained about the Cornell Expedition and asked Dittlinger if he would "show them where I used to collect along the banks of the Comal River."[2] Bradley, in his report, said: "Our visit to New Braunfels was made both more pleasant and more profitable by the kind hospitality of some old friends of Dr. Wheeler's, the Dittlingers, who showed us many courtesies, piloting us around to the most favorable collecting grounds." And the Comal River was indeed appreciated: "About 30 miles north of San Antonio at the old German community of New Braunfels, what is perhaps the finest of these rivers comes pouring out of the cliffs in a series of magnificent springs. It is the

Comal River and with all haste it empties its beautiful waters, after a very short course, into the Guadalupe. Calladia grow wild along the bank, and quite a jungle of semi-tropical plants, while its course is lined beneath with the most beautiful profusion of aquatic plants that I have ever seen. At these Comal springs we took our first real rest from the road and tarried several days."[3]

The drive across New Mexico and Arizona was slowed down because of excess rain and mired roads, but as Bradley said, "when we could not travel we could collect." And collecting was exceptional! Wheeler, writing to Bradley later, said that it was "wonderful tonic for both of us, and Ralph never tires of talking about his experiences."[4] He also told Bradley that he had missed perhaps the best part of the trip in late August when they "motored" to Sequoia National Forest, took the eleven mile trail to Alta Meadow (9,000 feet), and camped and collected for several days in that beautiful, alpine valley. Dr. Bequaert, who not long before had come to the United States from the Belgian Congo, wrote to Wheeler saying: "That trip through the continent was a wonderful experience for me. I learned a lot about America and Americans."[5] He apparently liked what he saw, for he stayed in this country, married an American girl, and moved to Harvard in 1923 to teach tropical entomology in the medical school. He was also an associate of the Museum of Comparative Zoology, and in 1945 he became curator of insects there. Though now retired, Dr. Bequaert is one of the most accomplished biologists of our day, being considered an expert by fellow workers in three different fields: Diptera (flies), Hymenoptera (specifically, wasps) and mollusks (land snails).

After 1917 Wheeler did not make any more major collecting trips within the continental boundaries of United States and Canada. Those few times he did collect in the west were usually done on brief vacations from New England winters or while he was delivering lectures. And, also in 1917, Wheeler published his last major paper on North American ants, "The Mountain Ants of Western North America." Now Wheeler was ready to enlarge his understanding of geographical distribution and other aspects of myrmecology, and he felt the lure of the tropics. Most of Wheeler's trips were to the American tropics, but he did make two extensive trips to Australia and one to the Mediterranean area. His first one made after coming to Harvard was in the winter of 1911–12 to Panama, Guatemala, and Costa Rica.

In early 1913 he spent several weeks in Cuba, but here he found the ant fauna poor—though he and Dr. Thomas Barbour did apparently collect special academic honors from the University of Havana. Wheeler did not get back to tropical America again until the summer of 1920 when he joined an old friend, William Beebe (1877–1962) of the New York Zoological Society, at his biological station, Kartabo, in the jungles of British Guiana. By 1920, Beebe and Wheeler had been friends for some years—no doubt dating back to Wheeler's years at the American Museum. Beebe had published several popular accounts of the maneuvers of army ants, and Wheeler had identified Beebe's specimens for him. Wheeler's appetite was whetted, and he had to go and see for himself.

To digress briefly: the attraction between the two men, Wheeler and Beebe, is an interesting one. Beebe, although he started out as an ornithologist, is considered today not so much as a scientist but as a man who, because of his unusual ability to write interestingly of his observations in the field, brought to the public the world of the outdoors, particularly the exotic world of the tropical jungles and ocean depths. His books, such as *Jungle Days, Edge of the Jungle, The Book of Bays, Galapagos, World's End* (Wheeler was with Beebe in the Galapagos briefly in April 1923), and many others, were widely read in their day. Beebe's abilities were appreciated by Wheeler, for Wheeler was, to some extent, trying to do the same thing—to observe the rich life of the tropics and to record it not only for scientists, but for nonscientists as well. In his anthology of the best natural history writings, *The Book of Naturalists,* published in 1964, Beebe had much to say about Wheeler and classified him as a scientist of first rank who could also write superbly.

The summer of 1920 in Kartabo was a rich one for Wheeler, for Beebe put the resources of the station at his disposal. Ants were everywhere, but two kinds were of particular interest: army ants and acacia ants. For studies of the acacias themselves, Professor I. W. Bailey, one of the botanists from Bussey, had accompanied Wheeler to British Guiana. Ralph Wheeler (then an upperclassman at Harvard) and another young man, Alfred Emerson, already at the station, helped with the field work. Emerson, in a letter to the authors in 1964, wrote of that summer:

Personally, I found Wheeler a most stimulating person and he did much to steer my professional life. I had already become interested in termites before he came to Kartabo, but his enthusiasm at some of my early discoveries was most encouraging to a young man starting to feel his way into scientific research. In several instances he advised me to make certain special studies in British Guiana that I followed out with fruitful results. I also helped him in his ant studies, largely by carrying equipment, or doing the muscle work involved in finding some army ant queens. All of this intimate contact with him was a rich experience for me. His recommendation was an important factor in securing the offer to come to the University of Chicago in 1929 after eight years of teaching at the University of Pittsburgh. He was also instrumental in getting me my first position as an Instructor at the University of Pittsburgh.[6]

Wheeler liked the biological station in British Guiana, but in later years he came to like another more—perhaps because he was instrumental in its founding. This was Barro Colorado in Panama. In April 1922, Wheeler had been appointed to the Board of Directors of the Gorgas Memorial Institute of Tropical and Preventive Medicine, located in Washington, D.C. In February of the following year, Wheeler, accompanied by Mrs. Wheeler and their daughter, Adaline, sailed to Balboa, Panama, where on February 18 he witnessed the laying of the cornerstone of the Gorgas Memorial Laboratory. Apparently the Institute had not decided at the time of Wheeler's trip to Panama whether to make one laboratory for both marine and terrestrial investigations or whether to make one in Ancon (marine) and one somewhere else in Panama for terrestrial and possibly fresh-water studies. With the flooding of the land to make the Panama Canal and the resulting formation of Gatun Lake, several square miles of virgin jungle, which had once been a hilltop, were left more or less undisturbed and could be obtained at little or no expense from the Canal Zone for a site for a tropical laboratory. In a letter dated March 21, 1923 to Thomas Barbour, Wheeler wrote in part:

Yesterday Zetek, Johannsen, Dr. Strong and I spent an hour on the island which is reached by launch in 20 minutes from Frijoles on the railroad, not far from Colon. In a small clearing, less than an acre in area, I took 19 species of ants, Zetek took 10 species of termites (1 new species) and both of us took a dozen species of myrmecophiles and termitophiles (2 new genera, one a beetle, very remarkable!). The vegetation is extraordinarily diverse. It is an ideal place for a lab. in every way . . . The ground is dry and rises in hills, one of which ("Gigante") is about 500 ft. above the level of Gatun Lake. A small bungalow with screened verandah for a lab., a good launch, a resident director (Zetek would be just the man), a cook and one or two competent negroes as assistants, the development of a few good trails across the island, and we should have an ideal zoological and botanical paradise. The King of all the tapirs lives on the island with many of his descendants, together with ocelots 9 ft. [*sic*] long and other beasts too numerous to mention.[7]

The site was available, but money was not forthcoming from government sources, and apparently neither was organization. So Wheeler appealed to his friends and associates for both. He wrote to his old friend, David Fairchild, suggesting that he take over the administration of the laboratory because of his previous experience with laboratories in the tropics. Wheeler, Fairchild, and Barbour not only contributed considerable amounts of their own cash to the enterprise, but persuaded many others to do likewise. One incident is described in one of David Fairchild's colorful letters (January 13, 1924):

Tom Barbour and Wigglesworth tried to go through Miami almost without stopping but the train didn't run so they had to stay over a day and I had the pleasure of introducing them to Barbour Lathrop and Allison Armour both so to speak stock holders of the Laboratory Company.

I also took them to the Aquarium and they talked with Allison the owner and with Capt. "Charlie Thompson" the Collector of the Aquarium who is leaving for the South Sea Islands via Panama on the 20th as collector for W. K. Vanderbilt. If Tom does not get Vanderbilt over to Barro Colorado and tie him to a tree with a boa constrictor until he has promised to become a Stock Holder he is not the Tom Barbour giant we both thought him to be at all.[8]

In the summer of 1924 Wheeler spent six weeks at the new laboratory on Barro Colorado in the company of David Fairchild and his son, Graham, Nathan Banks, curator of insects from the Museum of Comparative Zoology, George Wheeler, a student at Bussey, and several others. James Zetek, the director, was in and out helping to set up the station, and much of the time was spent helping him with that as well as collecting and observing. Wheeler wrote to Barbour, who had been there for similar reasons in the spring, saying that "Banks and I have now been on the island four days [letter written June 22, 1924] and are having a wonderful time. The number of

A group of workers at Barro Colorado Island, Panama, in 1924. Left to right: G. S. Dodds, James Zetek, Ignacio Molino, Nathan Banks, George Wheeler, Graham Fairchild, Frederick Burgess, David Fairchild, William Morton Wheeler. (Archives of the Museum of Comparative Zoology, Harvard University)

species and individuals of insects is simply inexhaustible. Banks is making very extensive collections, especially of Diptera, Coleoptera and Spiders. I have found many interesting ants. The trails are partly completed. The one running directly back from the lab. and across the island, though straight, is so precipitous and up and down that I find it impossible to use it. It will not be practicable during the wet season. The Indians who cut it were twice injured by slipping on the rocks, etc. The trail should have been made sinuous, if possible." What Wheeler did not say was that practically every able-bodied person on the island, whether hired to clear trails, to build shelters, or to study vertebrates, sooner or later ended up helping Wheeler study ants. Wheeler was a marvel in the field, and he seemed to gather everybody in. Wheeler appreciated the help he received, and acknowledged it, usually in the paper he wrote summarizing his work.

In a letter to the authors, dated March 17, 1969, Graham Fairchild recalled the summer of 1924 and Dr. Wheeler:

I first met him [Wheeler] on B.C.I. in 1924, and remember vividly helping to collect a colony of *Eciton hamatum*. A gallon can with cotton soaked in chloroform was jammed down over the colony at the base of a tree. I think the hero who did this was Ignacio Molino, Zetek's assistant. The rest of us stood well back, our boots smeared with vaseline and flit guns in our hands to protect ourselves and spray Molino. All hands turned to that night to sort through several quarts of ants to find the queen and assorted myrmecophiles. Uncle Bill [Wheeler] was pretty excited and as I remember quite puzzled at the relatively small size of the queen, which he spotted . . . I was 17 at the time, and had a friend of the same age with me. All of us slept on a row of cots in one big room in the only building on the Island then, and it was our duty to serve the assembled great men bananas as an eye opener. We learned a lot, as this dawn session was devoted largely to discussing the amatory adventures of absent colleagues. As I remember it, W.M.W. and Nathan Banks contributed the raciest items. The staff at Woods Hole and U.S.N.M. [U.S. National Museum] figured largely in these anecdotes.[9]

In the spring, before going to Panama, Wheeler had tried unsuccessfully to entice Raymond Pearl to join the group for the summer. "It makes me very sad," wrote Wheeler to Pearl on March 24, 1924, "that you will not be able to go to Barro Colorado this summer. I feel that the first few years of Barro Colorado will be the only delightful ones for people of our make-up. It will be like Woods Hole, which was also interesting during the early years of its development. As soon as the crowd begins to assemble down in the Canal and the rows begin to develop we shall have to look for some other place."[10] The crowds did come, but Wheeler did not return until 1935 when he made a spring trip to Guatemala and Panama, spending only a couple of weeks or so in each place. An amusing letter from Wheeler to Dr. Barbour's secretary describes the beginning of that trip:

It was very delightful to receive your and the "eateria's" congratulations on my 70th birthday. Mrs. W. and I have had such a hectic time that this is the first letter I have been able to write since reaching Guatemala. I was also much indebted to you for helping us catch our boat, the Santa Marta, in New Orleans. Owing to a freight wreck near Monmouth Junction, N.J., involving all four R.R. tracks, we left the Pa. station in N.Y. 1½ hours late. When the train of some 12 heavy Pullmans was plunging into the tunnel the engineer lost control of the locomotive, put on the emergency brakes and almost welded the train to the track, so that it took 5 locomotives to pull us back out of the tunnel and into Pa. Station again. When they decided to route us to

Washington we were 5 hours late in getting into our glorious capitol. Then while we were leaving Virginia we ran over two men in an auto and this delayed us another hour so that we were 6 hours late in reaching New Orleans and the Santa Marta, after waiting for us in vain till noon, had started down the river for Guatemala. A Fruit Co. official met us at the station and said we might perhaps catch the boat just as she was passing out into the Caribbean, since it takes her about 5 hours to get down the river. We therefore took an auto and drove 65 miles in 1½ hours along the Louisiana coast . . . reached a place called Burus in the wildest of wild coastal places, were met by a fast speed boat provided by the fruit company, hurried 25 miles down to where the Santa Marta was waiting for us and, after the speed-boat had broken down twice (once when she was about a block from the steamer), we were hurried up the ladder among the (it seemed to me) somewhat irritated acclamations of the passengers. Of course our nerves were completely disengeezelized, as Dr. Barbour would say. The passengers had been waiting impatiently for so long that they expected at least Franklin or Eleanor and were naturally disappointed to see two old things crawl up on the deck like a couple of frightened raccoons.

The voyage to Puerto Barios was delightful and the trip up to Guatemala City very interesting. When we were leaving the train someone was squeezing my arm and I found that it was Marston Bates who had come up from Honduras for a vacation. Mrs. W. went off to Antigua and Chichicastenango with my brother and sister-in-law and the other tourists, while Bates and I went on a week's collecting tour to Quezaltenango and Moca. At the last mentioned place we were cared for by Mr. Gordon Smith on his finca, which is the most beautiful in Guatemala. It is right at the base of the beautiful volcan de Atitlan."[11]

In a letter written to Tom Barbour on April 2, Wheeler said that he and Mrs. Wheeler were sailing for New York from Balboa on April 15. "I could have wished for more time on Barro Colorado, but as I have been asked to attend the meeting of the British Association in September and the meetings of the Zool. Congress in Lisbon somewhat later in the same month and shall therefore have to prepare at least two papers with the usual agony, I had best spend as much time in Cambridge and Colebrook as possible in the meantime."

During the last decade or so of his life, Wheeler was very close to three men in particular, George Howard Parker, Thomas Barbour, and David Fairchild. The first two he came to know better after he moved from Bussey in Jamaica Plain to the campus in Cambridge, and the last he found especially good company on field trips or in other informal situations. Although Fairchild and Wheeler had known each other since their first days at Naples together, they did not become close friends until they were together in the summer of 1924. In late August of that year, after Fairchild had left Barro Colorado, he wrote back to Wheeler: "I should have written you long ago and told you what I am telling all my friends here that you are just the most exciting thing in the whole range of jungle things. Joking aside dear Wheeler that stay with you was one of the most wonderful times I have had or ever expect to have in my life. Do not tie up with anything next summer until I see if we can't arrange to get together somewhere in Europe. I think that Marian and Mrs. Wheeler might hit it off splendidly. They both have real intellects you know, better ones a great deal than we have and they would let us gossip all we want to."

Marian Fairchild was indeed a stimulating person, and Wheeler obviously enjoyed her company as much as he did that of her husband. Mrs. Fairchild's father was Alexander Graham Bell, her son, Alexander Graham Bell Fairchild. And though the telephone is associated with his grandfather's name, plants with his father's name, Graham ("Sandy") Fairchild is an entomologist, specializing in horseflies. He has spent much of his career at the Gorgas Memorial Laboratory, not far from Barro Colorado Island.

David Fairchild and William Morton Wheeler did get together in the summer of 1925, and Wheeler found the trip a great relief after his strenuous spring. He had left New York on February 28 for Paris, where during a period of about six weeks he gave twelve lectures in French on evolution of the social insects. Dora and Adaline had accompanied him to France, but when his lectures were over, they went to England, and Morton, as David and Dora both called him, went down to Casablanca, Morocco, to meet the Fairchild family. As Wheeler explained it in his paper "The Ants of the Canary Islands": "During July and August, 1925, as the guest of Mr. Allison V. Armour on his yacht, the 'Utowana,' I had an opportunity to visit the Canary Islands. The visit was the more interesting and enjoyable on account of the companionship of my old friend, Dr. David Fairchild, of the Bureau of Plant Industry, who was making a detailed study of the agricultural and horticultural plants of the Islands."

The group on Allison Armour's yacht off Casablanca, August 1925. Left to right: William Morton Wheeler, Mrs. Jordan Mott, Graham Fairchild (standing), David Fairchild, Allison V. Armour (standing), and Jordan Mott. (From David Fairchild's *Exploring for Plants,* 1930)

David Fairchild wrote a book called *Exploring for Plants* (Macmillan, 1930) about his various trips with Allison Armour, and included in this book are many details of the adventures they shared with Wheeler. By mid-May, the Fairchilds had already been in Morocco for some time. Said Fairchild: "My old friend Morton Wheeler of the Bussey Institution had been at the Sorbonne, lecturing on ants, and from his letters I judged he was tired of the gray skies of Paris and needed a change as badly as we needed his company in Morocco. We had wired him to join us and when we reached Marrakesh we had word that he would take the next boat for Casa . . . Wheeler arrived on a French boat crowded with troops. I took the bus up to Casa to meet him and the next day we collected around the town and then took the bus back to Marrakesh. What a delight it was to have Wheeler, even though I felt afraid that I had misrepresented the character of the insect fauna when I wrote him to come, but Morocco had much more than ants to offer him, it had a society, a 'collectivity' as he might call it, which kept his wondering brain busy, and I have no doubt contributed to his later writings a touch that they could not otherwise have had."

Wheeler, Dr. and Mrs. Fairchild, and Graham traveled from Casablanca to Marrakesh, as described, then by car south to Mogador and then along the coast to Agadir. The latter generally was forbidden to visitors (even other Moroccans), but this biological party had been given a special travel permit so that they could study plants and insects there. David Fairchild was especially intrigued with Mogador, and said he thought it was the most romantic spot he had ever visited:

"Far away and long ago" are words that always send a sad but pleasant sense of longing through me, the sighing of the wind through the pines brings a similar feeling, and the name of Mogador does the same. Something of Poe's "Raven," "Lenore" and "nevermore" hang around my memory of the place. Rhymes kept forming themselves in my head all the time of our stay in the drowsy little town by the sea:

Camels walking on the shore
Sea and sand and nothing more.

and so on, indefinitely. To this Wheeler added, after a day spent back of the town with Graham:

Insect fauna mighty pore
On the coast by Mogador.

On the nineteenth of June the Utowana was due in Casablanca, and Morton and David and Graham met it there to begin their trip to the Canaries. "Those first mornings on the yacht, how can I describe them!" said Fairchild. "The library of floras of the eastern tropics had to be arranged, the microscope tables had to be equipped, the dark room needed all sorts of attention. Everything was *couleur de rose*. Each morning coffee was served in the library and we sat for a half hour in our djellabas, or Arab wrappers, and listened, as we sipped our coffee, to Wheeler's rendering of Bernard Shaw's Caesar and Cleopatra, which resulted in his acquiring the unearned title of Tatatita for the rest of the cruise." In another chapter, Dr. Fairchild described some of the problems of too much comfort:

Mr. Armour had put a special porthole leading onto the aft hatch so that the light in the laboratory was ideal, and I found that I had no trouble in microscopic work with magnifications under 500 diameters. The boat was so steady, never rolling enough to give us any real trouble, that even when there was a good deal of motion I should have been able, by bracing myself, to do microscopic work quite comfortably. The books in the library once in a while took to popping out, but this was before Mr. Armour devised a spring arrangement which kept them splendidly in place. Every detail of arrangement was perfect. The trouble was in the interior mechanism of my own make-up, which slowed down as the swell rose, until there came a time when anyone might have pitched the microscope overboard for all I cared. It was on such days as these that Wheeler would sit and devour book after book and smoke so many pipes full of tobacco and burn up so many matches in doing it, that he nearly ran the yacht short of safeties. It was at such times as these that I ceased to wonder that there were so few serious scientific books written on board ship.

Would that science had the assistance of a few such private yachts today—to test this hypothesis, of course!

In the Canaries the party visited four of the larger islands, and in spite of a summer drought, Wheeler was able to observe and collect most of the ants known to occur there. In a letter Fairchild wrote Barbour from Grand Canary (July 26, 1925), he described their excursion there: "Together we have scaled on muleback some of the most dizzy heights of these Canary Islands and slipping and sliding over the almost perpendicular steps our mules only by the grace of the gods that they respect have refrained from throwing us over into the 1000 feet deep Barrancos with which the islands are furrowed."

From the Canaries, they went to the Balearic Islands in the Mediterranean and visited Iviza, Majorca, and Minorca. In Majorca, Dr. Fairchild described a trip inland in these words:

As we were driving along up the steep winding road leading to the town of Soller, Wheeler spied a hole in the rocks. It had nothing particular about it to attract my attention but he wanted to stop, so stop we did, and Wheeler climbed up into the hole. There was a yell, ending in a word we had by this time all learned by heart, 'Vermileo!!!' He had found characteristic craters in the fine dust covering the bottom of the niche, and they were not those of the ordinary ant lion but of the rare vermileo. Out came the battery of cameras, the bottles and other collecting paraphernalia and when we left that hole in the wall Wheeler was walking on air, like a child with a new toy, and we had in the bottles every vermileo that had dared to build a crater in that dust, and a quart or so of the dust, too. Like the lampromyas found in the Gauncho caves of Teneriffe this *Vermileo vermileo* De Geer, is a "Leptid Dipteron," or fly, and because of its peculiar habits has a special interest, as I have already explained. Wheeler's care of his dust-filled jars in which were these two species of larvae, was the care of a mother for her baby or a child for her dolly. He got them to Boston and they hatched out in the Bussey Institution and I presume the beautiful flies their pupae produced are now somewhere in that Institution with horrible insect pins stuck through their thoraces, while a detailed description of them has already gone down into history.

But I have no inclination to belittle this discovery in any way, for it remains in my memory as one of the most splendid examples of a thing it were well for more people to understand, the fact that there is more delight in the discovery of a rare or new natural object than in almost any other activity of the human mind.

Furthermore has not Wheeler inscribed one of his newly discovered ants as *Solenopsis latro fairchildi* subsp. nov. in the immortal book of insect descriptions that now includes a half-million names? I could wish of course that it were a species instead of a subspecies, but since I shall probably never meet it I shall not run the risk of mistaking it for its close relative, a subspecies also, discovered in Portugal.

Wheeler's activities for the year 1925, which kept him away from home for almost nine months, had two unexpected consequences: the discovery in Paris of an unpublished manuscript of Réaumur on ants (subsequently published), and a "slight mental breakdown," as he called it. The latter did not come at once, but in early 1926 Wheeler did not feel well, and during February and March he was at a mental hospital near New York City. In between coppersmithing and the like, however, he read proof of his French book, *Les Sociétés d'Insectes*. From April first through the summer he spent most of his days at Colebrook relaxing.

George Howard Parker (1864–1955) wrote an autobiography called *My World Expands* (Harvard University Press, 1946) telling much about his days at Harvard, and including, of course, many reminiscences of Wheeler. At first these men may seem to have had little in common, for Dr. Parker was associated mainly with laboratory studies of the nervous system of animals, but in reality they were much alike. They had known each other at Woods Hole in its early days, and Parker helped the Wheelers find a house when they first came to Harvard in 1908. Both men seemed cool and austere upon first acquaintance, but this reserve disappeared when they were together or with others they liked. Both could display wit and charm in lectures or writings, as well as in the endless conversations they enjoyed with colleagues. Parker, though he was more of an experimental biologist, believed in simple apparatus, and once wrote what sounds like a "Wheelerism": "To Loeb the problem of the universe is soluble in a finger bowl; to Morgan in a milk-jar; and we must never forget that the importance of a result is often inversely proportional to the complication of the apparatus by which it is obtained."[12]

Parker enjoyed walking, and one of the stories in his book is about a walking trip in the Adirondacks in 1918. Three other men went with him; these were Wheeler, a Mr. F. W. Rogers, and Dr. E. G. Conklin (1863–1952), professor of biology at Princeton and summer resident at Woods Hole. According to Parker, at first they had trouble finding a guide for the hiking-camping excursion to Mt. Marcy, but eventually they did locate a man named Ed who started them out. He left them, however, before they reached the summit—saying he had to meet another party to guide.

The summit of Marcy was within easy sight and we said good-bye to our guide who disappeared very quickly leaving his pack of provisions with us and full directions for the summit. We pressed on valiantly and reached the top of Marcy about noon. And what a view we had! The Adirondacks spread out in all directions at our feet! The day as clear as anyone could wish. We saw, well marked, the beginning of the trail down to Placid. Having taken in all these matters we fell to our pack of midday lunch. And now we learned what a deceiver Ed had been. Three thin sandwiches for four hungry men and all the rest of the pack filled with paper, empty receptacles and the like! . . . We shared our meagre supply of buttered bread very evenly and a little before one o'clock began our long, long trek toward Placid. The trail though clear was none too easy, and weak from insufficient food we stumbled on in silence. We walked all the afternoon and into the evening. At last we gained a narrow road and where it crossed a small stream Wheeler stopped, took off his shoes, and bathed his swollen feet. Then, after an hour's rest, on again we went into the darkening night. It seemed as though we were doomed to put up supperless overnight somewhere on the open road.

However, between nine and ten they stumbled onto a farmhouse and persuaded the woman there to give them something to eat. She was reluctant at first, but finally did, and then sent them on to another farm where they got a place to sleep. Parker continued: "Wheeler and I got a room with a double bed and the same kind of accommodation fell to Rogers and Conklin. What the other two did I do not know but Wheeler and I took off our shoes and outer clothing and in a few minutes were in deep sleep. And what a sleep it was! I never before had come so near utter exhaustion. We left word to be wakened early that we might catch a bus which passed that way to Placid. Again we paid a good bill, caught the bus, and by breakfast time were in Placid. We learned afterwards, as I had suspected at the time, that we had slept in a tuberculosis camp. But in our dire straits such small matters counted as nothing." Parker said that when they parted they "swore eternal friendship like four shipwrecked mariners who had starved together on a raft, but had in the end been rescued."

Conklin (whose experiments with cell lineage later provided him with a basis for a keen interest in studying and lecturing on evolution) and Wheeler, who had long ago become well-acquainted at Woods Hole, had made a much longer trip together four years before this hike in the Adirondacks. In July of 1914, they had sailed together from San Francisco to Australia. Conklin was on his way to the international scientific meetings, but Wheeler was about to fulfill one of his dreams of studying and collecting the primitive ants of Australia. We know little about this trip, except for the collecting notes in Wheeler's resulting papers and the brief diary that he kept of the trip. Before leaving California, Conklin and Wheeler visited Stanford University and apparently spent several evenings at "the theater." They reached Papeete, Tahiti, on August 2, and went to a dance of the natives in town. On the next day they visited the local market, and Wheeler managed to collect a little. He also mentioned that there were rumors of a possible war between Germany and Russia. He neglected to say that the rumors were confirmed when the captain of the ship asked him to translate some German messages the wireless operator had picked up but could not understand.

Conklin's stay in Australia was not as long as Wheeler's, for he left September 26, and Wheeler did not leave until three months later. Shortly after Conklin sailed, however, Wheeler had a memorable experience. On September 28, he took a train from Sydney to Oatley, and started out from there on foot to collect and take pictures of ants and ant colonies. One of the sites that he decided to photograph was at the base of a bridge. As a result, he was arrested as a possible spy, photographed, and questioned. Apparently his fluency in German did not make him appear any friendlier to the jittery local Australians. But the good word of the Australian entomologists saved everyone from further embarrassment, and Wheeler was released.

Wheeler may have had to call for help in order to be recognized on his first trip to Australia, but his entrance into the island continent on his second trip in 1931 was an entirely different matter. One imagines that he may have wished now and then for some of the quiet of 1914, for crowds of university and museum people as well as reporters greeted his expedition wherever it went. And not only did the Australian press keep track of the party's progress, so did the Boston press. *The Boston Traveler* for August 8, 1931, devoted about half a page, with pictures, to a story headed "Harvard Party Will Seek Strange Mammals in Wilds of Australia." (Considering that at least half of the six people going were not particularly interested even in vertebrates, let alone mammals, this headline must have provoked a few scathing remarks!) To continue:

> A search amid tropical wildernesses for the wombat, the microscopic honey mouse, specimens of the kangaroo, a thousand other species of fauna little known and rare, to provide additions to the Harvard University and Australian museums is the objective of an unusual expedition which has left Boston, members of which will spend a year in travel and collecting in the wilds of Australia.
>
> The group, all Harvard men expect one, six in number and under the leadership of Dr. William Morton Wheeler, entomologist and the world's leading authority on social insects, are now en route to Sydney on the Pacific. The expedition is under the direction of the University Museum, Dr. Thomas Barbour, director."

Wheeler did not stay in Australia for a full year because of teaching duties at Harvard, but he was there during the first part of the trip (August through December). And fortunately this part is well documented in the informal letters exchanged between Wheeler and Barbour.[13]

"Dear Tom," wrote Wheeler on August 6, 1931, as his boat was approaching Pago Pago. "We had a glorious day in Honolulu. Two of my former students—F. X. Williams and Hagan, who wrote the paper on the embryology of the Polycteids which we took in Cuba

. . . —met us at the wharf with their cars and spent the entire day with us."

One of Wheeler's most interesting letters was written in Perth, in Western Australia, on September 5, 1931, and began as follows:

Dear Tom: We reached Perth safely yesterday morning after a very comfortable and interesting journey from Adelaide. The whole country was literally covered with white, yellow and pink flowers, not mixed but in stretches of many miles. We were given a civic reception by the Mayoress of Kalgoorlie, who drove us all over the town, served good whiskey to us in her room in the town hall and showed us the mines on the "Golden Mile" which has yielded £100,000,000 worth of gold during recent years. This she did because we had a stop of 3 hours in Kalgoorlie to change to the narrow gauge R.R. for Perth. On our arrival in a pouring rain and chilling cold—most unusual and unseasonable, of course, we were met by a large delegation of University professors, reporters, etc. In the evening the women's club of Perth entertained us like a lot of prima donnas at their joint, the Katta Karra Club. Le Loeuf (son of the zoo director at Sydney) and zoo director at Perth says marsupials of many kinds and echidnas are abundant near Perth and he is going to help us get some. A fine spiny-tailed lizard was sent us this morning. Everybody is most kind and we hope for great things.

One interesting episode occurred near Wiluna, a prosperous gold mining community some 350 miles north of Kalgoorlie. As the members of the expedition rested in the afternoon on the shores of Lake Violet, "a beautiful body of brackish water," Wheeler noted several *Bembix* digger wasps carrying damselflies into their nests. *Bembix* wasps occur over most of the globe and everywhere prey upon true flies rather than damselflies; in fact no wasps then or since have been found to specialize on damselflies. Wheeler conjectured that true flies were scarce at Wiluna and that the *Bembix* "had shifted their predatism to the very insects which were responsible for the dearth of Diptera."[14] Wheeler's unusual observations prompted the junior author to visit Lake Violet thirty-eight years later (to the day) in order to study these wasps in more detail. Wiluna is now nearly a ghost town, since the mines are no longer productive, and Lake Violet was bone dry when he visited it. Nevertheless the *Bembix* were busily provisioning their nests, not with damselflies (which require water to develop), but with true flies. But by digging in the sand it was possible to uncover many old nests, most of them filled with the remains of damselflies that had been consumed by past generations of wasp larvae. Thus Wheeler was correct that these wasps are able to switch from flies to damselflies—a most unusual trait for solitary wasps—and thus to take advantage of large, easy-to-catch prey when abundant and to persist through dry periods on flies when no damselflies are on the wing.

The Harvard collecting party spent September, October, and most of November in Western Australia, with varying degrees of success but always with enthusiasm and with regret that no botanists had accompanied them. The botany, said Wheeler, is vastly more interesting than the zoology. In his letter dated November 16, from Pemberton, Wheeler added a postscript: "This is the region of the tallest trees in the world, Karri and red-gum, up to 200 feet or more. A pure Karri forest is a wonderful sight, with the silvery, barkless trunks rising perfectly straight and the ground covered with bracken fern." From Perth the party went back across the continent on the long and dusty train ride to Sydney, stopping off in Adelaide and Melbourne mainly to visit other entomologists. Wheeler had to sail on December 26, but some of the younger men stayed on into the fall to collect in the eastern and northern parts of Australia.

Two of the young men, still associated with the Museum of Comparative Zoology, were William Schevill and Philip Darlington, and their recollections of this trip are interesting, too. Dr. Wheeler, at least after his sixty-fifth birthday, did not see the need for making his field work any more uncomfortable than necessary. Consequently his party stayed at hotels or tourist houses, where they could more easily clean up after a day of collecting, sleep in a comfortable bed (hopefully!), and have hot meals served to them. This was fine for Wheeler, who could collect ants even on a sidewalk, but for the men interested in more furtive beasts, this was certainly less than ideal. Nonetheless, the collectors managed to do well by all their special interests, and Dr. Barbour seemed to feel that the expedition was a great success, and that the expense of it had been worthwhile.

Although the Wheeler-Barbour letters were mainly about Australia and collecting, there were also comments about what was going on at home, and particularly the problems of the finishing of the new biology building. On October 12, Barbour wrote to Wheeler: "Dear Uncle Bill: You are lucky to be in Australia instead of trying to work here! The new Biological Building, according to the architects, is finished but the people who have moved in are

William Morton Wheeler collecting ants in a national park in New South Wales, Australia, in 1931. (Archives of the Museum of Comparative Zoology, Harvard University)

getting along without half the furniture they need. Two carloads of chairs and shelving are said to be 'lost' and apparently nothing can be done about it until they turn up! The students are marking time until the laboratories are in shape for them to begin work." Barbour included some gossip about happenings in the Museum, and then added: "Everyone awaits your letters with interest and we are all delighted to hear that you are having such a good time and such success. If you came back today your trip would have more than paid for itself in our estimation."

Wheeler's last years at Harvard were happy ones. He had left his administrative duties behind, and his teaching responsibilities were not heavy. He retired from teaching entirely in 1934, and about the same time resigned from many of the organizations to which he had belonged for many years. He resigned, for example, from the Board of Trustees of Wellesley College, on which he had served for eight years. He was freer to travel, and Mrs. Wheeler, now that the children were grown, could accompany him more frequently. Adaline and Ralph had finished their respective colleges (Wellesley and Harvard), and Ralph had gone on to medical school, receiving an

M.D. from Harvard in 1926 and a D.P.H. from Johns Hopkins in 1932. Ralph had a charming wife and a son, William Morton II, later a second son, Paul. The family albums for the early thirties contain many pictures of the doting grandfather and his namesake grandson.

The trips that Dr. and Mrs. Wheeler took together in the thirties were usually short and not very strenuous. One of their longer trips was in the spring of 1930 to California and Hawaii. Days on the train or aboard ship were relaxing, Dr. Wheeler often playing cards or enjoying his pipe, a book, or conversation with other passengers. In California they spent a week at Palm Springs resting, reading, and collecting ants in the nearby canyons. In Hawaii they visited friends and professional acquaintances and had dinner with "Mr. Dole" and a tour of his pineapple plantations. Back in California they spent a week in Yosemite, living in a cabin and hiking about looking for ant lions, some of which Wheeler took home alive for behavior and life history studies later recounted in *Demons of the Dust.*

When not traveling, Wheeler spent his days commuting between his home on West Cedar Street on Beacon Hill in Boston, and his office in Harvard's new biological laboratories. Here he, C. T. Brues, and F. M. Carpenter occupied adjacent rooms on the second floor. The huge new building provided facilities far beyond anything previously available, but some of the economies of construction had brought about one of Wheeler's occasional bursts of temper. Only a few of the offices had hot water, and Wheeler's was not one of them. It is said that he marched into G. H. Parker's office (Parker was then chairman of the department) and with appropriate invective and much pounding on the desk persuaded Parker of his need for hot water. Later, as he laughed over the episode with Tom Barbour on the steps of the Agassiz Museum, Wheeler remarked that for the first time in his life Parker had answered him: "Yes, sir!"

During these years Tom Barbour and "Uncle Bill" Wheeler grew especially close. Together they had helped found Barro Colorado Island Laboratory and had fought for the appointment of Raymond Pearl to replace Wheeler as dean of Bussey. They had both supported the Cambridge Entomological Club when lack of interest and money seemed to indicate an untimely demise. (It is now the oldest continuous entomological society in the United States and rapidly is approaching its hundredth birthday.) When Wheeler had

William Morton Wheeler, Thomas Barbour, and Henry B. Bigelow on the steps of the Museum of Comparative Zoology, June 1935. (Archives of the Museum of Comparative Zoology, Harvard University)

moved from Bussey to Cambridge, Barbour had made every effort to make him welcome and comfortable. They were both world travelers and top scientists in their own fields, yet knowledgeable in many facets of biology, considering themselves "naturalists of the old school." Because of the age and structure of the old Agassiz Museum building, smoking was not allowed inside, and some of the pleasantest memories of the students in the early 1930s are those spent sitting on the steps outside (whether they were smokers or not) listening to Wheeler, pipe in hand, Barbour, Henry Bigelow, and others telling stories—often in language most colorful if not exactly of the classroom type! Wheeler and Barbour, in fact, were notorious for "their command of the swear language," and sprinkled it liberally through their conversations, or if provoked, astonished their listeners with an unexpected vocabulary. As Mrs. C. T. Brues has told the authors, not only was Dr. Wheeler an accomplished linguist, his gutturals would have fooled a German, and "even his use of swear words so carefully phrased and effectively delivered were something."[15]

Thomas Barbour (1884–1947) had done his undergraduate and graduate studies at Harvard (in herpetology) because as a child he had visited the Museum of Comparative Zoology (the Agassiz Museum) and decided that that was where he wanted to be. In 1927 he was appointed its director, and under his guidance it became again a dynamic and modern institution. But it was one of its lesser known activities that became a favorite of Wheeler's. This was the "Eateria." Soon after Barbour became director, he found it convenient to bring his lunch and eat it in his back office. Soon other members of the staff joined him. In his book *Naturalist at Large,* Barbour described the Eateria in these words:

> Then it occurred to me that we had in Gilbert, working here in the Museum at odd tasks, a most courtly old-fashioned colored servant who, as he put it himself, had been left to the museum with Mr. Brewster's collection of birds. We installed an electric stove, proper sink and electric refrigerator, and screened these objects away in a corner of my office, which is a large one, so that they do not obtrude.
>
> William Morton Wheeler joined our groups and in time the Eateria became quite an institution. As long as Mr. Lowell was President of the University he not only called up frequently and said that he was coming to lunch but brought guests who he thought would be interested in learning about the Museum under entirely informal circumstances.[16]

Often the menu was enriched with exotic dishes. Dr. Barbour liked to hunt, and after one of his successful excursions, venison, duck, or even terrapin appeared. Frequently David Fairchild provided (in person or by airmail express) innumerable tropical fruits, such as white sapotas, agles, canistels, mangoes, and papayas. But when these men met the conversation was even more fascinating than the food: museum affairs, tropical paradises for observing nature, amateurs in biology, current books, most any subject directly or indirectly connected with biology.

Another subject of mutual interest to Barbour and Wheeler was the Boston Natural History Society, "a veritable Mother Hubbard of American natural history," as Wheeler once called it.[17] In his last book, *A Naturalist's Scrapbook,* Dr. Barbour devotes a chapter to the society, of which he was president at two different times, and talks of its history and accomplishments, as well as its difficulties.[18] Since its beginning in 1830, the society and its museum had been a center of scientific research, and its meetings well attended by many of Boston's most illustrious citizens. But, according to Dr. Barbour, as the Museum of Comparative Zoology grew, it gradually overshadowed the work of the Boston Society. Dr. Barbour felt that there was a place for both in the Boston area, with the Harvard museum concentrating on the scientific pursuits, and the Boston museum specializing in New England flora and fauna and in educating the public by working with the public schools and by offering adult education classes and good public exhibits. Dr. Wheeler had apparently agreed with these ideas, for in 1930, on the hundreth anniversary, Wheeler was (had been for many years) a vice-president and a member of the Board of Trustees of the Society. Today the Boston Society of Natural History no longer exists as such (it has been replaced by the Museum of Science, which is devoted entirely to public education). With it went, many felt, something of the rich tradition and cultured heritage that had made Boston "the hub of the universe." Perhaps it was just as well that Wheeler and Barbour did not live to see the end of the society or the takeover of the old museum building by a department store.

Wheeler had many pleasant hours with the men at the Agassiz Museum, but he also enjoyed his other intellectual associations in and out of Harvard. Among his special friends were L. J. Henderson and Hans Zinsser of the Harvard Medical School, Harlow Shapley, who is now an emeritus professor of astronomy at Harvard (but likes ants almost as much as stars),[19] and E. B. Holt, who had been a student of William James interested in social psychology and had taught psychology at Harvard from 1901 to 1919, but who had moved in later years to Princeton. Conklin, Pearl, Morgan, Jennings, Wilson, Donaldson, and many others who shared Wheeler's interest in the Marine Biological Laboratory at Woods Hole, The American Society of Naturalists, The American Philosophical Society, and The American Academy of Arts and Sciences, seldom missed a chance to attend an annual meeting, whether in Philadelphia, Baltimore, Washington, New York, or elsewhere. Another annual event Wheeler looked forward to was the mid-October weekend he spent at Cobalt, Connecticut, with his philosophical "cronies", including Alfred North Whitehead and Henderson. Henderson (1878–1947), although he taught biochemistry, had "never felt himself completely a medical student and lived in Cambridge, where began the friendships which were to lead for a time to an

Alfred North Whitehead, L. J. Henderson, H. O. Taylor, and William Morton Wheeler at Taylor's home in Cobalt, Connecticut, about 1935. (Archives of the Museum of Comparative Zoology, Harvard University)

interest in philosophy." In the 1920s he was making a physiological study of blood, but at the same time pursuing his philosophical interests. To quote again from an obituary in *Science:* "But several years before, about 1926, William Morton Wheeler had brought the Italian economist and sociologist, Pareto, to his attention. The Traité de Sociologie Generale had impressed Henderson as an accurate qualitative description and a workable classification of the uniformities found in all observed social systems. Once aware of its importance, he was impelled to master it by eager study, to lead others to an understanding of its implications and to test its validity.

After 1930 his efforts wholeheartedly served these ends. By 1935 his last published book, 'Pareto's General Sociology—A Physiologist's Interpretation' had appeared."[20] During this same time, Wheeler had also interested the Fairchilds in Pareto, and as a consequence Mrs. Fairchild was translating Pareto with Wheeler's help.

Wheeler's ability with language never ceased to astonish his friends. His friendship with C. K. Ogden, the British scholar and eccentric, was based on a mutual interest in language and its improvement. Wheeler was particularly fond of Greek poetry, and loved to read it aloud in the original. Ralph Wheeler recalls that late in the evening the cadences of Virgil and Homer would often boom through the house as his father relaxed to the music of those ancient bards. In the early 1930s Hans Zinsser was writing his *Rats, Lice and History,* and Wheeler spent many evenings with him checking Latin and Greek sources. In his preface, Zinsser acknowledges his gratitude to his "wise and kindly friend, Professor W. Morton Wheeler."[21] On the steps of the museum, Wheeler would often regale his listeners with tidbits from his conversations with this witty and voluble man.

An even more astonishing interest of Wheeler's was his familiarity with the Bible, particularly the Old Testament, which he had read in Greek. Wheeler and Parker, and various other friends at different times, held what they called a "Sunday School" on Tuesday for lunch at fairly regular intervals in one of the local Italian restaurants. Originally, according to Parker in his autobiography, these excursions had been made to get away from Cambridge and academic worries—a way of relaxing. "Wheeler, who was always given to scholarly performances, insisted that at these agreeable outings there should be food for the soul as well as for the body. He therefore declared for some good reading and proposed to this end the Book of Job. This we undertook with Wheeler as reader and I as listener and commentator. In the course of a few months, by reading some chapters each week, we completed the book to our great edification. To natural historians with a psychological trend this book of the Old Testament is a marvelously intimate portrayal of an abnormal personality who is forced to consider some of the most difficult problems of the human soul, the place and meaning of pain and affliction in this life of man. We read Job and pondered it with a deep sense of its significance." The fact that Job is considered by many scholars to have been taken from an old Greek tragedy may have added to the interest of these two men.

Something of Wheeler's great appetite for knowledge can be grasped by a visit to his study on the second floor of his house on West Cedar Street on Beacon Hill in Boston. It is still very much as he left it. The books on the shelves cover many subjects, but especially those relating to the more philosophical aspects of biology and sociology. Virtually all have marginal annotations, indicating that Wheeler had not only read them but had weighed every thought carefully. "N.B." [nota bene] appears beside passages he especially liked, while sections he did not agree with are marked "No!" or with some stronger word or a brief rebuttal. For example, in C. E. M. Joad's *Guide to Modern Thought* (1933) there is a section on "ectoplasm . . . the stuff of the medium's body which is temporarily dematerialized into a kind of amorphous, pulpy mass . . . I have myself seen ectoplasm issuing as a shapeless, fluid substance . . . from the medium's nose and ears." Wheeler's marginal comment: "Excellent for mending furniture, as E. B. Holt says." When Joad comments that "if I had not fallen out of the window at the age of five, I should not be afraid of heights now," Wheeler remarks: "He evidently fell on his head!"

Professor Wheeler rarely invited students to his home in his later years, and when he did entertain, the guest was usually a visiting scientist or a local kindred spirit such as E. B. Holt or Harlow Shapley—or occasionally some of Mrs. Wheeler's political allies. Students nevertheless found Wheeler far more approachable and willing to chat than he had been while dean of the Bussey. Here was a man who had been everywhere, done everything biological, a tradition with a pipe in his hand and a twinkle in his eye . . .

Probably there is not a man alive who does not hope that he will be alert and active up until the moment of death. Wheeler was one of those fortunate few who was. In 1936 the number of his publications was ten—not much different from the preceding years. In 1937 he left with Mrs. Wheeler in mid-February for a six-week trip to Tucson, Arizona, and then down through Guaymas, Mazatlan, and Guadalajara to Mexico City. On April 7, Wheeler wrote to Dr. Alfonso Dampf in Mexico City to say that he and Mrs. Wheeler were safely home again, and to thank him for his hospitality. And then in the evening of April 19, exactly a month after his seventy-second birthday, he dropped dead suddenly. As was reported in the *Boston Herald* the next morning: "the seventy-two year old professor emeritus at Harvard and world renowned figure in natural science, collapsed on the Harvard Square subway platform shortly after 8 o'clock last night and died within a few minutes. Police who tried for nearly three hours to identify the well-dressed man, finally succeeded in tracing the initials W.M.W. on his belt buckle."

Funeral services were held three days later in the chapel on the campus at Harvard, and, according to the newspaper reports, "throngs" of friends and associates were in attendance. In addition, scholars and scientists from all over the world sent messages of sympathy to Mrs. Wheeler. Both the *New York Times* and *New York Herald Tribune* carried long obituaries telling of his accomplishments. But perhaps the finest tribute to him—the one he would have liked the most—came from his old friend, David Fairchild. It was written in longhand in Florida to Thomas Barbour in Cambridge, and dated April 20, 1937:

Dear Tom, How thoughtful of you to persist until you got me on the phone! It just does not seem possible he's gone. The dreadful silence falls upon me that has fallen when other loved ones have departed but somehow Morton had so much of the undying love of the past about him that when he stops writing or talking then drops out of my life, more, so much more, than has ever left me when anyone else has died that I seem to lose the power of going on.

Back forty-two years come the memories of him and they haunt me by their unique intellectual character. For, somehow, our friendship had a sparkle and an intellectuality about it that kept Morton alive in my mind when I was away from him.

He was easily the most erudite mind that was ever my privilege to meet and when I think how many delightful hours we spent together, I feel I've no right to complain for how many others knew him as I did?

My sympathy dear Tom for you really drowns my own sorrow for to you Morton was more even than he was to Marian and me. I wish that there was something we could do to comfort you but know there isn't. I go to my slab house and stare at the motto there in the cement written in 1924 "Push On" and take up the burden of life again but things will never be the same now—never.

Thank you dear Tom. I do sympathize with you. Marian loved him too with an affection that was most unusual. I wired Dora and have written her. Poor girl!

Lots of love David.[22]

14 / The Measure of a Man

Upon his retirement in 1934, William Morton Wheeler received the following document from the Zoological Laboratory of Cambridge University, England, signed by the entire staff, including such distinguished figures as A. D. Imms, W. H. Thorpe, George Salt, S. M. Manton, and others:

We, the undersigned zoologists and entomologists from another Cambridge extend to you our cordial greetings. We ask you to accept this expression of our good wishes on your approaching retirement from your professorship at Harvard University. We recall with pleasure your early fundamental zoological studies on insect embryology, on Myzostoma, on the lamprey, and on other aspects of the science. Today we regard you as the world's master of myrmecology and wish to express our appreciation of your many contributions that have shed so much light on the evolution of the social Hymenoptera. Viewing your problems from the standpoint of a skilled investigator, endowed with philosophic insight, you have interpreted with conspicuous success some of the most complex phenomena in the realm of animal behaviour. We look upon you as one whose example has done much to inspire the highest standard in American Zoology. The long list, of now well known names, of those who came year after year to study in your Laboratory at the Bussey Institution affords ample testimony of your skilled and sympathetic guidance. In asking you to accept this slight testimony of your services to biology we feel that we are only voicing the opinions of your co-workers and admirers in many lands. We, in Cambridge, England, desire to convey, herewith, our expression of the hope that you may long enjoy a well-earned retirement.[1]

Wheeler's sudden death in 1937 provoked a number of eulogies.

Here is part of an appreciation published in *Science* by L. J. Henderson, Thomas Barbour, F. M. Carpenter, and Hans Zinsser:

In thought and feeling he was a practitioner *and* a theorist; a specialist of the first rank, and, in the ancient sense, a philosopher; a great professor, a man of vast encyclopedic learning and the least pedantic of men; a diagnostician of genius who could instantly recognize the significant patterns in things and events, but who confirmed his conclusions by meticulous and systematic observation and study. He always had and never lost satisfaction in the pursuit of minute detail and in the accumulation of facts, so that hard work was a necessity of his being, but he could set no limits, within his wide competency, to the scope of his thought or to its sources or to its reference . . .

A highly developed specialist in his own calling, Wheeler was more completely the intellectual man of the world than any but a very few of his contemporaries in this or any other country. One never left him without having learned something, and one walked down the hill after an evening with him with ever-renewed admiration and affection—and usually with a chuckle.

The death of a great naturalist, like that of a great physician, does more than put an end to a scientific career. It destroys an accumulation and synthesis of knowledge, skill, judgment and experience that cannot be transmitted and preserved, because it is as yet incommunicable . . .

His written contributions to his subject will perpetuate his scientific memory, and his less technical writings will be read with interest and amusement for a long time to come. But as a personality, Wheeler was one of the great experiences in the lives of his friends and, in this sense, he will not really die until all those who knew him are well gone.[2]

Wheeler was one of the most honored scientists of his generation. He held honorary degrees from Chicago (Sc.D., 1916), California (LL.D., 1928), Harvard (Sc.D., 1930),* and Columbia (Sc.D., 1933). He received the Leidy Medal of the Academy of Natural Sciences of Philadelphia, the Elliot Medal of the National Academy of Sciences, the cross of the French Legion of Honor, and numerous other awards. He was a member of the National Academy of Sciences, an honorary member of several foreign societies, and an active or

* The citation read by President Lowell at the awarding ceremony for the honorary degree from Harvard is especially apropos: "William Morton Wheeler. Eminent zoologist and Dean of the Bussey Institution. Profound student of the social life of insects, who has shown that they also can maintain complex communities without the use of reason."

honorary member of many scientific societies in this country. As a lecturer he was in constant demand (turning down the majority of requests, especially toward the end of his life), and publishers vied for his books, lengthy and technical though they were. He authored nine books during his lifetime, and two others appeared posthumously. His list of publications exceeds 450 titles, covering a span of more than half a century, and his books and papers have surely been cited in the bibliographies of other publications more than those of any other American entomologist.

Yet, only three decades after his death, Wheeler's image does not loom as large as it once seemed to. Is this a reflection of the present generation's unmindfulness of the past (something Wheeler himself could never be accused of), or are there more weaknesses in Wheeler's research and writings than were formerly apparent? In this final chapter we shall analyze what we consider his major strengths and weaknesses and try to arrive at a balanced view of his place in science.

To our minds there is no question that his major strength lay in his interest and success in communicating, not merely to students and colleagues, but to the educated public generally. His lectures, even to his classes, were erudite and well organized, filled with facts gleaned from his insatiable reading in many fields and sprinkled with quotations from both ancients and contemporaries. His public lectures also sparkled with wit and with humorous allusions to anyone he considered unreasonable—from Father Wasmann to Williams Jennings Bryan and Mary Baker Eddy. Many of his lectures were tried out first at seminars in the Bussey, then polished further for presentation to a wider audience; finally, replete with footnotes and illustrative material, most of them made their appearance in his books. The surprising thing is that he managed to pack so much entomological detail into lectures designed for a general audience. His own thoughts on this matter were well presented in the introduction to *Demons of the Dust* (1930), his last major book (other than the posthumous collection *Essays in Philosophical Biology*):

Some of my colleagues maintain that I have attempted the impossible, that there are only two ways of writing on such a subject as the one I have chosen. Either I should have written as an undefiled specialist in the refined jargon and with the ponderous documentation demanded by the ritual of my caste, or I should have turned to literature and presented the matter with the well-known devices and embellishments of rhetoric, which may perhaps delight, but are sure to mislead the uncritical reader. Any intermediate course, they claimed, can only lead to a compromise distasteful alike to the high-brow scientist and the thrill-seeking, movie-fed public . . .

If, as seems probable, I have really crashed between two stools in attempting to interest both the entomologist and the general reader . . . I shall not plead guilty but proceed to rationalize my conduct. There are, in fact, serious and perhaps insurmountable obstacles to popularizing or even semipopularizing entomology in its present state. Quite apart from the disgust experienced by many people at the mere sight of insects, the more we know about their structure and behavior, the more inhuman and demoniac they become. The number of their species, subspecies, varieties and individuals is appalling, and it would have been impossible, during the past two centuries, to weld together the inumerable ascertained and recorded minutiae of their classification, structure, development and distribution, to make the science of entomology, without the aid of an elaborate technical terminology, or symbolism. Though this terminology is awkward, tortuous and pedantic to a degree, it is, nevertheless, consistent and far more concise and regulated than the vague, emotion-soaked language of ordinary discourse . . .

Since science is a social undertaking, I would beg the reader to cooperate with me to the extent of consulting a good dictionary for the more general terms with which he may be unfamiliar. Many intelligent people spend hours communing with dictionaries and other works of reference for the purpose of solving cross-word puzzles. It is expecting too much to ask the general reader to do as much when he is hoping to obtain information from an entomological treatise?

As may be seen, Wheeler made no attempt to water down his material for the uninitiated; rather, he felt that his subject was sufficiently interesting and important to justify a good deal of effort on the part of the reader. This sometimes led to lengthy letters between him and his publishers regarding the use of common versus scientific names, lengthy quotes in their original languages, and so forth. For example, prior to the publication of *Demons of the Dust,* W. W. Norton, the publisher, wrote to Wheeler as follows: "Do you remember the flattering reference I made to your footnotes? I withdraw it and instead of saying that you write the most attractive footnotes I have ever read, you write the worst, if you insist on using these Latin ones! After all you are writing this book to be read, and how many people that pay $5 for it are going to be able to read these footnotes in Latin?"

Wheeler defended his footnotes, remarking that they were important in the history of the field and that one was "a beautiful piece of Latin." "I am now in a state of mind," Wheeler concluded, "in which the whole book seems to me very dull and stupid. I cannot, therefore, see the beastly thing in any proper perspective. There are so many dry and arid passages in the text that I have sometimes felt that the work is highly artistic, because I have carried through the dust and sand motif with a vengeance, and quite subconsciously, which I suppose is of the nature of true art!"[3]

Demons of the Dust was, by the way, a major departure from Wheeler's favorite subjects, and ants and other social insects, since it was an account of two kinds of insects that make pits in sandy soil and lie in wait for prey in the bottom of the pit: the ant lions and the worm lions. The worm lions (larvae of flies of the family Rhagionidae) make up the greater and more original part of the book. There is little in Wheeler's notes or diaries to suggest why he undertook this major diversion from his ants. He had, of course, published several papers on Diptera prior to 1900, and in 1918 he had described an interesting new worm lion he had collected on his 1917 trip through the Southwest and into California. In the summer of 1918 he was asked to talk to the Marine Biological Laboratory at Woods Hole and in accepting wrote to Frank R. Lillie that his lecture would be on "The Life-History and Behavior of the Worm-lion." "This is not on ants, you will be pleased to learn," he wrote. "I am growing tired of talking about them and absolutely nauseated writing about them."[4] In 1920 he gave a Sigma Xi lecture at Syracuse University on some of the early naturalists who had studied ant lions and worm lions (later to form the first chapter of *Demons of the Dust*). His interest in this group was doubtless revitalized when he and David Fairchild discovered several exciting aggregations of worm lions in the Canaries and the Balearic Islands in 1925 and 1926. Some of the specimens collected in the Balearics were kept alive until he reached home. They survived the winter and the following summer were moved to Colebrook, where they were reared in pans of sand on the windowsill and fed with termites; in late summer the last of them emerged as adult flies. Three years later, Wheeler returned to Yosemite and obtained more material of the species he had studied there earlier. All of this provided him with the makings of a book, which he subtitled *A Study of Insect Behavior* and dedicated to Thomas Barbour. The ant lions and worm lions he found to provide an unusually fine example of evolutionary convergence in behavior, since the two groups are quite unrelated and yet have many similarities in their predatory strategies. Both groups proved interesting, too, in that their behavior was considerably more plastic and adaptable than had been supposed.

William Beebe reviewed *Demons of the Dust* for the *New York Herald Tribune* on December 28, 1930, remarking that "Wheeler has no master in this country in scientific acumen, in sarcastic and in kindly humor, in knowledge of human nature and in diction which it is sheer joy to read." The *Tribune* printed several striking illustrations from the book. On March 14, 1931, the *Boston Evening Transcript* used the publication of *Demons of the Dust* as a point of departure for a review of Wheeler's scientific contributions, illustrating its half-page feature article with figures from *Ants, Social Life among the Insects,* and *Demons of the Dust.*

Wheeler's most popular book was *Foibles of Insects and Men,* published in 1928 by Alfred Knopf and dedicated "to my friend David Fairchild in memory of many happy excursions in the jungles of Panama and the mountains of Morocco." This book is a collection of essays first published elsewhere, including his early paper on *Leptothorax emersoni* (somewhat revised), his 1911 papers on parasitism and on the ant colony as an organism (both originally lectures at Woods Hole), and several others, including two amusing pieces first presented as lectures to the American Society of Naturalists. The book opened with an essay first given as a lecture to the Entomological Society of America in 1922 and then published by Raymond Pearl in his *Quarterly Review of Biology* in 1927. It was placed first in *Foibles* in accordance with the suggestion of H. L. Mencken to Alfred Knopf, who had asked him to read the manuscript. The essay was "The Physiognomy of Insects," a discussion of the functional significance of the many peculiar head shapes observed in insects, but with frequent references to types of body form in humans as well as human resemblances to insects. This essay is a major example of a device Wheeler used many times to capture the attention of his audience—analogies, often absurd, between people and insects.

This device is carried still further in the final essay in *Foibles,*

"The Termitodoxa, or Biology and Society" (published first in *Scientific Monthly,* 1920). In this, the best known of his essays, Wheeler tells of a letter he wrote to "the most wonderful of all insects," the queen of the African bellicose termite. His letter was answered by the king, Wee-wee, her majesty being too busy, having laid an "egg every three minutes for the past four years." After describing his own social system, Wee-wee turns to human society, and the result is an amusing satire on a variety of human foibles which is no less to the point now than it was in 1920. In his review of *Foibles* in the Sunday *New York Times,* Charles Johnston spoke of these essays as "liberally suffused with formic acid," yet not without pathos beneath the mockery. Most of the barbs were reserved for "the stuffy and the solemn, not excluding professors of biology."

Wheeler's lifelong respect for the nonprofessionals may well stem from his early association with such part-time naturalists as Shelley G. Crump (a storekeeper) and Frederick Rauterberg (a mail carrier), as well as his apprenticeship with Henry A. Ward, a man who devoted his life to bringing biology to the public. Later in life Wheeler was much at home in the company of such "popularizers" as William Beebe, Thomas Barbour, and David Fairchild. In several of his essays, he defended the amateur vigorously and expressed regret that the training of young biologists was left to professionals. "The typical professor," he wrote in "The Dry-Rot of Our Academic Biology," "has about the same liking for the amateur that the devil has for holy water, and the amateur habitually thinks of the professor in terms which I should not care to repeat . . . We have all known amateurs who could make an enthusiastic naturalist out of an indifferent lad in the course of an afternoon's ramble and, alas, professors who could destroy a dozen budding naturalists in the course of an hour's lecture."

"Surely we should realize, like the amateur," Wheeler wrote, "that the organic world is also an inexhaustible source of spiritual and esthetic delight. And especially in the college we are unfaithful to our trust if we allow biology to become a colorless, aridly scientific discipline, devoid of living contact with the humanities. Our intellects will never be equal to exhausting biological reality . . . We should all be happier if we were less completely obsessed by problems and somewhat more accessible to the esthetic and emotional appeal of our materials, and it is doubtful whether, in the end, the growth of biological science would be appreciably retarded."

Today these words sound strange indeed, for most of our biological laboratories have lost all contact with the amateur—one is almost tempted to say, with nature—and to mention science and the humanities in one sentence is heresy. One wonders, however, whether any science can survive in isolation from society. We could use a few Wheelers today.

Wheeler was first, last, and always a naturalist, with the naturalist's absorption in the drama of living things and man's role in that drama. Some of his comments on the role of naturalist are interesting in the light of various schisms in many biology departments today:

What was formerly called natural history is the perennial foundation of the biological sciences. It has given rise to all the theoretical branches and will no doubt give rise to others in the future, and all the practical applications of biology have their roots in ecology . . . It formulates most of the basic problems which the experimentalists and biometricians are endeavoring to solve . . . No matter how far the naturalist may specialize in his study of single groups of organisms or of the faunas and floras of particular regions or geological ages, he is always keenly aware both of the limitations of his specialty and of its relations to the whole realm of living things. Such modesty is not always apparent in the biologist in the strict sense, because he is not engaged in sympathetically exploring the contours of nature, but in determining the extent to which phenomena conform with his experimental, metrical and therefore highly rational procedure.

In one of his last publications, a review of Tarlton Rayment's *A Cluster of Bees,* Wheeler had more to say on this subject, in what is probably one of his most self-revealing series of comments:

The perusal of such contributions sets one to wondering whether the naturalist may not be after all an artist *manqué* or incompletely developed, and not a genuine "scientist." Undoubtedly the activities of the naturalist are to a considerable degree motivated and sustained by esthetic delight in the wonderful forms and colors of animate beings and the intricate temporal patterns, or configurations of their life histories, behavior and ecological settings. Unlike the "pure scientist" who is devoted to simplicity, the naturalist like the artist, feels at home in the profusion, multiplicity and diversity of natural phenomena.

Perhaps . . . the naturalist's preoccupation with taxonomy has its roots in an unavowed conviction that this "science" is really an art . . . The artist is quite as much concerned with truth as the scientist, though the naturalist seems to be more desirous of appreciating and understanding than of explaining the phenomena with which he deals. This may account for the non-mathematical and non-experimental spirit in which he approaches and handles his materials. It may also account for his somewhat unsympathetic attitude towards the bright high school boys now bombinating in the biological laboratories of our universities. He feels that not a few of these neophytes manifest a somewhat gangster attitude towards Nature, so eager are they to assault, scalp or rape her, and so unmindful of the old apothegm, *Natura non vincitur nisi parendo.*

The great naturalists of the past appear again and again in Wheeler's writings, and in the course of his career he did much to revive interest in figures such as Réaumur, Huber, Bonnet, Fabre, and others. As early as 1899 he had published on George Baur and Kaspar Wolff, and his tribute to Jean-Henri Fabre, published in *The Journal of Animal Behavior* upon the death of that great naturalist and writer, was one of his finest efforts. All of his major books from *Ants* to *Demons of the Dust* contained sections on the history of the subject, and two were primarily historical. These were his translations of Réaumur's long unpublished manuscript *The Natural History of Ants* (Alfred Knopf, 1926) and his joint effort with Thomas Barbour, *The Lamarck Manuscripts at Harvard* (Harvard University Press, 1933). In both instances the manuscripts were first presented in the original French, then in translation, and both books contained a valuable introduction. In his introduction to *The Natural History of Ants,* Wheeler expressed his thoughts on the neglect of the history of biology:

Zoologists of the present day find the study of the animal world so vast, intricate and absorbing and so dependent on elaborate and time-consuming methods of investigation that, like most scientists, they take little or no interest in the history of more than the past few decades of their specialties . . . Yet we can scarcely blame the scientists for this lack of historical interest. The very habit of impersonal estimation, which every investigator endeavours to cultivate, naturally leads him to eliminate from the history of his subject the personalities of even its great pioneers and to reduce them to mere names and dates and the barest mention of their most eminent achievements. To the recognition of this tendency to depersonalization we may add the rather obvious consideration that the lives of scientists are for the most part very drab . . . Hence the progressive disregard of the very persons who, within three centuries, have brought about a revolution in our thinking and activities so amazing that the previous development of our race, through thousands of centuries of post-Pliocene times, seems not unlike the monotonous stability of some animal species.

Nowhere is this lack of historical and biographical interest more noticeable than among entomologists . . . Not only do the personalities of the great entomologists of the past tend to fade into oblivion, but the records of their discoveries, apart from those in the field of taxonomy, are mostly forgotten and their abiding influence on the investigators of our time no longer perceived. Their writings are neglected, not because they are unreadable, for they are certainly no duller than the works of our present-day entomologists, but because specialization has become so intense . . .

Although a too constant preoccupation with historical details might tend to inhibit rather than to advance entomology, there are at least two reasons why a more intimate acquaintance with the lives, works and influence of the great pioneers of the science is eminently desirable. First, these pioneers are a permanent source of encouragment and inspiration. Can any entomologist, who meditates on the deeply pathetic lives of such indomitable investigators as Swammerdam and Fabre, fail to take heart and pursue with renewed zest the studies which in moments of depression seem so trivial and so exhausting in comparison with many other pursuits? And second, how can we hope to estimate our own efforts and those of our contemporaries in complete detachment from the achievements of former centuries?

Whether we know it or not, we are still following in the footsteps of the leading entomologists of the past. And among these leaders the greatest, I am convinced, is Réaumur, for of all the early investigators he combined the most comprehensive vision with the keenest powers of observation and the clearest conception of importance, both theoretical and practical, of the study of insects.

Wheeler discovered the unpublished Réaumur manuscript on ants when he was visiting professor at the University of Paris in 1925, although his interest in Réaumur began much earlier. In 1918 Wheeler and the Harvard historian of science George Sarton exchanged thoughts on Réaumur's achievements, and they continued to correspond on this subject until after the publication of the book, which Sarton reviewed in *Isis,* the journal of the history of science which he edited. Sarton especially liked the forty-two pages of notes which followed the text: "The ideal annotator of a specific scientific text is not so much the historian as the scientific expert, who alone can estimate the full meaning of every scientific fact involved. Or better still, the perfect annotation requires a combination of historical and scientific qualities which are but rarely united in a single head. The whole History of Science implies, as the name indicates, an intimate collaboration of historians and scientists, the former drawing the frames and placing every event in its proper perspective, the latter analyzing more deeply the technical contents of specific works."[5]

One of Wheeler's major disciples in this field was Donald Culross Peattie, who had been a student in one of Wheeler's classes in 1921 and who was a classmate of Ralph Wheeler's. Peattie's *An Almanac for Moderns* (1935) and his *Green Laurels* (1936) are filled with tributes to biologists of the past, and his debt to Wheeler is evident in the correspondence between the two as well as in the books themselves. Réaumur, Huber, Fabre, and others of Wheeler's favorites are vividly portrayed in Peattie's pages, and of Lamarck he wrote that "his ideas appear to be indispensable if we are ever to explain the origin of instinct."

Wheeler also authored a short history of the Bussey Institution in 1930 and one of entomology at Harvard in 1936. His reasons for writing the latter were expressed in a letter to Samuel Henshaw in 1933, three years after the publication of *A History of Applied Entomology* by his friend L. O. Howard of the U.S. Bureau of Entomology. "I am very much disappointed with some of the recent histories of entomology in the United States, especially with the works of Escherich and L. O. Howard . . . Howard and Escherich have quite distorted the early history of entomology in the United States, so far as Harvard University is concerned."[6]

Wheeler often expressed his feelings on this subject, and George Sarton also remarked in his review of the Réaumur translation: "When shall we be given a good history of entomology?" Such a history has never been undertaken, and now, three decades later, entomologists (and biologists generally) seem more unmindful of their past than ever. In a letter to Wheeler, Sarton characterized biologists as "the least curious of the past of all men of science."[7]

Wheeler, with much justification, regarded himself as a "spiritual great-great-grandson" of Réaumur. Réaumur (1683–1757) had strongly influenced Charles Bonnet in Geneva, and it was Bonnet who encouraged the blind François Huber to undertake his famous studies of the honeybee. Huber's son Pierre conducted brilliant studies on the Swiss ants, and it was the receipt of a copy of Pierre Huber's book that set Auguste Forel upon his career as "historian of the ants." Of Wheeler's debt to Forel we have spoken many times. Not only was Wheeler's approach to ant behavior and to systematics similar to that of Forel, but his interest in psychology may owe something to Forel. Wheeler's often diffuse style of writing and his enthusiasm for even the minutest details may even have resulted in part from his admiration of Forel's publications. In the introduction to *The Life of the Ant* (1930), Maurice Maeterlinck fittingly commented that "As an observer, Wheeler is no less scrupulous than Forel and Wasmann; but as a thinker he sees further and probes more deeply, and derives from what he has seen reflections and general ideas of greater scope than those of his colleagues."[8]

During his lifetime, Wheeler was commonly regarded as one of the world's foremost biologists, the greatest American zoologist since Agassiz, and so forth. Whether he was in truth a genius depends upon how one defines that much-abused word. If one accepts Thomas A. Edison's definition of genius as one percent inspiration and ninety-nine percent perspiration, then he surely qualifies. The touchstones of Wheeler's success were enthusiasm, self-discipline, and hard work. That one person could do so much reading (in several languages), so much writing, lecturing, instructing of graduate students, traveling, and collecting, and still find time to identify ants for almost everyone who asked him remains a mystery. In the midst of all his work on ants and social insects, and while he was still dean of the Bussey Institution, he found time to breed and study an unusual little bethylid wasp, *Scleroderma,* and to rear ant lions and worm lions! True, he had relatively little social life and no hobbies aside from his work, but the same could be said of many a less productive scientist.

We would say, however, that there was somewhat more than one percent inspiration in Wheeler's work (as there most assuredly was in Edison's!). After all, he was a pioneer in the field of insect embryology, in the study of several marine invertebrates, in the science of ethology, in the systematics and biology of ants, in the field of animal sociality. In each case he not only synthesized the published information in a useful manner, but completed a major body of original work. In many cases he improved upon the existing conceptual schemes or contributed concepts of his own. Yet he was not a highly original thinker, but rather an eclectic, a man with a voracious thirst for facts and ideas and a capacity for drawing upon his erudition for a series of brilliant essays, books, and research papers. Two of his major strengths, his sense of the history of his subject and his broad knowledge of the classics, may also have been his major weaknesses, for it can be argued that his failure to contribute important new ideas of permanent value stemmed from "a too constant preoccupation with historical details," to use his own words. And surely many of his essays are, by contemporary standards, overloaded with allusions to the past and with quotations in a variety of languages. Wheeler was, of course, being true to himself, and uninterested in reaching the "movie-fed public" on any other terms.

Evidently Wheeler himself regarded his essays as of less importance than his taxonomic studies. In the foreword of the posthumous *Essays in Philosophical Biology* (Harvard University Press, 1939), Thomas Barbour wrote as follows:

> Wheeler wrote most of these essays in a spirit of fun and spent extraordinarily little time in their preparation. I do not think that he himself ever really rated them very highly for he was unbelievably modest and ill equipped to appraise his own intellectual products. In talking of himself, as he often did, for we lunched together no less than 447 times between 1930 and 1937, and how many times before that I have no record, he always made it clear that he considered his taxonomic and ecological work of much more importance than anything else which he did and he rated artistry in taxonomy above all other of his powers. No one who knew him, or his work, can deny that his systematic work will last as long as there are entomologists and insects; on the other hand still less could one deny that his vast range of varied reading, his penetrating knowledge of the classics, and his extraordinary familiarity with modern foreign languages in their utmost refinements, finds an outlet and expression in these articles which, from the very deathlessness of their quality, demands preservation in a more permanent form than he gave them in their original publication.

Essays in Philosophical Biology contains a somewhat different selection than *Foibles of Insects and Men,* although four essays appear in both. *Foibles* has been long out of print, as have *Demons of the Dust, The Social Insects,* and most of the rest of Wheeler's publications. *Ants* (1910) has been reissued by the Columbia University Press (1960), and *Essays in Philosophical Biology* has recently been reprinted by Russell and Russell of New York (1967). These books and essays contain the essence of Wheeler, and despite their occasionally turgid style and frequent references to persons and concepts that no longer seem important to us, they remain readable and entertaining, as well as valuable syntheses of knowledge in Wheeler's time. While most of his writings are available in only the best libraries, he is still frequently quoted. For example, the final paragraphs of "The Dry-Rot of Academic Biology" were quoted recently in no less than three different books,* and "The Termitodoxa" was reprinted in its entirety in Beebe's *The Book of Naturalists* (1944).

In 1941 Clifton Fadiman published an anthology called *Reading I've Liked,* and included in this was C. K. Ogden's review of the *Encyclopaedia Britannica* originally published in 1926 in *The Saturday Review of Literature.* Ogden had scolded the encyclopaedia for its omissions, including "no mention of so profound and influential a thinker as Professor W. M. Wheeler, American's leading entomologist, and perhaps her leading sociologist as well." Fadiman pointed out in his introduction to this essay how shrewdly Ogden called the intellectual turn, and that while fifteen or twenty years later it was easier to estimate the importance of Wheeler and the other prominent men, "in 1926 it required rare erudition, judgement, and boldness to be as certain of their worth as was C. K. Ogden."[9]

To many of his contemporaries, Wheeler was very much a "philosophical biologist," a fairly rare breed, since most biologists seem content to leave philosophy to the physicists and the mathematicians. Here is a sample of his philosophy, taken from a talk to the American Association for the Advancement of Science in 1920, but just as appropriate to the contemporary scene, if not more so:

As the earth becomes more densely covered with its human populations, it becomes increasingly necessary to retain portions of it in a wild state, that is, free from the organizing mania of man, as national and city

* Joseph Wood Krutch, *Grand Canyon* (1957); L. S. Dillon and E. S. Dillon, *The Science of Life* (1964); H. E. Evans, *Life on a Little Known Planet* (1968).

parks or reservations to which we can escape during our holidays from the administrators, organizers, and efficiency experts and everything they stand for and return to a Nature that really understands the business of organization. Why may we not regard scientific research, artistic creation, religious contemplation, and philosophic speculation as the corresponding reservations of the mind, great world parks to which man must resort to escape from the deadening, overspecializing routine of his habits, mores, and occupations and enjoy veritable creative holidays of the spirit? These world parks are in my opinion the best substitute we are ever likely to have for the old theological Heaven, and they have the great advantage that some of us are privileged to return from them with discoveries and inventions to lighten the mental and physical burdens of those whose inclinations or limitations leave them embedded in routine.

In retrospect, it appears that Wheeler was wrong in rating his taxonomic work of more permanent value than his essays. It was in taxonomy, especially, that his work suffered from undue devotion to the past, for it was his admiration for Emery and Forel, in particular, that led him to cling to outmoded terms and to describe many trivial variants as varieties and subspecies. Every taxonomist commits errors as the result of drawing conclusions from too limited material, but Wheeler's insistence on applying a pentanomial system of nomenclature, while making little attempt to analyze the expected variation in series from one nest or one geographic area, led to his creation of an unusual number of names that have since been relegated to synonymy. If taxonomy was Wheeler's "knitting," as Mrs. Wheeler used to say, then it must be admitted that others have had to do a good deal of unravelling. This does not mean that his taxonomic work was not of major importance, for indeed he laid the foundations of myrmecology in North America and did much to improve knowledge of ants of other parts of the world. But far more important than his many descriptions of new ants was his enthusiasm for ants as living, social organisms and as exemplars of so many biological principles.

Any estimate of Wheeler's accomplishments must also take into consideration the excellence of his students, for in fact it would be difficult to point out another American entomologist whose students, taken as a group, have made such important contributions. Many have one or more books to their credit, and several are or were acknowledged leaders in their fields: George Salt in insect parasitology, P. J. Darlington, Jr. in zoogeography, F. M. Carpenter in insect paleontology, A. C. Kinsey in gall wasp systematics and later in human sexual behavior, to mention only a few. Several went on to make important contributions to myrmecology: W. S. Creighton, G. C. Wheeler, and Neal Weber; still others became leaders in the systematics of other groups: Z. P. Metcalf, T. B. Mitchell, R. F. Hussey, Howard Parshley, and others. Even after this categorization, one still has an imposing list of leftovers: William Mann, A. L. Melander, C. T. Brues, F. X. Williams, O. E. Plath, Marston Bates, and still others. The essence of Wheeler's success with students is perhaps best summarized by C. T. Brues in an obituary published in *Psyche:*

> Wheeler always dealt with his students as he would with colleagues. With his broad intellectual viewpoint he could do this with ease, and without apparent effort he would quickly stimulate these young men to accomplishments quite beyond their own expectations. He was always enthusiastically interested in his own work and however deeply immersed in it, was always ready to welcome the student who wandered into his laboratory at any time. Frequently, such conferences would turn to an account of what he was doing at the moment or to a critical review of some important book which he had just read. The immediate effect of such contacts was frequently disheartening in the extreme, as it emphasized the extent of any biological problem and the inadequate background of the young man who was attempting to solve it. However, the final result of a series of such meetings was highly salutary, and it gave to most of his students the impetus needed to complete their work well, and furthermore to prolong their studies after the inevitable doctor's thesis had been finished. This ability to instill his own ideals of research into the minds of younger men was a salient characteristic of his personality and it has done much to further the real advance of entomological investigation in many fields.[10]

Neither of the authors ever had the privilege of meeting Wheeler or of hearing him lecture—we were in high school and early college when he died. Our impressions of him have been formed after many conversations with his students, family, and friends, as well as after several years of study of his writings, both formal and informal. Wheeler was in every sense a self-made man, a person of modest origins whose active mind and inner drive kept him working long after others were resting and relaxing from an eight-hour day.

Even Colebrook, which symbolized escape for his family, was to him merely a transfer from his desk in the city to another in the country, though his chance to get outdoors and collect was undoubtedly his way of relaxing. His incessant energy took its toll on his nervous system, and he had neurotic tendencies throughout his life. His diaries and letters contain many references to the state of his health, and from time to time he had periods of depression, the worst being his collapse in 1926. But, as we have shown, even in the hospital he could not completely escape his absorption in his writing.

Wheeler's erudition and his devotion to his work made him appear aloof and unsympathetic to less capable or less experienced workers. More than one person has told us of his disconcertingly piercing blue eyes, and how after a foolish remark or an unsuccessful effort to be funny or profound, he would respond with a withering look, though seldom with a comment. Along with his self-assurance went something not always found in men absorbed in their work: a concern with his appearance. He was almost always well dressed, even in the field. As a young man, he took great interest in his mustache. When he went to work for Ward's he bought a new suit as soon as he could afford it, and when he went to Germany in 1893 he stopped over in London long enough to have a suit tailor-made. Wheeler had no church affiliations (he shed those when a very young man), yet he liked formality in the dinners arranged by Mrs. Wheeler for his friends, in the commencement exercises at Harvard, and in his chauffeur-driven car.

Wheeler was an incessant smoker, preferring a pipe but sometimes using cigars or even cigarettes when he ran out of pipe tobacco. On shipboard or at family gatherings he often played cards, but was said to be not always the best loser. (At Woods Hole, after losing at "Skat" again to Whitman, he would retire to his quarters to practice by himself!) Although basically a quiet, withdrawn person, in appropriate company he proved to have an unusual fund of lively stories, not always of a kind suitable for "polite society." In informal gatherings he was often the center of attention, and it was here that his numerous prejudices came to the surface: geneticists, psychologists, and even Jews (whom he enjoyed citing as examples of social parasitism).

His streak of antisemitism is curious, for one of Wheeler's closest friends at the German-English Seminary and during the years immediately following was a Jewish boy named Adolph Bernhard. In the Wheeler family papers are several of Bernhard's letters, although unfortunately none of Wheeler's to Bernhard. The two exchanged confidences about the future of the world, thoughts on women, on professors, on being Jewish, and on other subjects common to most friends age twenty. During World War I the German-Jewish geneticist Richard Goldschmidt, always a controversial figure, was at Bussey before being interned as a German national, and Wheeler and Goldschmidt did not get along for reasons we can only surmise. Perhaps some of Wheeler's prejudices began then. Certainly he was not very helpful in later years when asked to help bring foreign scientists to this country where they could get a new start. Yet he had many professional friends who were Jewish. Jacques Loeb and Wheeler, for example, had their disagreements, yet they patched their quarrels periodically and respected one another's abilities throughout their lives. Wheeler befriended and inspired Phil Rau, a Jewish storekeeper and amateur biologist, in his wasp behavior studies. To what extent Wheeler's biases were genuine and to what extent nurtured for conversation's sake is difficult to say.

To say that Wheeler was not without faults or above prejudice is only to say that he was human and that he was rather more complex and versatile than most of us. He was unique, a rare spirit who fortunately left abundant traces in his writings and in his students. That he no longer seems the giant he did to many of his contemporaries is largely a consequence of the fact that biology itself is expanding at such a rate that few individuals, past or present, assume large proportions. In that expansion Wheeler himself played a not unimportant role. Were he alive today, he would surely have something to say about our modern biology laboratories, insulated as they are from nature by barriers of sophisticated equipment and filled with faceless individuals engaged in trophallaxis with their computers. But such an analogy with social insects would have been made in fun, for Wheeler, despite his jibes at the experimentalists, held strong opinions on the importance of biology in its broadest sense. Science (like politics) would profit from a little more humor. And as to the role of biology in human affairs, Wheeler had it on the best authority:

It is my opinion . . . that if you will only increase your biological investigators a hundred fold, put them in positions of trust and responsibility much more often and before they are too old, and pay them at least as well as you are paying your plumbers and bricklayers, you may look forward to making as much social progress in the next three centuries as you have made since the Pleistocene. That some such opinion may also be entertained by some of your statesmen sometime before the end of the present geological age, is the sincere wish of

Yours truly,
Wee-Wee
43rd Neotenic King of the 8429th
Dynasty of the Bellicose Termites

Bibliography of William Morton Wheeler

Notes

Index

Bibliography of William Morton Wheeler

The following list is based on a similar one by C. T. Brues, which was published in *Psyche* in September 1937. We have made fairly numerous corrections and several additions.

1884

The Colugo and His Cousin. *Ward's Nat. Sci. Bull.* 3, no. 2 (July 1, 1884): 10–11.

1885

Catalogue of Specimens of Mollusca and Brachiopoda for Sale at Ward's Natural Science Establishment. Rochester, New York. 120 pp., 84 figs.

1886

A List of Trees Found in the City of Milwaukee. *Proc. Wisc. Pharm. Assoc.* 5:24–25.

1887

On the Distribution of Coleoptera along the Lake Michigan Beach of Milwaukee County. *Proc. Wisc. Nat. Hist. Soc.* April 1887, pp. 132–140.

1888

The Flora of Milwaukee Co., Wisconsin. *Proc. Wisc. Nat. Hist. Soc.,* April 1888, pp. 154–190.

(With G. W. Peckham and E. G. Peckham) The Spiders of the Sub-Family Lyssomanae. *Trans. Wisc. Acad. Sci., Arts, Lett.* 7:222–256. 2 pls.

Report of the Custodian. *Ann. Rep. Milw. Pub. Mus.,* no. 6, pp. 7–18.

In Memoriam: Thure Ludwig Theodor Kumlien (1819–1888). *Ann. Rep. Milw. Pub. Mus.,* no. 6, 10 unnumbered pp. following p. 35.

1889

The Embryology of *Blatta germanica* and *Doryphora decemlineata. J. Morph.* 3:291–386. 7 pls., 16 figs.

Homologues in Embryo Hemiptera of the Appendages to the First Abdominal Segment of Other Insect Embryos. *Am. Naturalist* 23:644–645.

Über drüsenartige Gebilde im ersten Abdominalsegment der Hemipteren-embryonen. *Zool. Anz.,* Jahrg. 12, pp. 500–504, 2 figs.

Report of the Custodian. *Ann. Rep. Milw. Pub. Mus.,* no. 7, pp. 7–24.

On Two New Species of Cecidomyid Flies Producing Galls on *Antennaria plantaginifolia. Proc. Wisc. Nat. Hist. Soc.,* April 1889, pp. 209–216.

Two Cases of Insect Mimicry. *Proc. Wisc. Nat. Hist. Soc.,* April 1889, pp. 217–221.

Flora of Milw. Co., 1st Suppl. *Proc. Wisc. Nat. Hist. Soc.,* April 1889, pp. 224–231.

1890

Descriptions of Some New North American Dolichopodidae. *Psyche* 5:337–343, 355–362, 373–379. 1 pl.

The Supposed Bot-Fly Parasite of the Box-Turtle. *Psyche* 5:403.

Review of E. B. Poulton's "The Colors of Animals, Their Meaning and Use, Especially Considered in the Case of Insects." *Science* 16:286.

Hydrocyanic Acid Secreted by *Polydesmus virginiensis* Drury. *Psyche* 5:442.

Review of R. H. Lamborn's "Dragon-Flies versus Mosquitoes." *Science* 16:284.

On the Appendages of the First Abdominal Segment of Embryo Insects. *Trans. Wisc. Acad. Sci., Arts, Lett.* 8:87–140. 3 pls.

Note on the Oviposition and Embryonic Development of *Xiphidium ensiferum* Scud. *Insect Life* 2:222–225.

Über ein eigenthümliches Organ in Locustidenembryo. *Zool. Anz.,* Jahrg. 13, pp. 475–480.

Report of the Custodian. *Ann. Rep. Milw. Pub. Mus.,* no. 8, pp. 7–24.

1891

The Embryology of a Common Fly. *Psyche* 6:97–99.

The Germ-Band of Insects. *Psyche* 6:112–115.

Neuroblasts in the Arthropoid Embryo. *J. Morphol.* 4:337–343. 1 fig.

Hemidiptera haeckelii. Psyche 6:66–67.

BIBLIOGRAPHY OF WILLIAM MORTON WHEELER

1892

Concerning the "Blood-Tissue" of the Insecta. *Psyche* 6:216–220, 233–236, 253–258. 1 pl.

A Dipterous Parasite of the Toad. *Psyche* 6:249.

1893

A Contribution to Insect Embryology: Inaugural Dissertation. *J. Morphol.* 8:1–160. 6 pls., 7 figs.

The Primitive Number of Malpighian Vessels in Insects. *Psyche* 6:457–460, 485–486, 497–498, 509–510, 539–541, 545–547, 561–564. 2 figs.

1894

Syncoelidium pellucidum, a New Marine Triclad. *J. Morphol.* 9:167–194. 1 pl.

Planocera inquilina, a Polyclad Inhabiting the Branchial Chamber of *Sycotypus canaliculatus,* Gill. *J. Morphol.* 9:195–201. 2 figs.

Protandric Hermaphroditism in Myzostoma. *Zool. Anz.* 17:177–182.

1895

The Behavior of the Centrosomes in the Fertilized Egg of *Myzostoma glabrum,* Leuckart. *J. Morphol.* 10:305–311. 10 figs.

Translation of Wilhelm Roux's *The Problems, Methods, and Scope of Developmental Mechanics. Biol. Lectures,* Marine Biol. Lab., Woods Hole, pp. 149–190. Boston: Ginn & Co.

1896

The Sexual Phases of *Myzostoma. Mitth. Zool. Station zu Neapel* 12:227–302. 3 pls.

The Genus *Ochthera. Entomol. News* 7:121–123. 2 figs.

Two Dolichopodid Genera New to America. *Entomol. News* 7:152–156. 1 fig.

A New Genus and Species of Dolichopodidae. *Entomol. News* 7:185–189. 1 fig.

A New Empid with Remarkable Middle Tarsi. *Entomol. News* 7:189–192. 3 figs.

An Antenniform Extra Appendage in *Dilophus tibialis* Loew. *Arch. Entwicklungsmech. Organ.* 3:261–268. 1 pl.

1897

A Genus of Maritime Dolichopodidae New to America. *Proc. Calif. Acad. Sci.* ser. 3, 1:145–152. 1 pl.

The Maturation, Fecundation, and Early Cleavage of *Myzostoma glabrum* Leuckart. *Arch. Biol.* 15:1–77. 3 pls.

Two Cases of Mimicry. *Chicago Univ. Record.* 2:1.

[Marine Fauna of San Diego Bay, California.] *Science,* n.s. 5:775–776. (Paper read before the Zoological Club of the University of Chicago, April 14, 1897.)

1898

A New Genus of Dolichopodidae from Florida. *Zool. Bull.* 1:217–220. 1 fig.

Review of Bürger and Carrière's "The Embryonic Development of the Wall-Bee (*Chalicodoma murana* Fabr.)." *Am. Naturalist* 32:794–798.

Review of A. S. Packard's *A Text Book of Entomology, Including the Anatomy, Physiology, Embryology, and Metamorphoses of Insects. Science,* n.s. 7:834–836.

A New Peripatus from Mexico. *J. Morphol.* 15:1–8. 1 pl., 1 fig.

1899

George Baur's Life and Writings. *Am. Naturalist.* 33:15–30. 1 pl.

The Life-History of *Dicyema. Zool. Anz.* 22:169–176.

Anemotropism and Other Tropisms in Insects. *Arch. Entwicklungsmech. Organ.* 8:373–381.

The Prospects of Zoological Study in Texas. *University of Texas Record* 1:335–339.

New Species of Dolichopodidae from the United States. *Proc. Calif. Acad. Sci.* ser. 3, 2:1–84, 4 pls.

The Development of the Urinogenital Organs of the Lamprey. *Zool Jahrb. Abth. Anat. und Ontogenie* 13:1–88. 7 pls.

J. Beard on the Sexual Phases of *Myzostoma. Zool. Anz.* 22:281–288.

Caspar Friedrich Wolff and the *Theoria Generationis. Biol. Lectures,* Marine Biol. Lab., Woods Hole, pp. 265–284. Boston: Ginn & Co.

1900

The Free-Swimming Copepods of the Woods Hole Region. *U.S. Fish Comm. Bull. for 1899,* pp. 157–192. 30 figs.

On the Genus *Hypocharassus,* Mik. *Entomol. News* 11:423–425.

The Study of Zoology. *Univ. of Texas Record,* vol. 2, no. 2, pp. 125–135.

Review of E. Korschelt and K. Heider's *Text-book of Embryology of the Invertebrates,* vols. II and III. *Science,* n.s. 11:148–149.

The Female of *Eciton sumichrasti* Norton, with Some Notes on the Habits of Texan Ecitons. *Am. Naturalist* 34:563–574. 4 figs.

The Habits of *Myrmecophila nebrascensis* Bruner. *Psyche* 9:111–115. 1 fig.

A Singular Arachnid (*Koenenia mirabilis* Grassi) Occurring in Texas. *Am. Naturalist* 34:837–850. 4 figs.

A New Myrmecophile from the Mushroom Gardens of the Texan Leaf-Cutting Ant. *Am. Naturalist* 34:851–862. 6 figs.

A Study of Some Texan Ponerinae. *Biol. Bull.* 2:1–31. 10 figs.

The Habits of *Ponera* and *Stigmatomma*. *Biol. Bull.* 2:43–69. 8 figs.

1901

(With W. H. Long.) The Males of Some Texan Ecitons. *Am. Naturalist* 35:157–173. 3 figs.

Impostors among Animals. *Century Mag.* 62:369–378. 8 figs.

The Compound and Mixed Nests of American Ants. *Am. Naturalist* 35:431–448, 513–539, 701–724, 791–818. 20 figs.

Microdon Larvae in *Pseudomyrma* Nests. *Psyche* 9:222–224. 1 fig.

The Parasitic Origin of Macroërgates among Ants. *Am. Naturalist* 35:877–886. 1 fig.

An Extraordinary Ant-Guest. *Am. Naturalist* 35:1007–1016. 2 figs.

Notices biologique sur les fourmis Mexicaines. *Ann. Soc. Entomol. Belg.* 45:199–205.

1902

A New Agricultural Ant from Texas, with Remarks on the Known North American Species. *Am. Naturalist* 36:85–100. 8 figs.

A Consideration of S. B. Buckley's "North American Formicidae." *Trans. Tex. Acad. Sci. for 1901* 4:17–31.

(With A. L. Melander.) Empididae. *Biol. Centrali-Americana: Diptera (supp.)*, pp. 366–376.

A Neglected Factor in Evolution. *Science,* n.s. 15:766–774.

'Natural History,' 'Oecology,' or 'Ethology?' *Science,* n.s. 15:971–976.

Formica fusca Linn. subsp. *subpolita* Mayr. var. *perpilosa* n. var. *Mem. Rev. Soc. Cient. "Antonia Alzate," Mexico* 17:141–142.

New Agricultural Ants from Texas. *Psyche* 9:387–393.

Translation of Carlo Emery's "An Analytical Key to the Genera of the Family Formicidae for the Identification of the Workers." *Am. Naturalist* 36:707–725.

Review of P. Bachmetjew's *Temperaturverhältnisse bei Insekten. Am. Naturalist* 36:401–405.

Review of J. H. Comstock and V. L. Kellogg's *The Elements of Insect Anatomy. Science,* n.s. 16:351–352.

An American *Cerapachys,* with Remarks on the Affinities of the Cerapachyinae. *Biol. Bull.* 3:181–191. 5 figs.

The Occurrence of *Formica cinerea* Mayr and *Formica rufibarbis* Fabricius in America. *Am. Naturalist* 36:947–952.

1903

Review of James Mark Baldwin's *Development and Evolution. Psychol. Rev.* 10:70–80.

Erebomyrma, a New Genus of Hypogaeic Ants from Texas. *Biol. Bull.* 4:137–148. 5 figs.

(With J. F. McClendon.) Dimorphic Queens in an American Ant (*Lasius latipes* Walsh). *Biol. Bull.* 4:149–163. 3 figs.

Ethological Observations on an American Ant (*Leptothorax emersoni* Wheeler). *J. Psychol. Neurol.* 2:1–31. 1 fig.

A Revision of the North American Ants of the Genus Leptothorax Mayr. *Proc. Acad. Nat. Sci. Phila.* 55:215–260. 1 pl.

Review of G. N. Calkin's *The Protozoa. Am. Naturalist* 37:214–216.

Review of T. W. Headley's *Problems of Evolution. Psychol. Rev.* 10:193–199.

A Decad of Texan Formicidae. *Psyche* 10:93–111. 10 figs.

The North American Ants of the Genus Stenamma Sensu Stricto. *Psyche* 10:164–168.

How Can Endowments Be Used Most Effectively for Scientific Research? *Science,* n.s. 17:577–579.

The Origin of Female and Worker Ants from the Eggs of Parthenogenetic Workers. *Science,* n.s. 18:830–833.

Review of "Report on the Collections of Natural History made in the Antarctic Regions during the Voyage of the Southern Cross, London, 1902." *Bull. Am. Geog. Soc.* 35:572–573.

Some Notes on the Habits of *Cerapachys augustae*. *Psyche* 10:205–209. 1 fig.

Extraordinary Females in Three Species of Formica, with Remarks on Mutation in the Formicidae. *Bull. Am. Mus. Nat. Hist.* 19:639–651. 3 figs.

Some New Gynandromorphous Ants: With a Review of the Previously Recorded Cases. *Bull. Am. Mus. Nat. Hist.* 19:653–683. 11 figs.

Review of C. W. Dodge's *General Zoology: Practical, Systematic, and Comparative*. *Science,* n.s. 18:824–825.

1904

Translation of August Forel's "Ants and Some Other Insects: An Inquiry into the Psychic Powers of These Animals with an Appendix on the Peculiarities of Their Olfactory Sense." *Monist* 14, nos. 1 and 2 (October 1903 and January 1904): 33–66, 177–193. 3 figs. (Reprinted, Religion of Science Library, no. 56, Chicago, 1904.)

Three New Genera of Inquiline Ants from Utah and Colorado. *Bull. Am. Mus. Nat. Hist.* 20:1–17. 2 pls.

Review of E. E. Austen's "A Monograph of the Tsetse-Flies (Genus *Glossina,* Westwood) Based on the Collection in the British Museum." *Bull. Am. Geog. Soc.* 35:573–575.

Woodcock Surgery. *Science,* n.s. 19:347–350.

The Obligations of the Student of Animal Behavior. *Auk* 21:251–255.

A Crustacean-Eating Ant (*Leptogenys elongata* Buckley). *Biol. Bull.* 6:251–259. 1 fig.

The American Ants of the Subgenus *Colobopsis*. *Bull. Am. Mus. Nat. Hist.* 20:139–158. 7 figs.

Dr. Castle and the Dzierzon Theory. *Science,* n.s. 19:587–591.

The Ants of North Carolina. *Bull. Am. Mus. Nat. Hist.* 20:299–306.

On the Pupation of Ants and the Feasibility of Establishing the Guatemalan Kelep, or Cotton-Weevil Ant, in the United States. *Science,* n.s. 20:437–440.

Social Parasitism among Ants. *Am. Mus. J.* 4:74–75.

A New Type of Social Parasitism among Ants. *Bull. Am. Mus. Nat. Hist.* 20:347–375.

The Phylogeny of the Termites. *Biol. Bull.* 8:29–37.

Some Further Comments on the Guatemalan Boll Weevil Ant. *Science,* n.s. 20:766–768.

Ants from Catalina Island, California. *Bull. Am. Mus. Nat. Hist.* 20:267–271.

1905

An Interpretation of the Slave-Making Instincts in Ants. *Bull. Am. Mus. Nat. Hist.* 21:1–16.

Ethology and the Mutation Theory. *Science,* n.s. 21:535–540.

The Ants of the Bahamas: With a List of the Known West Indian Species. *Bull. Am. Mus. Nat. Hist.* 21:79–135. 1 pl., 23 figs.

Some Remarks on Temporary Social Parasitism and the Phylogeny of Slavery among Ants. *Biol. Centralbl.* 25:637–644.

New Species of *Formica. Bull. Am. Mus. Nat. Hist* 21:267–274.

Ants from Catalina Island, California. *Bull. S. Calif. Acad. Sci.* 4:60–62.

The Structure of Wings. *Bird Lore* 7:257–262.

A New *Myzostoma,* Parasitic in a Starfish. *Biol. Bull* 8:75–78. 1 fig.

How the Queens of the Parasitic and Slave-Making Ants Establish Their Colonies. *Am. Mus. J.* 5:144–148.

The North American Ants of the Genus *Dolichoderus. Bull. Am. Mus. Nat. Hist.* 21:305–319. 2 pls., 3 figs.

The North American Ants of the Genus *Liometopum. Bull. Am. Mus. Nat. Hist.* 21:321–333. 3 figs.

An Annotated List of the Ants of New Jersey. *Bull. Am. Mus. Nat. Hist.* 21:371–403. 4 figs.

Ants from the Summit of Mount Washington. *Psyche* 12:111–114.

Worker Ants with Vestiges of Wings. *Bull. Am. Mus. Nat. Hist.* 21:405–408. 1 pl.

Dr. O. F. Cook's "Social Organization and Breeding Habits of the Cotton-Protecting Kelep of Guatemala." *Science,* n.s. 21:706–710.

1906

The Habits of the Tent-Building Ant (*Crematogaster lineolata* Say). *Bull. Am. Mus. Nat. Hist.* 22:1–18. 6 pls., 3 figs.

On the Founding of Colonies by Queen Ants, with Special Reference to the Parasitic and Slave-Making Species. *Bull. Am. Mus. Nat. Hist.* 22:33–105. 7 pls., 1 fig.

On Certain Tropical Ants Introduced into the United States. *Entomol. News* 17:23–26.

The Queen Ant as a Psychological Study. *Pop. Sci. Monthly* 68:291–299. 7 figs.

The Kelep Excused. *Science,* n.s 23:348–350.

Pelastoneurus nigrescens Wheeler, a synonym of *P. dissimilipes* Wheeler: A Correction. *Entomol. News* 17:69.

New Ants from New England. *Psyche* 13:38–41. 1 pl.

Fauna of New England: List of the Formicidae. *Occas. Papers Boston Soc. Nat. Hist.* 7:1–24.

A New Wingless Fly *(Pulciphora borinquensis)* from Porto Rico. *Bull. Am. Mus. Nat. Hist.* 22:267–271. 1 pl.

The Ants of Japan. *Bull. Am. Mus. Nat. Hist.* 22:301–328. 1 pl., 2 figs.

The Ants of the Grand Cañon. *Bull. Am. Mus. Nat. Hist.* 22:329–345.

The Ants of the Bermudas. *Bull. Am. Mus. Nat. Hist.* 22:347–352. 1 fig.

Concerning *Monomorium destructor* Jerdon. *Entomol. News* 17:265.

An Ethological Study of Certain Maladjustments in the Relations of Ants to Plants. *Bull. Am. Mus. Nat. Hist.* 22:403–418. 6 pls., 1 fig.

The Expedition to Colorado for Fossil Insects. *Am. Mus. J.* 6:199–203. 5 figs.

1907

A Collection of Ants from British Honduras. *Bull. Am. Mus. Nat. Hist.* 23:271–277. 2 pls.

The Polymorphism of Ants, with an Account of Some Singular Abnormalities Due to Parasitism. *Bull. Am. Mus. Nat. Hist.* 23:1–93. 6 pls.

Notes on a New Guest-Ant, *Leptothorax glacialis,* and the Varieties of *Myrmica brevinodis* Emery. *Bull. Wisc. Nat. Hist. Soc.* 5:70–83.

On Certain Modified Hairs Peculiar to the Ants of Arid Regions. *Biol. Bull.* 13:185–202. 14 figs.

The Fungus-growing Ants of North America. *Bull. Am. Mus. Nat. Hist.* 23:669–807. 5 pls., 31 figs.

The Origin of Slavery among Ants. *Pop. Sci. Monthly* 71:550–559.

Pink Insect Mutants. *Am. Naturalist* 41:773–780.

1908

The Ants of Porto Rico and the Virgin Islands. *Bull. Am. Mus. Nat. Hist.* 24:117–158. 2 pls., 4 figs.

Comparative Ethology of the European and North American Ants. *J. Psychol. Neurol.* 13:404–435. 2 pls., 6 figs.

The Ants of Jamaica. *Bull. Am. Mus. Nat. Hist.* 24:159–163.

Ants from Moorea, Society Islands. *Bull. Am. Mus. Nat. Hist.* 24:165–167.

Ants from the Azores. *Bull. Am. Mus. Nat. Hist.* 24:169–170.

Vestigial Instincts in Insects and Other Animals. *Am. J. Psychol.* 19:1–13.

The Ants of Texas, New Mexico, and Arizona: I. *Bull. Am. Mus. Nat. Hist.* 24:399–485. 2 pls.

Honey Ants: With a Revision of the American *Myrmecocysti. Bull. Am. Mus. Nat. Hist.* 24:345–397. 28 figs.

The Polymorphism of Ants. *Ann. Entomol. Soc. Am.* 1:39–69. 1 pl.

The Ants of Casco Bay, Maine, with Observations on Two Races of *Formica sanguinea* Latreille. *Bull. Am. Mus. Nat. Hist.* 24:619–645.

A European Ant (*Myrmica levinodis*) Introduced into Massachusetts. *J. Econ. Entomol.* 1:337–339.

Studies on Myrmecophiles: I. *Cremastochilus*. *J. N.Y. Entomol. Soc.* 16:68–79. 3 figs.

Studies on Myrmecophiles: II. *Hetaerius*. *J. N.Y. Entomol. Soc.* 16:135–143. 1 fig.

Studies on Myrmecophiles: III. *Microdon*. *J. N.Y. Entomol. Soc.* 16:202–213. 1 fig.

The Ants of Isle Royale, Michigan. *Rep. Mich. Geol. Surv., 1908,* pp. 325–328.

1909

A Small Collection of Ants from Victoria, Australia. *J. N.Y. Entomol. Soc.* 17:25–29.

Ants Collected by Professor Filippo Silvestri in Mexico. *Boll. Lab. Zool. Gen. Agrar. Portici* 3:228–238.

Review of P. Deegener's *Die Metamorphose der Insekten*. *Science,* n.s. 29:384–387.

Predarwinian and Postdarwinian Biology. *Pop. Sci. Monthly* 74:381–385.

Ants Collected by Professor Filippo Silvestri in the Hawaiian Islands. *Boll. Lab. Zool. Gen. Agrar. Portici.* 3:269–272.

Ants of Formosa and the Philippines. *Bull. Am. Mus. Nat. Hist.* 26:333–345.

A Decade of North American Formicidae. *J. N.Y. Entomol. Soc.* 17:77–90.

A New Honey Ant from California. *J. N.Y. Entomol. Soc.* 17:98–99.

Review of A. D. Hopkins' "The Genus *Dendroctonus*." *J. Econ. Entomol.* 2:471–472.

Observations on Some European Ants. *J. N.Y. Entomol. Soc.* 17:172–187. 2 figs.

1910

Ants: Their Structure, Development, and Behavior. Columbia University Biological Series no. 9. New York: Columbia University Press. 663 pp., 286 figs. (2d printing, 1926; 3d printing, 1960).

Two New Myrmecophilous Mites of the Genus *Antennophorus*. *Psyche*. 17:1–6. 2 pls.

Review of W. Dwight Pierce's "A Monographic Revision of the Twisted Winged Insects Comprising the Order Strepsiptera Kirby." *J. Econ. Entomol.* 3:252–253.

Small Artificial Ant-Nests of Novel Patterns. *Psyche*. 17:73–75. 1 fig.

Review of H. Friese's "Die Bienen Afrikas nach dem Stande unserer heutigen Kenntnisse." *Science,* n.s. 31:580–582.

The Effects of Parasitic and Other Kinds of Castration in Insects. *J. Exp. Zool.* 8:377–438. 8 figs.

Colonies of Ants (*Lasius neoniger* Emery) Infested with *Laboulbenia formicarum* Thaxter. *Psyche* 17:83–86.

An Aberrant Lasius from Japan. *Biol. Bull.* 19:130–137. 2 figs.

Three New Genera of Myrmicine Ants from Tropical America. *Bull. Am. Mus. Nat. Hist.* 28:259–265. 3 figs.

A New Species of *Aphomomyrmex* from Borneo. *Psyche* 17:131–135. 1 fig.

A Gynandromorphous Mutillid. *Psyche* 17: 186–190. 1 fig.

The North American Forms of *Lasius umbratus* Nylander. *Psyche* 17: 235–243.

The North American Forms of *Camponotus fallax* Nylander. *J. N.Y. Entomol. Soc.* 18:216–232.

The North American Ants of the Genus *Camponotus* Mayr. *Ann. N.Y. Acad. Sci.* 20:295–354.

Family Formicidae. In J. B. Smith's *Annual Report of the New Jersey State Museum, Including a Report of the Insects of New Jersey, 1909,* pp. 655–663. Trenton: MacCrellish & Quigley.

1911

The Ant-Colony As an Organism. *J. Morphol.* 22:307–325.

Additions to the Ant-Fauna of Jamaica. *Bull. Am. Mus. Nat. Hist.* 30:21–29.

Review of K. Escherich's "Termitenleben auf Ceylon." *Science,* n.s. 33:530–534.

On *Melanetaerius infernalis* Fall. *Psyche* 18:112–114. 1 fig.

Two Fungus-Growing Ants from Arizona. *Psyche* 18:93–101. 2 figs.

A New *Camponotus* from California. *J. N.Y. Entomol. Soc.* 19:96–98.

Three Formicid Names Which Have Been Overlooked. *Science* n.s. 33:858–860.

Ants Collected in Grenada, W. I., by Mr. C. T. Brues. *Bull. Mus. Comp. Zool.* 54:167–172.

Review of v. Kirchner's *Blumen und Insekten. Science,* n.s. 34:57–58.

A List of the Type Species of the Genera and Subgenera of Formicidae. *Ann. N.Y. Acad. Sci.* 21:157–175.

Literature for 1910 on the Behavior of Ants, Their Guests and Parasites. *J. Anim. Behav.* 1:413–429.

Notes on the Myrmecophilous Beetles of the Genus *Xenodusa,* with a Description of the Larva of *X. cava* Leconte. *J. N.Y. Entomol. Soc.* 19:163–169. 1 fig.

Pseudoscorpions in Ant Nests. *Psyche* 18:166–168.

Descriptions of Some New Fungus-Growing Ants from Texas, with Mr. C. G. Hartman's Observations on Their Habits. *J. N.Y. Entomol. Soc.* 19:245–255. 1 pl.

An Ant-Nest Coccinellid (*Brachyacantha quadripunctata* Mels.). *J. N.Y. Entomol. Soc.* 19:169–174. 1 fig.

Miastor Larvae in Connecticut. *J. N.Y. Entomol. Soc.* 19:201.

Lasius (Acanthomyops) claviger in Tahiti. *J. N.Y. Entomol. Soc.* 19:262.

A Desert Cockroach. *J. N.Y. Entomol. Soc.* 19:262–263.

Three New Ants from Mexico and Central America. *Psyche* 18:203–208.

Insect Parasitism and Its Peculiarities. *Pop. Sci. Monthly* 79:431–449.

1912

The Ants of Guam. *J. N.Y. Entomol. Soc.* 20:44–48.

New Names for Some Ants of the Genus *Formica*. *Psyche* 19:90.

Notes on a Mistletoe Ant. *J. N.Y. Entomol. Soc.* 20:130–133.

Notes about Ants and Their Resemblance to Man. *Nat. Geog. Mag.* 23:731–766. 34 figs.

Additions to Our Knowledge of the Ants of the Genus *Myrmecocystus* Wesmael. *Psyche* 19:172–181. 1 fig.

The Male of *Eciton vagans* Olivier. *Psyche* 19:206–207.

Review of J. H. Comstock's *The Spider Book*. *Science*, n.s. 36:745–756.

1913

Notes on the Habits of Some Central American Stingless Bees. *Psyche* 20:1–9.

A Giant Coccid from Guatemala. *Psyche* 20:31–33. 1 fig.

Review of F. W. L. Sladen's *The Humble Bee: Its Life History and How to Domesticate it*. *Science*, n.s. 37:180–182.

A Revision of the Ants of the Genus *Formica* (Linné) Mayr. *Bull. Mus. Comp. Zool.* 53:379–565. 10 figs.

Observations on the Central American Acacia Ants. *Trans. 2nd Intern. Entomol. Congr., Oxford, England, 1912,* pp. 109–139.

Zoological Results of the Abor Expedition, 1911–1912, XVII; Hymenoptera II: Ants (Formicidae). *Rec. Indian Mus.* 8:233–237.

Corrections and Additions to "List of Type Species of the Genera and Subgenera of Formicidae." *Ann. N.Y. Acad. Sci.* 23:77–83.

Ants Collected in Georgia by Dr. J. C. Bradley and Mr. W. T. Davis, *Psyche* 20:112–117.

The Ants of Cuba. *Bull. Mus. Comp. Zool.* 54:477–505.

Ants Collected in the West Indies. *Bull. Mus. Nat. Hist.* 32:239–244.

A Solitary Wasp (*Aphilanthops frigidus* F. Smith) That Provisions Its Nest with Queen Ants. *J. Anim. Behav.* 3:374–387.

1914

The Ants of the Baltic Amber. *Schrift. Physik-ökonom. Gesellsch. Königsberg* 55:1–142, 66 figs.

(With W. M. Mann.) The Ants of Haiti. *Bull. Am. Mus. Nat. Hist.* 33:1–61. 27 figs.

Gynandromorphous Ants Described During the Decade 1903–1913. *Am. Naturalist* 48:49–56.

Ants Collected by Mr. W. M. Mann in the State of Hidalgo, Mexico. *J. N.Y. Entomol. Soc.* 22:37–61.

Review of O. M. Reuter's *Lebensgewohnheiten und Instinkte der Insekten bis zum Erwachen der sozialen Instinkte*. *Science*, n.s. 39:69–71.

Formica exsecta in Japan. *Psyche* 21:26–27.

Notes on the Habits of Liomyrmex. *Psyche* 21:75–76.

Ants and Bees As Carriers of Pathogenic Microörganisms. *Am. J. Trop. Dis. Prevent. Med.* 2:160–168.

The American Species of *Myrmica* Allied to *M. rubida* Latreille. *Psyche* 21:118–122. 1 fig.

New and Little Known Harvesting Ants of the Genus *Pogonomyrmex*. *Psyche* 21:149–157.

1915

(With F. X. Williams.) The Luminous Organ of the New Zealand Glow-Worm. *Psyche* 22:36–43. 1 pl.

A New Linguatulid from Ecuador. In *Rep. First Harvard Exped. to South America (1913)*, pp. 207–208. Cambridge: Harvard University Press. 1 pl.

Neomyrma versus *Oreomyrma:* A Correction. *Psyche* 22:50.

Some Additions to the North American Ant-Fauna. *Bull. Am. Mus. Nat. Hist.* 34:389–421.

The Australian Honey-Ants of the Genus *Leptomyrmex* Mayr. *Proc. Am. Acad. Arts Sci.* 51:255–286. 12 figs.

Paranomopone: A New Genus of Ponerine Ants from Queensland. *Psyche* 22:117–120. 1 pl.

Hymenoptera. In "Scientific Notes on an Expedition into the North-Western Regions of South Australia." *Trans. Roy. Soc. S. Australia* 39:805–823. 3 pls.

A New Bog-Inhabiting Variety of *Formica fusca* L. *Psyche* 22:203–206.

Two New Genera of Myrmicine Ants from Brazil. *Bull. Mus. Comp. Zool.* 59:483–491. 2 figs.

On the Presence and Absence of Cocoons among Ants, the Nest-Spinning Habits of the Larvae and the Significance of the Black Cocoons among Certain Australian Species. *Ann. Entomol. Soc. Am.* 8:323–342. 5 figs.

The Bussey Institution. *Harvard Univ., Annual Reports, 1914–1915,* pp. 133–137.

1916

The Marriage-Flight of a Bull-Dog Ant (*Myrmecia sanguinea* F. Smith). *J. Anim. Behav.* 6:70–73.

Formicoidea. In *The Hymenoptera of Connecticut. Conn. Nat. Hist. Geol. Surv.,* bull. 22, pp. 577–601.

Prodiscothyrea: A New Genus of Ponerine Ants from Queensland. *Trans. Roy. Soc. S. Australia* 40:33–37. 1 pl.

The Australian Ants of the Genus *Onychomyrmex*. *Bull. Mus. Comp. Zool.* 60:45–54. 2 pls.

Ants Collected in British Guiana by the Expedition of the American Museum of Natural History during 1911. *Bull. Am. Mus. Nat. Hist.* 35:1–14.

(With W. M. Mann.) The Ants of the Phillips Expedition to Palestine during 1914. *Bull. Mus. Comp. Zool.* 60:167–174. 1 fig.

Ants Collected in Trinidad by Professor Roland Thaxter, Mr. F. W. Urich, and Others. *Bull. Mus. Comp. Zool.* 60:323–330. 1 fig.

Jean-Henri Fabre. *J. Anim. Behav.* 6:74–80.

Four New and Interesting Ants from the Mountains of Borneo and Luzon. *Proc. New Engl. Zool. Club* 6:9–18. 4 figs.

Review of H. St. J. K. Donisthorpe's *British Ants: Their Life-History and Classification. Science,* n.s. 43:316–318.

Some New Formicid Names, *Psyche* 23:40.

Notes on Some Slave-Raids of the Western Amazon Ant (*Polyergus breviceps* Emery). *J. N.Y. Entomol. Soc.* 24:107–118.

The Australian Ants of the Genus *Aphaenogaster* Mayr. *Trans. Roy. Soc. S. Australia* 40:213–223. 2 pls.

Note on the Brazilian Fire-Ant, *Solenopsis saevissima* F. Smith. *Psyche* 23:142–143.

An Anomalous Blind Worker Ant. *Psyche* 23:143–145. 1 fig.

Questions of Nomenclature Connected with the Ant Genus *Lasius* and Its Subgenera. *Psyche* 23:168–173.

Two New Ants from Texas and Arizona. *Proc. New Engl. Zool. Club* 6:29–35. 2 figs.

A Phosphorescent Ant. *Psyche* 23:173–174.

An Indian Ant Introduced into the United States. *J. Econ. Entomol.* 9:566–569. 1 fig.

Ants Carried in a Floating Log from the Brazilian Mainland to San Sebastian Island. *Psyche* 23:180–183.

The Bussey Institution. *Harvard Univ., Annual Reports, 1915–1916,* pp. 123–129.

1917

The Mountain Ants of Western North America. *Proc. Am. Acad. Arts Sci.* 52:457–569.

The Australian Ant-Genus *Myrmecorhynchus* (Ern. André) and Its Position in the Subfamily Camponotinae. *Trans. Roy. Soc. S. Australia* 41:14–19. 1 pl.

A New Malayan Ant of the Genus *Prodiscothyrea. Psyche* 24:29–30.

A List of Indiana Ants. *Proc. Indiana Acad. Sci.,* 1917, pp. 460–466.

The North American Ants Described by Asa Fitch. *Psyche* 24:26–29.

The Ants of Alaska. *Bull. Mus. Comp. Zool.* 61:15–22.

The Phylogenetic Development of Subapterous and Apterous Castes in the Formicidae. *Proc. Nat. Acad. Sci.* 3:109–117. 3 figs.

The Synchronic Behavior of Phalangidae. *Science,* n.s. 45:189–190.

Jamaican Ants Collected by Prof. C. T. Brues. *Bull. Mus. Comp. Zool.* 61:457–471, 2 pls., 3 figs.

The Temporary Social Parasitism of *Lasius subumbratus* Viereck. *Psyche* 24:167–176.

Notes on the Marriage Flights of Some Sonoran Ants. *Psyche* 24:177–180.

The Pleometrosis of *Myrmecocystus. Psyche* 24:180–182.

1918

The Ants of the Genus *Opisthopsis* Emery. *Bull. Mus. Comp. Zool.* 62:341–362. 3 pls.

The Australian Ants of the Ponerine Tribe Cerapachyini. *Proc. Am. Acad. Arts Sci.* 53:215–265. 17 figs.

Ants Collected in British Guiana by Mr. C. William Beebe. *J. N.Y. Entomol. Soc.* 26:23–28.

A Great Opportunity for Applied Science. *Harvard Alumni Bull.* 20:264–266.

A Study of Some Ant Larvae, with a Consideration of the Origin and Meaning of the Social Habit among Insects. *Proc. Am. Philos. Soc.* 57:293–343. 12 figs.

Vermileo comstocki, sp. nov., an Interesting Leptid Fly from California. *Proc. New Engl. Zool Club* 6:83–84.

Quick Key to a Knowledge of Common Insects: Review of F. E. Lutz's *Field Book of Insects. Am. Mus. J.* 18:381–382. 8 pls.

Introduction. In Phil and Nellie Rau's *Wasp Studies Afield,* pp. 1–8. Princeton: Princeton University Press.

The Bussey Institution. *Harvard Univ., Annual Reports, 1917–1918,* pp. 118–122.

1919

Two Gynandromorphous Ants. *Psyche* 26:1–8, 2 figs.

The Parasitic Aculeata: A Study in Evolution. *Proc. Am. Philos. Soc.* 58:1–40.

The Ants of Borneo. *Bull. Mus. Comp. Zool.* 63:43–147.

A New Subspecies of *Aphaenogaster treatae* Forel. *Psyche* 26:50.

The Ant Genus *Lordomyrma* Emery. *Psyche* 26:97–106. 4 figs.

A New Paper-Making *Crematogaster* from the Southeastern United States. *Psyche* 26:107–112. 2 pls., 1 fig.

The Ants of Tobago Island. *Psyche* 26:113.

The Ants of the Genus *Metapone* Forel. *Ann. Entomol. Soc. America* 12:173–191. 7 figs.

The Ants of the Galapagos Islands. *Proc. Calif. Acad. Sci.* 2:259–297.

The Ants of Cocos Island. *Proc. Calif. Acad. Sci.* 2:299–308.

A Singular Neotropical Ant (*Pseudomyrma filiformis* Fabricius). *Psyche* 26:124–131. 3 figs.

The Phoresy of *Antherophagus. Psyche* 26:145–152. 1 fig.

The Bussey Institution. *Harvard Univ., Annual Reports, 1918–1919,* pp. 107–108.

1920

The Termitodoxa: Or Biology and Society. *Sci. Monthly* 10:113–124.

The Subfamilies of Formicidae, and Other Taxonomic Notes. *Psyche* 27:46–55. 3 figs.

(With F. M. Gaige.) *Euponera gilva* (Roger), a Rare North American Ant. *Psyche* 27:69–72.

Charles Gordon Hewitt. *J. Econ. Entomol.* 13:262–263. (Obituary.)

(With I. W. Bailey.) The Feeding Habits of Pseudomyrmine and Other Ants. *Trans. Am. Philos. Soc.* 22:235–279. 5 pls., 6 figs.

Review of C. L. Bouvier's *La Vie Psychique des Insectes. Science,* n.s. 52:443–446.

The Bussey Institution. *Harvard Univ., Annual Reports, 1919–1920,* pp. 145–149.

1921

A New Case of Parabiosis and the "Ant Gardens" of British Guiana. *Ecology* 2:89–103. 3 figs.

The Organization of Research. *Science,* n.s. 53:53–67.

Chinese Ants. *Bull. Mus. Comp. Zool.* 64:529–547.

Observations on Army Ants in British Guiana. *Proc. Am. Acad. Arts Sci.* 56:291–328. 10 figs.

Professor Emery's Subgenera of the Genus *Camponotus* Mayr. *Psyche* 28:16–19.

A Study of Some Social Beetles in British Guiana and of Their Relations to the Ant-Plant *Tachigalia. Zoologica* 3:35–126. 5 pls. 12 figs.

The Tachigalia Ants. *Zoologica* 3:137–168. 4 figs.

Notes on the Habits of European and North American Cucujidae. *Zoologica* 3:173–183.

On Instincts. *J. Abnorm. Psychol.* 15:295–318.

Chinese Ants Collected by Prof. C. W. Howard. *Psyche* 28:110–115. 2 figs.

(With L. H. Taylor.) *Vespa arctica* Rohwer, a Parasite of *Vespa diabolica* De Saussure. *Psyche* 28:135–144. 3 figs.

The Bussey Institution. *Harvard Univ., Annual Reports, 1920–1921,* pp. 185–188.

1922

Ants of the Genus *Formica* in the Tropics. *Psyche* 19:174–177.

The Ants of Trinidad. *Am. Mus. Novitates,* no. 45, pp. 1–16. 1 fig.

A New Genus and Subgenus of Myrmicinae from Tropical America. *Am. Mus. Novitates,* no. 46, pp. 1–6. 2 figs.

(With the collaboration of J. C. Bequaert, I. W. Bailey, F. Santschi, and W. M. Mann.) Ants of the American Museum Congo Expedition: A Contribution to the Myrmecology of Africa. *Bull. Am. Mus. Nat. Hist.,* vol. 45. 45 pls., 103 figs. I. On the Distribution of the Ants of the Ethiopian and Malagasy Regions, pp. 13–37; II. The Ants Collected by the American Museum Congo Expedition, pp. 39–269; VII. Keys to the Genera and Subgenera of Ants, pp. 631–710; VIII. A Synonymic List of the Ants of the Ethiopian Region, pp. 711–1004; IX. A Synonymic List of the Ants of the Malagasy Region, pp. 1005–1055.

Observations on *Gigantiops destructor* Fabricius and Other Leaping Ants. *Biol. Bull.* 42:185–201. 3 figs.

Neotropical Ants of the Genera *Carebara, Tranopelta,* and *Tranopeltoides,* New Genus. *Am. Mus. Novitates,* no. 48, pp. 1–14. 3 figs.

(With J. W. Chapman.) The Mating of *Diacamma. Psyche* 29:203–211. 4 figs.

Formicidae from Easter Island and Juan Fernandez. In *The Natural History of Juan Fernandez and Easter Island,* ed. Dr. Carl Skottsberg, III, 317–319. Uppsala: Almquist & Wiksells.

The Bussey Institution. *Harvard Univ., Annual Reports, 1921–1922,* pp. 142–145.

Social Life among the Insects. *Sci. Monthly* 14 (1922):497–525; 15 (1922): 67–88, 119–131, 235–256, 320–337, 385–404, 527–541; 16 (1923):5–33, 159–176, 312–329. 113 figs.

1923

The Dry-Rot of Our Academic Biology. *Science,* n.s. 57:61–71.

(With W. M. Mann.) A Singular Habit of Sawfly Larvae. *Psyche* 30:9–13. 1 fig.

Report on the Ants Collected by the Barbados-Antigua Expedition from the University of Iowa in 1918. *Univ. Iowa Studies Nat. Hist.* 10:3–9.

Chinese Ants Collected by Professor S. F. Light and Professor A. P. Jacot. *Am. Mus. Novitates,* no. 69, pp. 1–6.

Formicidae: Wissenschaftliche Ergebnisse der Schwedischen entomologischen Reise des Herrn Dr. A. Roman in Amazonas 1914–1915. *Arkiv Zool.* 15:1–6.

Social Life among the Insects. New York: Harcourt, Brace & Co. 375 pp., 116 figs.

Ants of the Genera *Myopias* and *Acanthoponera. Psyche* 30:175–192. 5 figs.

The Occurrence of Winged Females in the Ant Genus *Leptogenys* Roger, with Descriptions of New Species. *Am. Mus. Novitates,* no. 90, pp. 1–16. 5 figs.

The Bussey Institution. *Harvard Univ., Annual Reports, 1922–1923,* pp. 170–174.

1924

Two Extraordinary Larval Myrmecophiles from Panama. *Proc. Nat. Acad. Sci.* 10:237–244. 3 figs.

A Gynandromorph of *Tetramorium guineense* Fabr. *Psyche* 31:136–137. 1 fig.

Hymenoptera of the Siju Cave, Garo Hills, Assam. *Rec. Indian Mus.* 26:123–125.

On the Ant-Genus *Chrysapace* Crawley. Psyche 31:224–225.

The Formicidae of the Harrison Williams Galapagos Expedition. *Zoologica* 5:101–122. 9 figs.

Ants of Krakatau and Other Islands in the Sunda Strait. *Treubia* 5:1–20. 1 map.

Courtship of the *Calobatas:* The Kelep Ant and the Courtship of Its Mimic *Cardiacephala myrmex. J. Hered.* 15:484–495. 8 figs.

The Bussey Institution. *Harvard Univ., Annual Reports, 1923–1924,* pp. 168–171.

1925

A New Guest-Ant and Other New Formicidae from Barro Colorado Island, Panama. *Biol. Bull.* 49:150–181. 8 figs.

(With J. W. Chapman.) The Ants of the Philippine Islands: Part I. Dorylinae and Ponerinae. *Philipp. J. Sci.* 28:47–73. 2 pls.

Neotropical Ants in the Collections of the Royal Museum of Stockholm: Part I. *Arkiv Zool.* 17A, no. 8, pp. 1–55.

Zoological Results of the Swedish Expedition to Central Africa 1921. Insecta 10, Formicidae. *Arkiv Zool.* 17A, no. 25, pp. 1–3.

The Finding of the Queen of the Army Ant *Eciton hamatum* Fabricius. *Biol. Bull.* 49:139–149. 8 figs.

L'Évolution des Insectes Sociaux. *Rev. Sci.* 63:548–557. 7 figs.

Carlo Emery. *Entomol. News* 36:313–320. (Obituary.)

The Bussey Institution. *Harvard Univ., Annual Reports, 1924–1925,* pp. 160–165.

1926

Les Sociétés d' Insectes: leur Origine, leur Évolution. Paris: Gaston Doin & Co. 468 pp., 61 figs.

The Natural History of Ants. New York: A. Knopf. 280 pp., 4 pls. (From an unpublished manuscript of Réaumur, translated and annotated.)

Social Habits of Some Canary Island Spiders. *Psyche* 33:29–31.

A New Word for an Old Thing. *Quart. Rev. Biol.* 1:439–443. (Review of J. B. Watson's *Behaviorism.*)

Emergent Evolution and the Social. *Science,* n.s. 64:433–440.

Ants of the Balearic Islands. *Folia Myrmecol. Termitol.* 1:1–6.

The Bussey Institution. *Harvard Univ., Annual Reports, 1925–1926,* pp. 168–172.

1927

The Occurrence of *Formica fusca* in Sumatra. *Psyche* 34:40–41.

Burmese Ants Collected by Professor G. E. Gates. *Psyche* 34:42–46.

Chinese Ants Collected by Professor S. F. Light and Professor N. Gist Gee. *Am. Mus. Novitates,* no. 255, pp. 1–12.

The Physiognomy of Insects. *Quart. Rev. Biol.* 2:1–36. 42 figs.

Ants collected by Professor F. Silvestri in Indochina. *Boll. Lab. Zool. Gen. Agrar. Portici* 20:83–106. 9 figs.

Ants of the Genus *Amblyopone* Erichson. *Proc. Am. Acad. Arts Sci.* 62:1–29. 8 figs.

A Few Ants from China and Formosa. *Am. Mus. Novitates,* no. 259, pp. 1–4.

The Ants of the Canary Islands. *Proc. Am. Acad. Arts Sci.* 62:93–120. 3 pls.

The Ants of Lord Howe Island and Norfolk Island. *Proc. Am. Acad. Arts Sci.* 62:121–153. 12 figs.

Carl Akeley's Early Work and Environment. *Nat. Hist.* 27:133–141. 5 figs.

The Occurrence of the Pavement Ant (*Tetramorium caespitum L.*) in Boston. *Psyche* 34:164–165.

Conserving the Family: A Review of three books on Human Reproduction and the Family. *J. Hered.* 18:119–120.

Emergent Evolution and the Social. Psyche Miniatures, General Series, no. 11. London: Kegan Paul, Trench, Trubner & Co. 57 pp.

The Bussey Institution. *Harvard Univ., Annual Reports, 1926–1927,* pp. 169–174.

1928

Foibles of Insects and Men. New York: A. Knopf. 217 pp., 56 figs.

The Social Insects: Their Origin and Evolution. New York: Harcourt, Brace & Co. 378 pp., 79 figs.

Ants Collected by Professor F. Silvestri in China. *Boll. Lab. Zool. Gen. Agrar. Portici* 22:3–38. 3 figs.

The Evolution of Ants. In *Creation by Evolution,* ed. Frances Mason, pp. 210–224. New York: Macmillan.

A New Species of *Probolomyrmex* from Java. *Psyche* 35:7–9. 1 fig.

Ants of Nantucket Island, Mass. *Psyche* 35:10–11.

Mermis Parasitism and Intercastes among Ants. *J. Exp. Zool.* 50:165–237. 17 figs.

Ants Collected by Professor F. Silvestri in Japan and Korea. *Boll. Lab. Zool. Gen. Agrar. Portici* 21:96–125.

Emergent Evolution and the Development of Societies. New York: W. W. Norton. 80 pp.

Zatapinoma, a New Genus of Ants from India. *Proc. New Engl. Zool. Club* 10:19–23. 1 fig.

The Bussey Institution. *Harvard Univ., Annual Reports, 1927–1928,* pp. 175–187.

1929

(With J. C. Bequaert.) Amazonian Myrmecophytes and Their Ants. *Zool. Anz.* 82 (Wasmann-Festband):10–39. 7 figs.

Two Interesting Neotropical Myrmecophytes (*Cordia nodosa* and *C. alliodora*). *IV. Intern. Congr. Entomol. Ithaca, N.Y., 1928* 2:342–353. Naumburg: G. Pätz.

Present Tendencies in Biological Theory. *Sci. Monthly* 28:97–109.

The Identity of the Ant Genera *Gesomyrmex* Mayr and *Dimorphomyrmex* Ernest André. *Psyche* 36:1–12. 1 fig.

Three New Genera of Ants from the Dutch East Indies. *Am. Mus. Novitates,* no. 349, pp. 1–8. 3 figs.

Ants Collected by Professor F. Silvestri in Formosa, the Malay Peninsula, and the Philippines. *Boll. Lab. Zool. Gen. Agrar. Portici* 24:27–64. 7 figs.

Two Neotropical Ants Established in the United States. *Psyche* 36:89–90.

Note on *Gesomyrmex*. *Psyche* 36:91–92.

The Ant Genus *Rhopalomastix*. *Psyche* 36:95–101. 2 figs.

A *Camponotus* Mermithergate from Argentina. *Psyche* 36:102–106. 1 fig.

Some Ants from China and Manchuria. *Am. Mus. Novitates,* no. 361, pp. 1–11.

Review of H. Friedmann's *The Cowbirds: A Study in the Biology of Social Parasitism*. *Science,* n.s. 70:70–73.

The Entomological Discoveries of John Hunter in "Exercises in Celebration of the Bicentenary of the Birth of John Hunter." *New Engl. J. Med.* 200: 810–823. 6 figs.

Is *Necrophylus Arenarius* Roux the Larva of *Pterocroce Storeyi* Withycombe? *Psyche* 36:313–320. 1 fig.

The Bussey Institution. *Harvard Univ., Annual Reports, 1928–1929,* pp. 163–166.

1930

Societal Evolution. In *Human Biology and Racial Welfare,* ed. E. V. Cowdry, pp. 139–155. New York: Hoeber.

History of the Bussey Institution. In *The Development of Harvard University Since the Inauguration of President Eliot, 1869–1929,* ed. S. E. Morison, pp. 508–517. Cambridge: Harvard University Press.

The Ant *Prenolepis imparis* Say. *Ann. Entomol. Soc. Am.* 23:1–26. 4 figs.

A Second Note on *Gesomyrmex*. *Psyche* 37:35–40. 1 fig.

Two New Genera of Ants from Australia and the Philippines. *Psyche* 37:41–47. 2 figs.

Two Mermithergates of *Ectatomma*. *Psyche* 37:48–54. 1 fig.

Formosan Ants Collected by Dr. R. Takahashi. *Proc. New Engl. Zool. Club* 11:93–106. 2 figs.

A New *Emeryella* from Panama. *Proc. New Engl. Zool. Club* 12:9–13. 1 fig.

A New Parasitic *Crematogaster* from Indiana. *Psyche* 37:55–60.

Review of Auguste Forel's *The Social World of the Ants Compared with That of Man*. *J. Soc. Psychol.* 1:170–177.

Philippine Ants of the Genus *Aenictus* with Descriptions of the Females of Two Species. *J. N.Y. Entomol. Soc.* 38:193–212. 7 figs.

(With P. J. Darlington, Jr.) Ant-Tree Notes from Rio Frio, Colombia. *Psyche* 37:107–117. 1 pl.

Demons of the Dust: A Study in Insect Behavior. New York: W. W. Norton. 378 pp., 49 figs.

1931

New and Little-Known Ants of the Genera *Macromischa, Croesomyrmex* and *Antillaemyrmex*. *Bull. Mus. Comp. Zool.* 72: 3–34.

A List of the Known Chinese Ants. *Peking Nat. Hist. Bull.* 5:53–81.

What is Natural History? *Bull. Boston Soc. Nat. Hist.,* no. 59, pp. 3–12.

Concerning Some Ant Gynandromorphs. *Psyche* 38:80–85.

Neotropical Ants of the Genus *Xenomyrmex* Forel. *Rev. Entomol.* 1:129–139. 2 figs.

Hopes in the Biological Sciences. *Proc. Am. Philos. Soc.* 70:231–239.

The Ant *Camponotus* (*Myrmepomis*) *sericeiventris* Guérin and Its Mimic. *Psyche* 38:86–98. 2 figs.

A Cuban Vermileo. *Psyche* 38:165–169.

1932

An Extraordinary Ant-Guest from the Philippines, *Aenictoteras chapmani,* gen. et sp. nov. In *Livre Centenaire Soc. Entomol. France,* pp. 301–310. 2 figs.

Ants of the Marquesas Islands. *Bernice P. Bishop Mus. Honolulu,* bull. 98, pp. 155–163.

Ants from the Society Islands. *Bernice P. Bishop Mus. Honolulu,* bull. 113, pp. 13–19.

A List of the Ants of Florida with Descriptions of New Forms. *J. N.Y. Entomol. Soc.* 40:1–17.

How the Primitive Ants of Australia Start Their Colonies. *Science,* n.s. 76:532–533.

Some Attractions of the Field Study of Ants. *Sci. Monthly* 34:397–402. 2 figs.

An Australian *Leptanilla. Psyche* 39:53–58. 1 fig.

1933

Colony-Founding among Ants: With an Account of Some Primitive Australian Species. Cambridge: Harvard University Press. 179 pp., 29 figs.

(And T. Barbour, eds.) *The Lamarck Manuscripts at Harvard.* Cambridge: Harvard University Press. 202 pp., 4 pls.

Mermis Parasitism in Some Australian and Mexican Ants. *Psyche* 40:20–31. 3 figs.

(With R. Dow.) Unusual Prey of *Bembix. Psyche* 40:57–59.

Formicidae of the Templeton Crocker Expedition. *Proc. Calif. Acad. Sci.,* Ser. 4, 21:57–64.

New Ants from China and Japan. *Psyche* 40:65–67.

A Second Parasitic *Crematogaster. Psyche* 40:83–86.

Translation of Maurice Bedel's "My Uncles, Louis Bedel and Henri d'Orbigny." *Quart. Rev. Biol.* 8:325–330. 1 fig.

A New Species of *Ponera* and Other Records of Ants from the Marquesas Islands. *Bernice P. Bishop Mus. Honolulu,* bull. 114, pp. 141–144.

An Ant New to the Fauna of the Hawaiian Islands. *Proc. Hawaiian Entomol. Soc.* 8:275–278. 1 fig.

A New *Myrmoteras* from Java. *Proc. New Engl. Zool. Club* 13:73–75. 1 fig.

Three Obscure Genera of Ponerine Ants. *Am. Mus. Novitates,* no. 672, pp. 1–23. 8 figs.

1934

Some Aberrant Species of *Camponotus (Colobopsis)* from the Fiji Islands. *Ann. Entomol. Soc. Am.* 27:415–424. 5 figs.

Ants from the Islands off the West Coast of Lower California and Mexico. *Pan-Pacific Entomol.* 10:132–144.

A Second Revision of the Ants of the Genus *Leptomyrmex* Mayr. *Bull. Mus. Comp. Zool.* 77:69–118. 16 figs.

Revised List of Hawaiian Ants. *Occas. Papers Bernice P. Bishop Mus. Honolulu* vol. 10, no. 21, pp. 1–21.

(With W. S. Creighton.) A Study of the Ant Genera *Novomessor* and *Veromessor*. *Proc. Am. Acad. Arts Sci.* 69:341–387. 2 pls., 1 fig.

Animal Societies (Biology and Society). *Sci. Monthly* 39:289–301.

Formicidae of the Templeton Crocker Expedition, 1933. *Proc. Calif. Acad. Sci.* 21:173–181. 1 fig.

Contributions to the Fauna of Rottnest Island, Western Australia: No. IX —The Ants. *J. Roy. Soc. W. Australia* 20:137–163.

An Australian Ant of the Genus *Leptothorax* Mayr. *Psyche* 41:60–62.

A Specimen of the Jamaican Vermileo. *Psyche* 41:236–237.

Foreword. In O. E. Plath's *Bumblebees and Their Ways,* pp. vii–x. New York: Macmillan.

Neotropical Ants Collected by Dr. Elisabeth Skwarra and Others. *Bull. Mus. Comp. Zool.* 77:157–240. 6 figs.

Some Ants from the Bahama Islands. *Psyche* 41:230–232.

1935

Two New Genera of Myrmicine Ants from Papua and the Philippines. *Proc. New Engl. Zool Club* 15:1–9. 2 figs.

Insects. In "Observations on the Behavior of Animals during the Total Solar Eclipse of August 31st, 1932." *Proc. Am. Acad. Arts Sci.* 70:36–45.

Review of F. Maidl's *Die Lebensgewohnheiten und Instinkte der Staatenbilden den Insekten. Biol. Abstr.* vol. 9, no. 9, #20116.

Ants of the Genera *Belonopelta* Mayr and *Simopelta* Mann. *Rev. Entomol.* 5:8–19. 4 figs.

The Australian Ant Genus *Mayriella* Forel. *Psyche* 42:151–160. 1 fig.

Check List of the Ants of Oceania. *Occas. Papers Bernice P. Bishop Mus. Honolulu,* vol. 11, no. 11, pp. 1–56.

New Ants from the Philippines. *Psyche* 42:38–52. 3 figs.

Myrmecological Notes. *Psyche* 42:68–72.

Ants of the Genus *Acropyga* Roger, with Description of a New Species. *J. N.Y. Entomol. Soc.* 43:321–329. 1 fig.

1936

Binary Anterior Ocelli in Ants. *Biol. Bull.* 70:185–192, 3 figs.

Entomology at Harvard University. In "Notes Concerning the History and Contents of the Museum of Comparative Zoölogy." Cambridge: Museum of Comparative Zoölogy, pp. 22–32.

Ants from Hispaniola and Mona Island. *Bull. Mus. Comp. Zool.* 80:195–211.

Notes on Some Aberrant Indonesian Ants of the Subfamily Formicinae. *Tijdschr. Entomol.* 79:217–221.

Review of Thomas Elliott Snyder's *Our Enemy the Termite. Psyche* 43:27–28.

The Australian Ant Genus *Froggattella. Am. Mus. Novitates,* no. 842, pp. 1–11.

A Singular *Crematogaster* from Guatemala. *Psyche* 43:40–48. 1 fig.

Ecological Relations of Ponerine and Other Ants to Termites. *Proc. Am. Acad. Arts Sci.* 71:159–243. 9 figs.

A Notable Contribution to Entomology. *Quart. Rev. Biol.* 11:337–341. (Review of T. Rayment's *A Cluster of Bees.*)

Ants from the Society, Austral, Tuamotu, and Mangareva Islands. *Occas. Papers Bernice P. Bishop Mus. Honolulu,* vol. 12, no. 18, pp. 1–17.

1937

Additions to the Ant-Fauna of Krakatau and Verlaten Islands. *Treubia* 16:21–24.

Ants Mostly from the Mountains of Cuba. *Bull. Mus. Comp. Zool.* 81:439–465.

Mosaics and Other Anomalies among Ants. Cambridge: Harvard University Press. 95 pp., 18 figs.

1938

Ants from the Caves of Yucatan. *Carnegie Inst. Wash. Publ.,* no. 491, pp. 251–255.

1939

(Edited by G. H. Parker.) *Essays in Philosophical Biology.* Cambridge: Harvard University Press. 261 pp., 2 pls., 2 figs.

1942

(Revised and edited by J. C. Bequaert.) Studies of Neotropical Ant-Plants and Their Ants. *Bull. Mus. Comp. Zool.* 90:1–262. 57 pls.

Notes

The following notes contain a list of sources for information other than those in Professor Wheeler's published papers, which are listed in the bibliography. Most of the unpublished material, cited as "the Wheeler family papers," is now in the Boston home of Professor Wheeler's daughter.

Prologue: Ants, Elephants, and Men

1. Two good sources, among many, of information on the life of Barnum are: P. T. Barnum, *Barnum's Own Story: The Autobiography of P. T. Barnum* (New York: Dover, 1961); Irving Wallace, *The Fabulous Showman: The Life and Times of P. T. Barnum* (New York: A Knopf, 1959).
2. Bergh's biography is called *Angel in Top Hat,* written by Zulma Steele and published by Harper in New York in 1942.
3. The standard reference on the life of Louis Agassiz is Edward Lurie, *Louis Agassiz: A Life in Science* (Chicago: University of Chicago Press, 1960).
4. Roswell Ward, "Henry A. Ward," *Rochester Historical Society Publications* 24 (1948):1–297.
5. Holmes' poem "Farewell to Agassiz" was printed first in 1865 in *The Atlantic Monthly* 16:584–585.
6. David Starr Jordan, *The Days of a Man,* 2 vols. (Yonkers, N.Y.: World, 1922).
7. Whittier's poem "A Prayer of Agassiz" was published in 1873. The authors found it in the library of the Museum of Comparative Zoology at Harvard University in *Tribune, Popular Science,* published in Boston by Henry L. Shepard and Company in 1894, but it has been reprinted in many places over the years.
8. R. C. Andrews, "Akeley of Africa," *Beyond Adventure: The Lives of Three Explorers* (New York: Duell, Sloan, and Pearce, 1954), pp. 91–142. Wheeler's years at Ward's establishment are described in his 1927 paper about Carl Akeley (see bibliography).
9. E. J. Dornfeld, "The Allis Lake Laboratory, 1886–1893," *Marquette Medical Review* 21 (1956):113–142.

10. F. R. Lillie, "Charles Otis Whitman," *Journal of Morphology* 22 (1911): xv–lxxvii.

11. The quotations about natural history, Agassiz, and amateurs were taken from Wheeler's "The Dry-Rot of Our Academic Biology," first published in 1923 (see bibliography).

12. L. J. Henderson, T. Barbour, F. M. Carpenter, and Hans Zinsser, "William Morton Wheeler," *Science* 85 (1937):533–535.

13. William Beebe, *The Book of Naturalists* (New York: A. Knopf, 1944).

14. David Fairchild, *The World Grows Round My Door* (New York: Scribner's, 1947).

Chapter 1: Bright Beginnings

1. The Wheeler genealogy is based on notes found in the family papers in Boston, old newspaper clippings in the "Wheeler Scrapbook" at the Flower Memorial Library in Watertown, New York, and Jefferson County deeds and census reports in Watertown, New York.

2. Olive Wheeler's work was described in the *Annual Report* of the Milwaukee Public Museum for 1904; her 1893 letter is in the Wheeler family papers.

3. Anonymous, *History of Milwaukee, Wisconsin* (Chicago: Western Historical, 1881).

4. E. R. Leland, "In Memoriam: Prof. Peter Engelmann," *Transactions of the Wisconsin Academy of Sciences, Arts, and Letters* 2 (1871):258–263.

5. The 1871 report on the German-English Academy is bound with several other miscellaneous papers entitled *Engelmann's Schule: Seminar: Milwaukee Souvenir*. The single volume is deposited in the collections of the Wisconsin Historical Society in Madison.

6. W. T. Hornaday, "The King of Museum-Builders," *The Commercial Travelers' Home Magazine* 6 (1896):147–159.

7. These remarks are from Wheeler's tribute to Akeley (see note 8, prologue).

8. *Ward's Natural Science Bulletin*. Volume 1 began in 1881. All the early copies were studied by the authors, but not all were numbered and can be cited. Especially helpful, however, were vol. 3, no. 1 (January 1, 1884), and one dated July 1, 1884.

9. Roswell Ward, "Henry A. Ward," *Rochester Historical Society Publications* 24 (1948):1–297.

10. Blake McKelvey, *Rochester the Flower City, 1855–1890* (Cambridge: Harvard University Press, 1949).

11. Charles H. Lyttle, "Reverend Newton M. Mann," *Meadville Theological School, Quarterly Bulletin* 24 (1929):19–28.

12. Letter of Lewis Akeley about his brother, Carl, quoted from a letter to Wheeler from Mrs. Mary Akeley, January 21, 1929, is now in the Wheeler family papers.

Chapter 2: The World of Science Opens

1. "Die Milwaukee Bai" is included in *Engelmann's Schule: Seminar: Milwaukee Souvenir.* See note 5, chapter 1. There are five stanzas in all; only the first and last are quoted here.

2. The newspaper story on George W. Peckham was published in the *Milwaukee Journal* on March 1, 1892.

3. R. A. Muttkowski, "George Williams Peckham, M.D., LL.D. (1845–1914)," *Entomological News* 25 (1914):145–148.

4. These remarks are from Wheeler's 1927 paper on Carl Akeley (see bibliography).

5. F. R. Lillie, "Charles Otis Whitman," *Journal of Morphology* 22 (1911): xv–lxxvii.

6. D. P. Costello, "Henry van Peters Wilson, 1863–1939," *National Academy of Sciences, Biographical Memoirs* 35 (1961):351–383.

7. E. J. Dornfeld, "The Allis Lake Laboratory, 1886–1893," *Marquette Medical Review* 21 (1956):113–142.

8. J. H. Gerould, "William Patten," *Science* 76 (1932):481–482.

9. Wheeler's *Annual Reports* of the Milwaukee Museum were in vols. 6 (1887–1888), 7 (1888–1889), and 8 (1889–1890).

10. F. R. Lillie, *The Woods Hole Marine Biological Laboratory* (Chicago: University of Chicago Press, 1944).

Chapter 3: Two Universities Are Born

1. Anonymous, "Clark University," *Science* 13 (1889):462–465.

2. A. E. Tanner, *History of Clark through Interpretation of the Will of the Founder* (unpublished manuscript in the archives of Clark University).

3. Lorine Pruette, *G. Stanley Hall: A Biography of a Mind* (New York: Appleton, 1926).

4. The quotation from William James was taken from E. L. Thorndike, "Granville Stanley Hall, 1846–1924," *National Academy of Sciences, Biographical Memoirs* 12 (1928):135–180.

5. Johannsen and Butt's book was published in New York by McGraw-Hill (1941).

6. Waddington's book, *The Principles of Embryology,* was published in London by Allen and Unwin in 1956.

7. L. N. Wilson, *G. Stanley Hall, A Sketch* (New York: Stechert, 1914).

8. F. P. Mall's letter is in the Presidential Papers in the archives of the University of Chicago.

9. F. R. Sabin, *Franklin Paine Mall, Anatomist* (Baltimore: Johns Hopkins, 1934).

10. Ray Ginger, *Altgeld's America: 1890–1905* (New York: Funk and Wagnalls, 1958).

11. R. J. Storr, *Harper's University, the Beginnings: A History of the University of Chicago* (Chicago: University of Chicago Press, 1966).

12. E. G. Conklin, "Henry Herbert Donaldson, 1857–1938," *National Academy of Sciences, Biographical Memoirs* 20 (1939):229–243.

13. William Burrows, "Edwin Oakes Jordan, 1866–1936," *National Academy of Sciences, Biographical Memoirs* 20 (1938):197–228.

14. W. D. Howells, *Letters of an Altrurian Traveller* (1893–94), facsimile, with an introduction by C. M. and R. Kirk (Gainesville, Fla.: Scholar's Facsimiles and Reprints, 1961).

Chapter 4: European Interlude

1. Whitman's letter is in the Wheeler family papers.

2. Boveri's letter is in the Wheeler family papers.

3. Fritz Baltzer, *Theodor Boveri: Life and Work of a Great Biologist,* trans. Dorothea Rudnick (Berkeley: University of California Press, 1967).

4. The quotation from von Ubisch is taken from Jane Oppenheimer's appreciative paper, "Theodor Boveri: the Cell Biologists' Embryologist," *Quarterly Review of Biology* 38 (1963):245–249.

5. Amalie Grob's letters are in the Wheeler family papers.

6. Most of the Jordan-Wheeler correspondence for 1893–94 is in the archives of the University of Chicago, the remainder in the Wheeler family papers.

7. A brief sketch of Escherich's life will be found in an anonymous article entitled "Karl Escherich zum 80 Geburtstag," published in *Zeitschrift für Angewandte Entomologie* 33 (1952):1–11. Inserted between the front cover and the first page is an unnumbered page with an unsigned notice of Escherich's sudden death on November 22, 1951.

8. E. Ray Lankester's obituary for Dohrn was in *Nature* 81 (1909):429–431.

9. The quotation about Lo Bianco is from C. A. Kofoid, *The Biological Stations of Europe* (Washington, D.C.: Government Printing Office, 1910).

10. David Fairchild's *The World Was My Garden* was published in 1938 in New York by Scribner's.

11. "Prof. E. van Beneden (1845–1910)," *Nature* 83 (1910):344–345. The obituary is signed with the initials A. S.

Chapter 5: Woods Hole and Chicago

1. Wheeler's letter to Jordan is in the archives of the University of Chicago.

2. H. H. Newman, "History of the Department of Zoology in the University of Chicago," *Bios* 19 (1948):214–239.

3. A. G. Pohlman and Phil Rau, "Charles Henry Turner, Ph.D.," *Academy of Science, St. Louis, Transactions* 24, no. 9 (1923):1–54.

4. D. H. Weinrich, "C. E. McClung," *Science* 103 (1946):551–552.

5. Ralph Emerson's autobiography was published privately in 1949 by his daughter, Mary E. Lathrop, at Ithaca, N.Y.

6. *Love Bound and Other Poems* was published in 1879 by University Press, Cambridge, Massachusetts.

7. Wheeler's letter to Mrs. Emerson is in the Wheeler family papers.

Chapter 6: Texas: Enter the Ant

1. The Whitman letter to Harper is in the Presidential Papers in the archives of the University of Chicago.
2. The Wheeler-Jordan correspondence for 1899–1900 is in the archives of the University of Chicago.
3. In the library of the University of Texas is much helpful material, including the *Biennial Reports* of the Texas Board of Regents, *The University Records, and The Alcalde,* as well as student registers and course catalogues.
4. Brues's article was published in *The Alcalde* (Texas alumni magazine) of December 1937.
5. Information on Augusta Rucker may be found in the first two numbers for the year 1919 of *The Alcalde.*
6. Hartman's essay is in *Biological Contributions,* University of Texas Publication 5914, ed. M. R. Wheeler (Austin, 1959), pp. 13–24.
7. C. T. Brues's obituary for H. E. Walter is in *Science* 102 (1945):554–555.
8. Wheeler wrote of his first attraction to ants in his paper on ant larvae, published in 1918 (see bibliography).
9. S. W. Geiser, *Naturalists of the Frontier,* 2d ed. (Dallas, Texas: Southern Methodist Press, 1948).
10. Mrs. Comstock's autobiography was edited by G. W. Herrick and R. G. Smith and published in 1953 by Comstock Publishing Associates (Cornell University Press), Ithaca, N.Y.
11. Helen N. Steven's biography of Adele Fielde was published in New York City by the Fielde Memorial Committee in 1918.
12. Mabel Abbott, *The Life of William T. Davis* (Ithaca, N.Y.: Cornell University Press, 1949).
13. The Wheeler-Bumpus correspondence for 1902–03 is in the archives of the American Museum of Natural History.

Chapter 7: In and Out of New York

1. I. E. Manchester, *The History of Colebrook and Other Papers* (Colebrook, Conn.: Sesqui-Centennial Committee, 1935); R. W. Rockwell, *The Rockwell Family in One Line of Descent* (Pittsfield, Mass.: privately published, 1924).
2. Ralph Emerson's 1911 letter to his wife about Wheeler is in the Wheeler family papers.
3. American Museum of Natural History, *Annual Reports,* nos. 33–40 (1901–1908); *The American Museum Journal,* vol. 7 (1907), under "Museum News Notes," carried information on some of Wheeler's activities for the year.
4. The Wheeler-Bumpus correspondence for 1904–1908 is in the archives of the American Museum of Natural History.
5. T. D. A. Cockerell, "William Morton Wheeler," *Scientific Monthly* 44 (1937):569–571.
6. T. D. A. Cockerell, "Recollections of a Naturalist, V: Fossil Insects," *Bios* 8 (1937):51–56.

7. C. D. Michener, obituary of Cockerell in the *Journal of the New York Entomological Society* 56 (1948):171–174.

8. W. A. Weber, "Theodore Dru Alison Cockerell, 1866–1948," *University of Colorado Studies,* Series in Bibliography, no. 1. (Boulder: University of Colorado Press, 1965).

9. J. Ewan's *Rocky Mountain Naturalists* was published by the University of Denver Press in 1950. The last chapter (pp. 95–116) is about Cockerell.

10. Wheeler's letter about Cockerell was dated November 26, 1918, and written to Dr. Henry E. Crampton at the American Museum of Natural History. A copy is in the Wheeler family papers.

11. T. D. A. Cockerell, "Recollections of a Naturalist, I," *Bios* 6 (1935):372–385.

12. W. M. Mann, *Ant Hill Odyssey* (Boston: Little, Brown, 1948).

13. H. C. Bumpus, Jr., *Hermon Carey Bumpus: Yankee Naturalist* (Minneapolis: University of Minnesota Press, 1947).

14. Barbour's 1907 letter to Henshaw is in the archives of the Museum of Comparative Zoology.

15. The summary of Bumpus's work for the National Parks is based on A. D. Mead's obituary for Bumpus in *Science* 99 (1944):28–30.

16. *Entomological Society of America, Annals* 1 (1908):1–37 (Constitution and Proceedings).

Chapter 8: "Ants Are to be Found Everywhere"

1. T. D. A. Cockerell's review of *Ants* was in *Science* 31 (1910):860–862.

2. E. O. Wilson's 1953 paper on polymorphism is in the *Quarterly Review of Biology* 28:136–156.

3. A copy of Wheeler's letter to Parker, dated March 25, 1910, is in the Wheeler family papers.

4. The Museum of Comparative Zoology published W. S. Creighton's *The Ants of North America* in the *Bulletin* as vol. 104.

5. Cockerell's letter to Wheeler, dated November 23, 1923, is in the Wheeler family papers.

6. F. M. Carpenter's "The Fossil Ants of North America" was in the *Bulletin* of the Museum of Comparative Zoology, vol. 70, pp. 1–66.

7. A copy of Wheeler's letter to Rohwer, dated December 2, 1909, is in the Wheeler family papers.

8. For the expression "the weird and wonderful pentanomial system" and other thoughts on ant taxonomy, see W. L. Brown, Jr., *A Century of Progress in the Natural Sciences, 1853–1953,* published in 1955 by the California Academy of Sciences, San Francisco, pp. 569–572.

9. Wheeler's letter to Hocking, dated May 22, 1917, is in the Wheeler family papers.

10. The quotation from Wheeler's 1935 letter to Creighton is in Creighton's paper "On Formicid Nomenclature," *Journal of the New York Entomological Society* 46 (1938):1–9.

Chapter 9: Professor at the Bussey

1. Wheeler's remarks about the economic entomologist were in a letter to T. D. A. Cockerell, dated April 7, 1917. A copy of this letter is in the Wheeler family papers.

2. The archives in Widener Library, Harvard University have been a rich source of information on Bussey with *Annual Reports,* catalogues (listing courses and students) and student doctoral theses.

3. E. A. Steinhaus's book, *Principles of Insect Pathology,* was published in 1949 in New York by McGraw-Hill.

4. James Chapman and Ethel Chapman, *Escape to the Hills* (Lancaster, Pa.: Jacques Cattell, 1947).

5. A copy of Wheeler's 1912 lecture (unpublished) is in the Wheeler family papers.

6. The Wheeler-Melander letters of 1915, or carbon copies thereof, are in the Wheeler family papers.

7. A copy of Wheeler's letter to Cockerell, dated June 18, 1915, is in the Wheeler family papers.

8. F. X. Williams's letter to the authors was dated January 13, 1964. The following reference gave much useful information on Williams's life: E. C. Zimmerman, "Francis Xavier Williams, 1882–1967," *Pan-Pacific Entomologist* 45 (1969):135–146.

9. W. M. Mann, *Ant Hill Odyssey* (Boston: Little, Brown, 1948).

10. Gordon MacCreagh's *White Waters and Black* was originally published in 1926 by Appleton-Century-Crofts and is now available in a paperback reprint (Doubleday Anchor, 1961).

11. A. C. Kinsey's letter to Wheeler, dated November 4, 1926, is in the Wheeler family papers.

12. Bentley Glass's review of the Kinsey report is in the *Quarterly Review of Biology* 23 (1948):39–42.

13. J. G. Myers, *Insect Singers* (London: Routledge, 1929).

14. T. B. Mitchell's *Bees of the Eastern United States* was published by the North Carolina Agricultural Experiment Station as Technical Bulletin 41 (1960, 1962).

15. The recollections of P. W. and Anna R. Whiting were sent to the authors with a letter dated March 16, 1967.

16. Herbert Friedmann's comments are in a letter to the authors dated May 23, 1967.

Chapter 10: Student of Animal Behavior

1. W. H. Thorpe, *Learning and Instinct in Animals* (Cambridge: Harvard University Press, 1956).

2. Quotation about G. Stanley Hall is from E. L. Thorndike, "Granville Stanley Hall, 1846–1924," *National Academy of Sciences, Biographical Memoirs* 12 (1928):135–180.

3. Quotation about von Uexküll is from H. S. Jennings, "The Work of J. von Uexküll on the Physiology of Movements and Behavior," *Journal of Comparative Neurology and Psychology* 19 (1909):313–336.

4. W. J. V. Osterhout, "Jacques Loeb, 1859–1924," *National Academy of Sciences. Biographical Memoir* 13 (1930):318–401.

5. Loeb's remarks on tropisms were taken from his 1918 work entitled *Forced Movements, Tropisms, and Animal Conduct* (Philadelphia: Lippincott).

6. Fraenkel and Gunn's book *The Orientation of Animals* was published in England by Oxford in 1940. In 1961 Dover reprinted it in paperback.

7. Jennings's talk, entitled "Tropisms," was included in the *Comptes Rendus* of the Congress, published in Geneva, Switzerland (pp. 307–324).

8. A. Kühn, *Die Orientierung der Tier in Raum* (Jena: Gustav Fischer, 1919).

9. F. R. Lillie, "Charles Otis Whitman," *Journal of Morphology* 22 (1911): xv–lxxvii.

10. C. O. Whitman's essay "Animal Behavior" was published as the sixteenth Woods Hole Biological Lecture (Boston: Ginn and Company), pp. 285–338.

11. E. H. Hess, *New Directions in Psychology* (New York: Holt, Rinehart, and Winston, 1962).

12. K. Z. Lorenz's remarks regarding Whitman are from *Physiological Mechanisms in Animal Behavior* (New York: Academic Press, 1950).

13. D. Spalding, "Instinct, with Original Observations on Young Animals," 1872, reprinted *British Journal of Animal Behavior* 2 (1954):2–11.

14. The quotation from Lorenz regarding Heinroth's founding of ethology is from the discussion following a paper by N. Tinbergen included in *Group Processes,* edited by Bertram Schaffner and published in New York by the Josiah Macy, Jr. Foundation in 1955.

15. R. R. F. Sheehan, "Conversations with Konrad Lorenz," *Harper's Magazine* 236 (May 1968):69–77.

16. W. H. Thorpe, "Ethology as a New Branch of Biology," in Buzzati-Traverso, ed., *Perspectives in Marine Biology* (Berkeley: University of California Press, 1956).

17. G. P. Baerends, "Ethological Studies of Insect Behavior," *Annual Review of Entomology* 4 (1959):207–234.

18. W. Craig, "Appetites and Aversions as Constituents of Instincts," *Biological Bulletin* 34 (1918):91–107.

19. E. Mayr, *Animal Species and Evolution* (Cambridge: Harvard University Press, 1963), chapter 3.

20. The quotation from Baldwin was taken from Wheeler's 1903 review of Baldwin's book.

21. G. G. Simpson, "The Baldwin Effect," *Evolution* 7 (1953):110–117.

22. E. Mayr, "Behavior and Systematics," in G. G. Simpson and Anna Roe, eds., *Behavior and Evolution* (New Haven, Conn.: Yale University Press, 1958), pp. 341–362. See also the discussion in H. E. Evans, *The Comparative Ethology and Evolution of the Sand Wasps* (Cambridge: Harvard University Press, 1966).

23. A. C. Hardy, *The Living Stream* (London: Collins, 1965). Chapters six and seven relate most closely to the ideas Wheeler expressed.

Chapter 11: Colleagues, Credos, and Crises

1. Anna Whiting's paper was published in the *Quarterly Review of Biology* 42 (1967):334–406.
2. E. B. Wilson's book *The Cell in Development and Inheritance* was published in 1900 in New York by Macmillan.
3. A. H. Sturtevant, "Thomas Hunt Morgan, 1866–1945," *National Academy of Sciences, Biographical Memoir* 33 (1959):283–325.
4. Originals or carbon copies of the Wheeler-Loeb letters are in the Wheeler family papers.
5. Originals or carbon copies of the Wheeler-Morgan letters are in the Wheeler family papers.
6. A. H. Sturtevant's letter to the authors was dated April 13, 1967.
7. L. C. Dunn's *A Short History of Genetics* was published in 1965 in New York by McGraw-Hill.
8. Wheeler's 1931 remarks on eugenics were in his essay "Hopes in the Biological Sciences" (see bibliography).
9. H. S. Jennings, "Raymond Pearl, 1879–1940," *National Academy of Sciences, Biographical Memoir* 22 (1943): 295–347.
10. Originals or carbon copies of the Wheeler-Pearl-Mencken letters are in the Wheeler family papers.
11. The 1928 Agassiz letter, undated, is in the Wheeler family papers.
12. Copies of the East letters are in the Wheeler family papers.
13. A copy of Wheeler's letter to James is in the Wheeler family papers.

Chapter 12: Student of Insect Sociality

1. D. S. Jordan's letter to Wheeler, dated April 25, 1928, is in the Wheeler family papers.
2. H. S. Jennings's letter to Wheeler, dated September 29, 1926, and a copy of Wheeler's letter to Jennings, dated October 5, 1926, are in the Wheeler family papers.
3. C. L. Prosser, "Levels of Biological Organization and their Physiological Significance," in J. A. Moore, ed., *Ideas in Modern Biology* (Garden City, N.Y.: Natural History Press, 1965), pp. 357–390.
4. T. Dobzhansky's book *The Biology of Ultimate Concern* was published in 1967 in New York by New American Library.
5. E. O. Wilson, "The Superorganism Concept and Beyond," in *L'effet de Groupe chez les Animaux,* Colloques internationaux Centre National de la Recherche Scientifique (Paris, France), no. 173 (1967), pp. 1–11.
6. J. Ishay and R. Ikan, "Food Exchange between Adults and Larvae in Vespa orientalis F.," *Animal Behaviour* 16 (1968):298–303.
7. P. W. Whiting's review was in *The Journal of Heredity* 29 (1938):189–193.
8. Nevin Weaver, "Physiology of Caste Determination," *Annual Review of Entomology* 11 (1966):79–102.
9. The *Boston Post's* story was dated September 26, 1932.
10. A copy of Wheeler's letter to J. H. Vincent, dated November 22, 1932, is in the Wheeler family papers.

Chapter 13: Distant Places and Home Port

1. A copy of Wheeler's letter to A. L. Melander, dated June 18, 1915, is in the Wheeler family papers.

2. A copy of Wheeler's 1917 letter to Dittlinger is in the Wheeler family papers.

3. J. C. Bradley's report of the 1917 trip comprised the first fifty-four pages of *Scientific Monthly* 8 (1919).

4. A copy of Wheeler's letter to Bradley, dated October 18, 1917, is in the Wheeler family papers.

5. Bequaert's letter to Wheeler, dated September 23, 1917, is in the Wheeler family papers.

6. A. E. Emerson's letter to the authors is dated January 23, 1964.

7. Wheeler's letters to Thomas Barbour about Barro Colorado are in the archives of the Museum of Comparative Zoology.

8. David Fairchild's letter to Wheeler is in the Wheeler family papers.

9. Graham Fairchild's letter to the authors is dated March 17, 1969.

10. A copy of Wheeler's letter to Pearl is in the Wheeler family papers.

11. Wheeler's letter describing his 1935 trip to Central America was addressed to Barbour's secretary, Miss H. M. Robinson, and is now in the archives of the Museum of Comparative Zoology.

12. The quotation about Loeb and Morgan is included in A. S. Romer, "George Howard Parker, 1864–1955," *National Academy of Sciences, Biographical Memoir* 39 (1967):359–390.

13. The Wheeler-Barbour correspondence about the 1931 trip to Australia is in the archives of the Museum of Comparative Zoology.

14. Wheeler's observations on the Wiluna *Bembix* were published in 1933 in an article titled "Unusual Prey of Bembix" (see bibliography).

15. Mrs. C. T. Brues's letter to the authors is dated October 1, 1968.

16. T. Barbour, *Naturalist at Large* (Cambridge: Harvard University Press, 1943).

17. The "Mother Hubbard" remark was contained in a letter to Pearl, dated February 25, 1932, and is in the Wheeler family papers.

18. T. Barbour, *A Naturalist's Scrapbook* (Cambridge: Harvard University Press, 1946).

19. Harlow Shapley has published an autobiography in which he tells of his interest in ants and his contacts with Wheeler: *Through Rugged Ways to the Stars* (New York: Scribner's, 1969).

20. R. M. Ferry, "Lawrence Joseph Henderson, 1878–1942," *Science* 95 (1942):316–318.

21. H. Zinsser, *Rats, Lice, and History* (Boston: Little, Brown, 1934).

22. Fairchild's letter to Barbour is in the archives of Widener Library at Harvard University.

Chapter 14: The Measure of a Man

1. A copy of the document from the Zoological Laboratory of Cambridge University is in the archives of the Museum of Comparative Zoology.

2. The tribute to Wheeler by Henderson, Barbour, Carpenter, and Zinsser was printed in *Science* 85 (1937):533–535.

3. The Wheeler-Norton correspondence, most of which occurred in September 1930, is in the Wheeler family papers.

4. A copy of Wheeler's letter to Lillie, dated May 8, 1918, is in the Wheeler family papers.

5. G. Sarton, review of *René Antoine Ferchault de Réaumur* in *Isis* 9 (1927): 445–447.

6. A copy of Wheeler's letter to Henshaw, dated December 26, 1933, is in the Wheeler family papers.

7. Sarton's letter to Wheeler, dated December 3, 1918, is in the Wheeler family papers.

8. Maurice Maeterlinck's *The Life of the Ant* was published in English in 1930 by Doubleday in New York.

9. Clifton Fadiman's anthology, *Reading I've Liked,* was published in 1941 by Simon and Schuster in New York.

10. C. T. Brues published the obituary to Wheeler in *Psyche* 44 (1937):61–96.

Index

INDEX